CARDIOVASCULAR AND RESPIRATORY BIOENGINEERING

CARDIOVASCULAR AND RESPIRATORY BIOENGINEERING

Edited by

NENAD FILIPOVIC
Faculty of Engineering, Center for Bioengineering, University of Kragujevac;
BioIRC, Bioengineering Research and Development Center, Kragujevac, Serbia

Academic Press is an imprint of Elsevier
125 London Wall, London EC2Y 5AS, United Kingdom
525 B Street, Suite 1650, San Diego, CA 92101, United States
50 Hampshire Street, 5th Floor, Cambridge, MA 02139, United States
The Boulevard, Langford Lane, Kidlington, Oxford OX5 1GB, United Kingdom

ISBN 978-0-12-823956-8

For information on all Academic Press publications
visit our website at https://www.elsevier.com/books-and-journals

Publisher: Mara Conner
Acquisitions Editor: Carrie Bolger
Editorial Project Manager: Howie M. De Ramos
Production Project Manager: Prasanna Kalyanaraman
Cover Designer: Miles Hitchen

Typeset by STRAIVE, India

Contents

7. Lab-on-a-chip for lung tissue from in silico perspective

Milica Nikolic

8. Chaotic mixing and its role in enhancing particle deposition in the pulmonary acinus: A review

Akira Tsuda and Frank S. Henry

9. Three-dimensional reconstruction and modeling of the respiratory airways, particle deposition, and drug delivery efficacy

Nenad Filipovic and Akira Tsuda

10. Tissue engineering—Electrospinning approach

Marko N Živanović and Nenad Filipovic

11. Application of numerical methods for the analysis of respiratory system

Aleksandra Vulović and Nenad Filipovic

12. Artificial intelligence approach toward analysis of COVID-19 development—Personalized and epidemiological model

Tijana Šušteršič and Anđela Blagojević

13. Economic analysis for in silico clinical trials of biodegradable and metallic vascular stents

Marija Gacic

Contributors

Branko Arsić Bioengineering Research and Development Center, BioIRC; Faculty of Science, University of Kragujevac, Kragujevac, Serbia

Anđela Blagojević University of Kragujevac, Faculty of Engineering; Bioengineering Research and Development Center (BioIRC), Kragujevac, Serbia

Smiljana Djorovic Faculty of Engineering, University of Kragujevac; Bioengineering Research and Development Center, Kragujevac, Serbia

Tijana Djukic Bioengineering Research and Development Center (BioIRC), Kragujevac, Serbia

Nenad Filipovic Bioengineering Research and Development Center (BioIRC), Kragujevac, Serbia

Marija Gacic University of Kragujevac, Institute for Information Technology; BIOIRC Bioengineering Research and Development Center, Kragujevac, Serbia

Frank S. Henry Department of Mechanical Engineering, Manhattan College, Riverdale, NY, United States

Dalibor D. Nikolic Bioengineering Research and Development Center (BioIRC), Kragujevac, Serbia

Milica Nikolic Steinbeis Advanced Risk Technologies Institute doo Kragujevac, SARTIK; Institute for Information Technologies Kragujevac, Kragujevac, Serbia

Igor Saveljic Bioengineering Research and Development Center (BioIRC), Kragujevac, Serbia

Tijana Šušteršič University of Kragujevac, Faculty of Engineering; Bioengineering Research and Development Center (BioIRC), Kragujevac, Serbia

Akira Tsuda Tsuda Lung Research, Shrewsbury, MA, United States

Aleksandra Vulović Bioengineering Research and Development Center (BioIRC), Kragujevac, Serbia

Marko N Živanović Department of Science, University of Kragujevac, Institute of Information Technologies Kragujevac, Kragujevac, Serbia

Preface

This book is intended for pregraduate and postgraduate students as well as for researchers in the domains of bioengineering, biomechanics, biomedical engineering, and medicine.

The book can be useful for researchers in various fields related to bioengineering and other scientific fields, including medical applications. It provides basic information about how a bioengineering (or medical) problem can be modeled, which computational model can be used, and what is the background of the applied computer models.

Different examples in the area of cardiovascular and respiratory systems give readers an overview of typical problems that can be modeled, followed by a complete theoretical background with numerical method behind.

There are different examples of application: electromechanical ventricle modeling, carotid artery disease, stent mechanical testing, ECG simulation for cardiac disease, aorta stenosis, lung tissue, particle deposition, pulmonary acinus, respiratory airways, drug efficacy, tissue engineering, electrospinning, AI, COVID-19, and economic analysis of in silico clinical trials.

This book will not only be useful for lecturers of bioengineering courses at universities, but it will also be very helpful for researchers, medical doctors, and clinical researchers.

In Chapter 1, computational modeling of electromechanical coupling of the left ventricle is presented. Chapter 2 introduces a deep learning approach in the stratification of patients with carotid artery disease. Simulation of stent mechanical testing is described in Chapter 3. Chapter 4 presents a basic numerical and experimental approach for ECG simulation of cardiac hypertrophic conditions. The simulation of carotid artery plaque development and treatment is presented in Chapter 5. Myocardial work and aorta stenosis simulation are described in Chapter 6. Chapter 7 gives numerical and experimental examples for lab-on-a-chip for lung tissue. Chapter 8 provides a review of the chaotic mixing and its role in enhancing particle deposition in the pulmonary acinus. Three-dimensional reconstruction and modeling of the respiratory airways, particle deposition, and drug delivery efficacy are introduced in Chapter 9. In Chapter 10, the basic tissue engineering—electrospinning approach is given. Chapter 11 focuses on the application of numerical methods for the analysis of the respiratory system. Artificial intelligence approach toward analysis of COVID-19 development—personalized and epidemiological model is presented in Chapter 12. Finally, economic analysis of in silico clinical trials is given in Chapter 13.

Nenad Filipovic
Faculty of Engineering, University of Kragujevac, Serbia

CHAPTER

1

Computational modeling of electromechanical coupling of left ventricle

Nenad Filipovic

Bioengineering Research and Development Center (BioIRC), Kragujevac, Serbia

1 Introduction

It is very important to use detailed and complex model, with high-resolution anatomically accurate model of whole heart electrical activity, which requires extensive computation times, dedicated software, and even the use of supercomputers (Gibbons Kroeker et al., 2006; Pullan et al., 2005; Trudel et al., 2004).

We recently developed methodology for a real 3D heart model, using a linear elastic and orthotropic material model based on Holzapfel experiments. Using this methodology, we can accurately predict the transport of electrical signals and the displacement field within the heart tissue (Kojic, Milosevic, Simic, Milicevic, et al., 2019).

Muscles in the body are activated by electrical signals transmitted from the nervous system to muscle cells, thus affecting the change of the cell membrane potentials. Additionally, calcium current and concentration inside the cell are the main causes of generating active stress within muscle fibers.

In order to simulate an electrical model of the heart, it requires an adequate model of the transfer between cardiac electrical activity and the ECG signals measured on the torso surface. Also, it requires solving a mathematically inverse problem for which no unique solution exists. Clinical validation in humans is very limited since simultaneous whole heart electrical distribution recordings are inaccessible for both practical and ethical reasons (Trudel et al., 2004).

https://doi.org/10.1016/B978-0-12-823956-8.00009-2

The rapid development of information technologies, simulation software packages, and medical devices in recent years provides the opportunity for collecting a large amount of clinical information. Creating comprehensive and detailed computational tools has become essential to process specific information from the abundance of available data. From the point of view of physicians, it becomes of paramount importance to distinguish "normal" phenotypes from the appearance of the phenotype in a specific patient in order to estimate its disease progression, therapeutic responses, and future risks. Recent computational models have significantly improved integrative understanding of the heart muscles behavior in HCM (hypertrophic) and DCM (diluted) cardiomyopathies. The development of novel integrative modeling approaches could be an effective tool in distinguishing the type and severity of symptoms in, for example, multigenic disorder patients and assessing the degree to which normal physical activity is impaired.

On the other hand, patient-specific modeling presents many new challenges, including (1) the lack of details regarding physical and biological properties of the human heart; (2) the need for a subject-specific estimation of parameters from limited, noisy data, typically obtained using noninvasive measurements; (3) the need to perform numerous large-scale computations in a clinically useful time frame; and (4) the need to store and share model metadata that can be reused without compromising patient confidentiality. Despite the difficulties, multiscale models of the heart can include a level of detail sufficient to achieve predictions that closely follow observed transient responses providing solid evidence for prospective clinical applications.

However, regardless of the substantial scientific effort by multiple research labs and significant amount of grant support, currently there is only one commercially available software package regarding multiscale and whole heart simulations called the SIMULIA Living heart model (Baillargeon et al., 2014). It includes dynamic, electromechanical simulation, refined heart geometry, a blood flow model, and a complete characterization of cardiac tissues including passive and active characteristics, its fibrous nature, and the electrical pathways. This model is targeted for use in personalized medicine, but active material characterization is based on a phenomenological model introduced by Guccione et al. (Guccione et al., 1993; Guccione & McCulloch, 1993). Therefore, SIMULIA cannot directly and accurately translate the changes in the contractile protein functional characteristics observed in numerous cardiac diseases. These changes are caused by mutations and other abnormalities at the molecular and subcellular level. The limited use of SIMULIA software in a small number of applications in clinical practice is a great example of today's struggles in developing higher-level multiscale human heart models. On the other hand, it is a motivation for developing a new generation of multiscale program packages that can trace the effects of mutations from the molecular to organ scale.

In silico clinical trials are a new paradigm for the development and testing of a new drug and medical device. The SILICOFCM project (H2020 project SILICOFCM, 2018–2022) is the multiscale modeling of familial cardiomyopathy which considers a comprehensive list of patient-specific features as genetic, biological, pharmacologic, clinical, imaging, and cellular aspects. The biomechanics of the heart is a key part of the in silico clinical platform. We have built this platform using state-of-the-art finite element modeling for macro-simulation of the fluid-structure interaction with micro-modeling on the molecular level for drug interaction with the cardiac cells.

2 Method

2.1 Fluid-solid coupling

The blood is considered as an incompressible homogenous viscous fluid. The fundamental laws of physics which include balance of mass and balance of linear momentum are applicable here. These laws are expressed by the continuity equation and the Navier-Stokes equations (Kojic, Milosevic, Simic, Geroski, et al., 2019).

We present here the final form of these equations to emphasize some specifics related to blood flow. The incremental-iterative balance equation of a finite element for a time step 'n' and equilibrium iteration 'i' has a form

$$\begin{bmatrix} \frac{1}{\Delta t}\mathbf{M} + {}^{n+1}\widetilde{\mathbf{K}}_{vv}^{(i-1)} & \mathbf{K}_{vp} \\ \mathbf{K}_{vp}^{T} & \mathbf{0} \end{bmatrix} \begin{Bmatrix} \Delta\mathbf{V}^{(i)} \\ \Delta\mathbf{P}^{(i)} \end{Bmatrix}_{blood} = \begin{Bmatrix} {}^{n+1}\mathbf{F}_{ext}^{(i-1)} \\ 0 \end{Bmatrix} - \begin{bmatrix} \frac{1}{\Delta t}\mathbf{M} + {}^{n+1}\mathbf{K}^{(i-1)} & \mathbf{K}_{vp} \\ \mathbf{K}_{vp}^{T} & 0 \end{bmatrix} \begin{Bmatrix} {}^{n+1}\mathbf{V}^{(i-1)} \\ {}^{n+1}\mathbf{P}^{(i-1)} \end{Bmatrix} + \begin{Bmatrix} \frac{1}{\Delta t}\mathbf{M}^{n}\mathbf{V} \\ 0 \end{Bmatrix} \tag{1}$$

where ${}^{n+1}\mathbf{V}^{(i-1)}$ ${}^{n+1}\mathbf{P}^{(i-1)}$ are the nodal vectors of blood velocity and pressure, with the increments in time step $\Delta\mathbf{V}^{(i)}$ and $\Delta\mathbf{P}^{(i)}$ (the index 'blood' is used to emphasize that we are considering blood as the fluid); Δt is the time step size and the left upper indices 'n' and '$n+1$' denote the start and end of time step. Note that the vector ${}^{n+1}\mathbf{F}_{ext}^{(i-1)}$ of external forces includes the volumetric and surface forces. In the assembling of these equations, the system of equations of the form (1) is obtained, with the volumetric external forces and the surface forces acting only on the fluid domain boundary (the surface forces among the internal element boundaries cancel).

The solid domain for the left ventricle was defined with a nonlinear finite element equation taking into account the Holzaphel and Hunter model (Kojic, Milosevic, Simic, Geroski, et al., 2019). The balance of linear momentum is derived from the fundamental differential equations of balance of forces acting at an elementary material volume. In dynamic analysis, we include the inertial forces. Then by applying the principle of virtual work

$$\mathbf{M}\ddot{\mathbf{U}} + \mathbf{B}^{w}\dot{\mathbf{U}} + \mathbf{K}\mathbf{U} = \mathbf{F}^{ext} \tag{2}$$

Here the element matrices are $\mathbf{M}$ is the mass matrix; $\mathbf{B}^{w}$ is the damping matrix in cases when the material has a viscous resistance; $\mathbf{K}$ is the stiffness matrix; and $\mathbf{F}^{ext}$ is the external nodal force vector which includes body and surface forces acting on the element.

About half of the cardiomyopathies are caused by genetic malformations with mutations in sarcomeric proteins (Vikhorev & Vikhoreva, 2018). In addition to significant changes at the level of molecular mechanisms within cardiomyocytes, significant changes are also observed at the macroscopic level in terms of changes in blood pressure, the left ventricular mass index, wall thickness, left ventricular diameter, left ventricular volume, fractional shortening, and ejection fraction. A change in these parameters induces many other physiologically important features and finally on status of suffering patients. Many drugs are created to counteract these changes by reducing the wall thickness, increasing the left ventricular volume, or increasing the ejection fraction.

2.2 Nonlinear material model of the left ventricle

In the heart cycle, we have two repeated regimes: systole and diastole. In the systole regime, the left ventricle (LV) contract and pump blood to the arterial system and the right ventricle (RV) contract and pump blood to the lung. At the same time, the atria expand while the blood comes from the veins to the right atrium and from the lung to the left atrium. In the diastole regime, the atria contract and blood flows from the atria into the ventricles which expand. Expansion is a consequence of the blood loading which enters the ventricles. At the end of diastole, there is maximum deformation and maximum passive stress within the tissue of ventricles. The active stress is generated during systole within the muscle cells and, together with the passive stress, provides the mechanical forces to overcome the resistance to blood flow to the arterial system (from the LV) and to the lung (from the RV). It can be observed that the resistance to blood flow to the arterial system is higher than to the lung. This directly indicates that overall loading is higher on the walls of the LV. That is a reason why it is very important to study processes in the LV. Heart dysfunction and heart failure are often related to the tissue of the left ventricle. Mechanical characteristics of the LV tissue represent one of the fundamental components of heart behavior, and have been under investigation over centuries.

The structural morphology of the left ventricle tissue is very complex. The microstructural composition has been intensively studied and it has been described in the medical and engineering literature (e.g., Sommer et al., 2015a; Holzapfel & Ogden, 2009; McEvoy, Holzapfel, & McGarry, 2018). The outline of the specificities which are most important for our computational modeling has been presented here. In Fig. 1 is shown a schematic representation of the ventricle according to Bovendeerd et al. (1992), indicating that the ventricle can geometrically be approximated by a thick shell. The wall is composed of layers (or sheets) of parallel muscle cells (myocytes) which have a fibrous character and occupy about 70% of the volume.

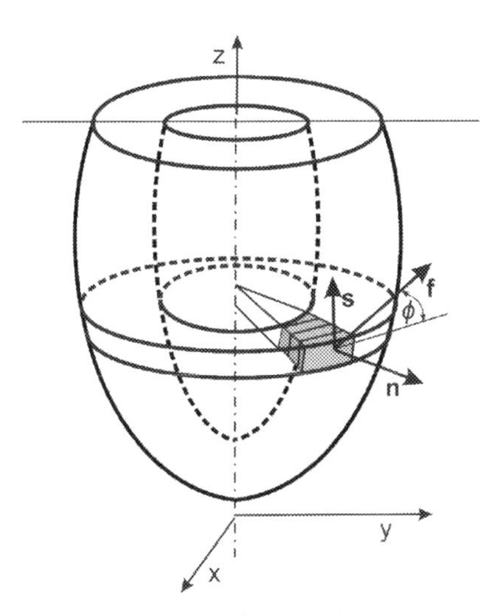

FIG. 1 Schematic representation of geometry of the left ventricle with material element and local unit vectors of fiber, sheet, and normal directions.

The remaining 30% consist of various interstitial components, where 2%–5% represent the collagen network for lateral connections.

Long muscle cells form fibers where the muscle activation occurs producing active stress along directions of fibers. The material can be considered orthotropic with the orthotropic unit vectors **f** and **s** for the fiber and sheet directions, respectively, lying in the tangential plane of the sheet surface. The third direction is defined by the unit vector **n** normal to the sheet plane, as shown in Fig. 1. As stated above, the fibers have a helicoidal character with the angle ϕ changing over the wall thickness.

2.2.1 Biaxial tests

The mechanical behavior of the ventricle as a thick shell structure is mainly characterized by the mechanical properties of the sheet layers. Due to these circumstances, experimental and theoretical investigations are focused on the kinematics of deformation and constitutive laws of the sheet layers, as outlined in the Introduction. Mechanical investigations are usually performed as biaxial loading and shear on a sample in a sheet plane. In Fig. 2 are shown the average constitutive curves obtained by using samples of 26 human ventricles subjected to biaxial loading in the sheet plane.

Tests were performed by loading and unloading into fiber (MFD—according to the notation in Sommer et al. (2015b)) and sheet (CFD) directions up to three levels of maximum stretch (1.05, 1.075,1.1) and maintaining the constant ratios between strain $e_2 = \lambda_2 - 1$ in the sheet direction and fiber strain $e_1 = \lambda_1 - 1$; here λ_2 and λ_1 are stretches. It can be seen that the constitutive curves are highly nonlinear, with hyperelastic characteristics usual for biological materials. The stress-stretch relationship depends on the stretch level to which the material is stretched before unloading and on the stretch ratio. The material displays a hysteretic character, with hysteresis and therefore the dissipation energy per unit volume is more pronounced at higher level of stretch. In Fig. 3 are shown average constitutive curves with the curves obtained using mean values of the loading and unloading paths. They can be used

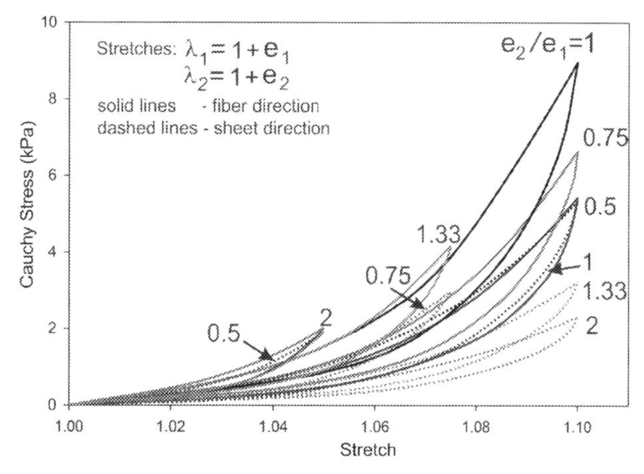

FIG. 2 Average constitutive curves (26 samples) for human left ventricle tissue subjected to biaxial loading for various ratios of sheet strain e_2 and fiber strain e_1, according to Sommer et al. (2015b).

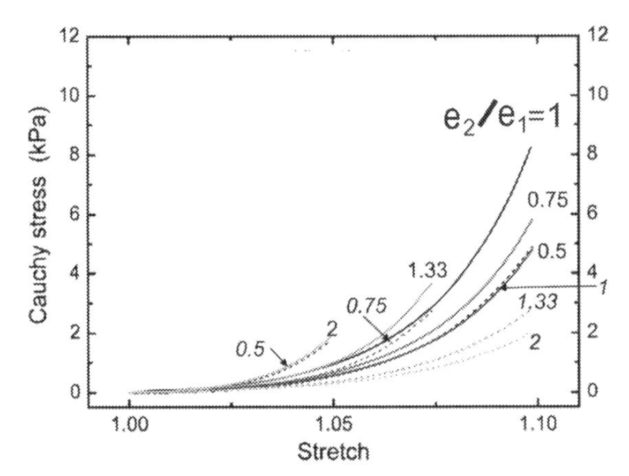

FIG. 3 Average constitutive curves for the human left ventricle tissue subjected to biaxial loading, with mean values of the loading and unloading, and for several ratios of sheet strain e_2 and fiber strain e_1, according to Sommer et al. (2015b).

in our computational procedure and also in applications of the analytical forms of the constitutive laws.

Here we add constitutive curves for the left ventricle, according to Stevens et al. (2003). These curves include the stress-strain relationship for the sheet-normal direction which will be used in our computational model (Fig. 4).

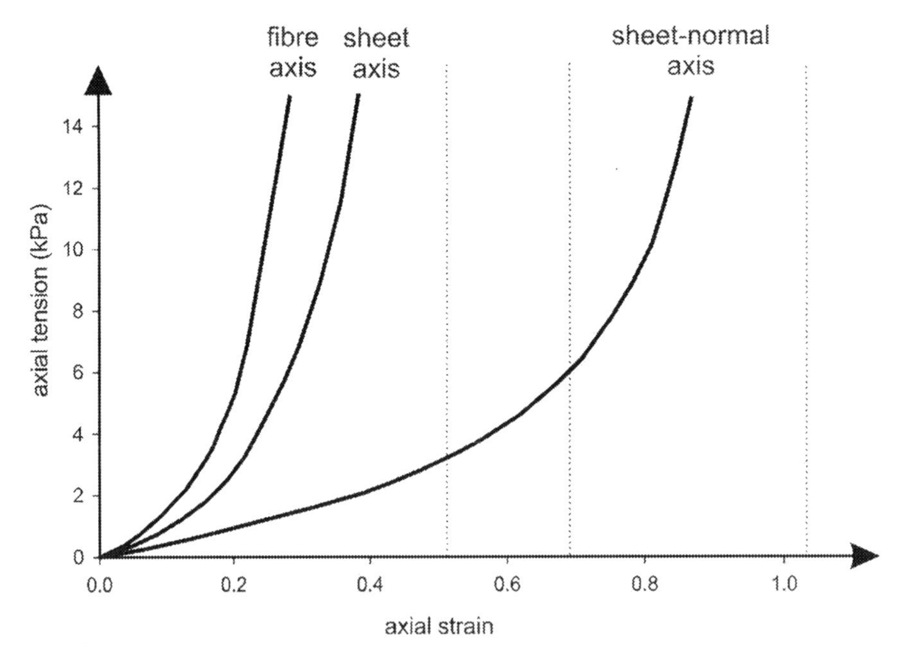

FIG. 4 Constitutive curves obtained by uniaxial tension in the three material directions of the left ventricle tissue, according to Stevens et al. (2003).

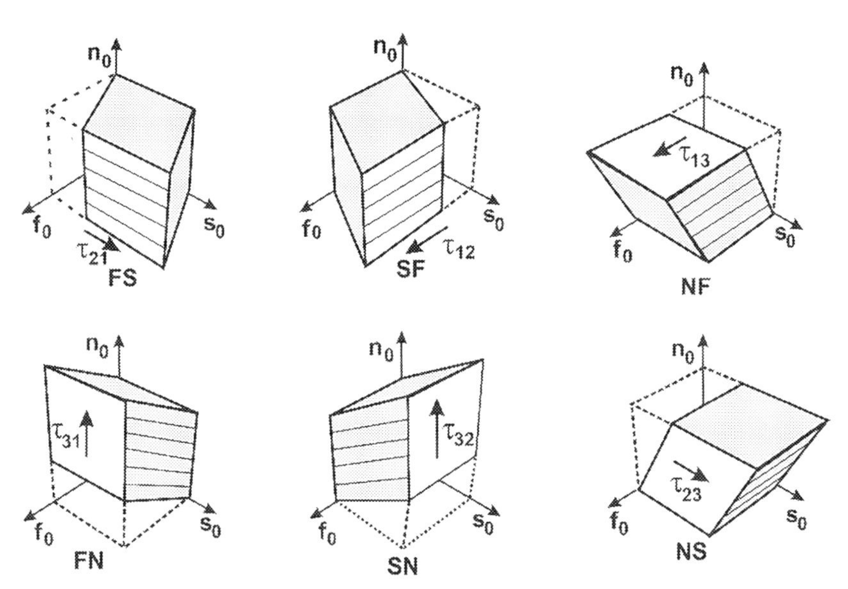

FIG. 5 Shear modes used in triaxial shear tests (H2020 project SILICOFCM, 2018–2022).

2.2.2 Shear tests

In Sommer et al. (2015b) are reported results of triaxial shear tests. Six modes of shear deformation are generated in planes corresponding to the f-s-n material coordinate system in the undeformed material, with initial unit vectors $\mathbf{f}_0$, $\mathbf{s}_0$, $\mathbf{n}_0$ as shown in Fig. 5. Shear stresses corresponding to each mode are expressed in terms of the 'amount of shear' $\Delta L/L$ where ΔL is the displacement in direction of the shear stress and L is the sample dimension. This amount of shear represents a part of the usually used engineering shear strain of a continuum. The specimen was loaded in cycles, where loading is increased and then decreased, further followed by loading in the opposite direction. Also, tests were performed with different load levels. Results are shown in Fig. 6 displaying the anisotropic and hysteretic character of the material under shear. The shear stresses are the largest on planes with normal **f** (FS and FN) and smallest on planes with normal **n** (NF and NS).

2.3 Electrophysiology of the left ventricle

Cardiac cells are filled and surrounded by ionic solution, mostly sodium $Na+$, potassium $K+$, and calcium Ca^{2+}. These charged atoms move between the inside and the outside of the cell through proteins called ion channels. Cells are connected through gap junctions which form channels that allow ions to flow from one cell to another.

An accurate numerical model is needed for a better understanding of heart behavior in cardiomyopathy, heart failure, cardiac arrhythmia, and other heart diseases. These numerical models usually include drug transport, electrophysiology, and muscle mechanics (Fitzhugh, 1961).

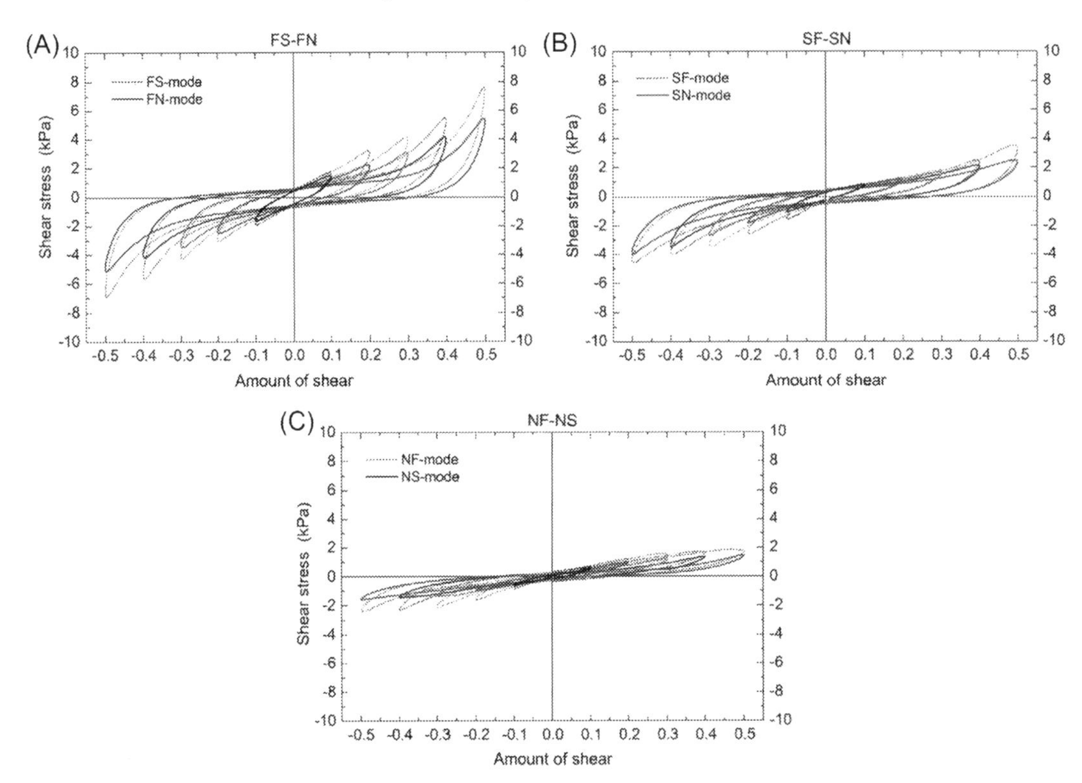

FIG. 6 Shear constitutive curves, relation between shear stress and amount of shear, for the six planes: (A) FS-FN, (B) SF-SN, and (C) NF-NS according to H2020 project SILICOFCM (2018–2022).

We have presented the heart geometry and seven different regions of the model where we included the: (1) sinoatrial node, (2) atria, (3) atrioventricular node, (4) His bundle, (5) bundle fibers, (6) Purkinje fibers, and (7) ventricular myocardium (Fig. 7).

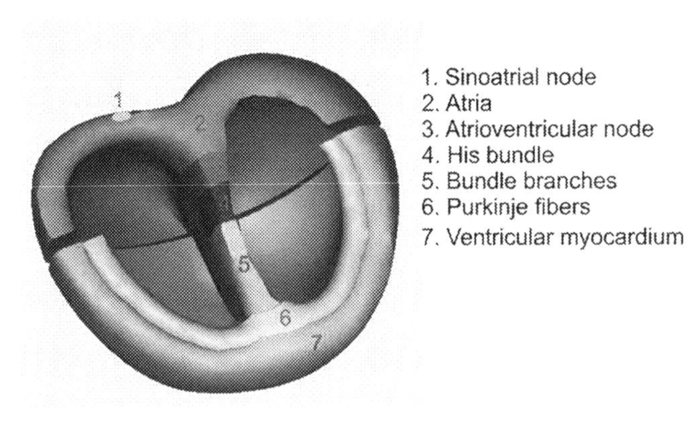

FIG. 7 Heart geometry and seven different regions of the model.

In this study, we used the monodomain model of the modified FitzHugh-Nagumo model of the cardiac cell (Nagumo et al., 1962; Sovilj et al., 2013; Wang & Rudy, 2006).

$$\frac{dV}{dt} = -kc_1(V_m - B)\left(-\frac{V_m - B}{A} + a\right)\left(-\frac{V_m - B}{A} + 1\right) - kc_2R(V_m - B)$$
$$\frac{dR}{dt} = -ke\left(\frac{(V_m - B)}{A} - R\right) \tag{3}$$

where V_m is the membrane potential, R is the recovery variable, a is relating to the excitation threshold, e is relating to the excitability, A is the action potential amplitude, B is the resting membrane potential, and c_1, c_2, and k are the membrane-specific parameters.

The monodomain model (Sovilj et al., 2013; Wang & Rudy, 2006) with incorporated modified FitzHugh-Nagumo equations is

$$\frac{\partial V_m}{\partial t} = \frac{1}{\beta C_m}(\nabla.(\sigma \nabla V_m) - \beta(I_{ion} - I_s)), D = \frac{\sigma}{\beta C_m} \tag{4}$$

where β is the membrane surface-to-volume ratio, C_m is the membrane capacitance per unit area, σ is the tissue conductivity, I_{ion} is the ionic transmembrane current density per unit area, and I_s is the stimulation current density per unit area.

Parameters for the monodomain model with modified FitzHugh-Nahumo equations are presented in Table 1.

The 12-Lead ECG became a standard in clinical practice since the American Heart Association published its recommendation in 1954. It consists in recording signals from 10 electrodes respecting the following placement (Fig. 8):

- V1: 4th intercostal space to the right of the sternum;
- V2: 4th intercostal space to the left of the sternum;
- V3: midway between V2 and V4;

TABLE 1 Parameters for monodomain model with modified FitzHugh-Nahumo equations.

Parameter	SAN	Atria	AVN	His	BNL	Purkinje	Ventricles
a	−0.60	0.13	0.13	0.13	0.13	0.13	0.13
b	−0.30	0	0	0	0	0	0
c_1 ($\mathrm{As\,V^{-1}\,m^{-3}}$)	1000	2.6	2.6	2.6	2.6	2.6	2.6
c_2 ($\mathrm{As\,V^{-1}\,m^{-3}}$)	1.0	1.0	1.0	1.0	1.0	1.0	1.0
D	0	1	1	1	1	1	1
e	0.066	0.0132	0.0132	0.005	0.0022	0.0047	0.006
A (mV)	33	140	140	140	140	140	140
B (mV)	−22	−85	−85	−85	−85	−85	−85
k	1000	1000	1000	1000	1000	1000	1000
σ ($\mathrm{mS\,m^{-1}}$)	0.5	8	0.5	10	15	35	8

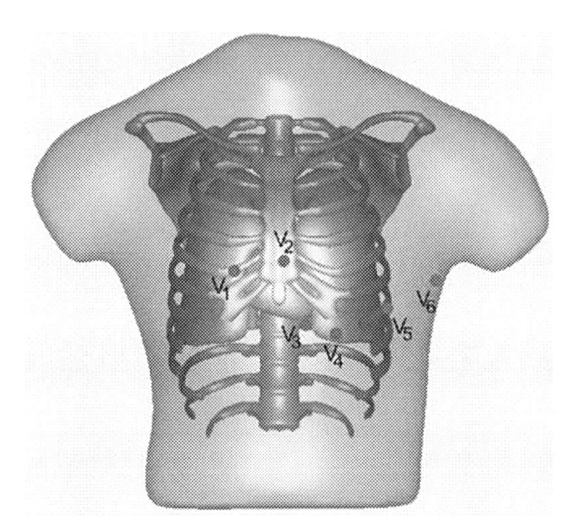

FIG. 8 Six electrodes (V1–V6) which are positioned at the chest to model the precordial leads.

- V4: 5th intercostal space at the midclavicular line;
- V5: anterior axillary line at the same level as V4;
- V6: midaxillary line at the same level as V4 and V5;

Computer simulations were conducted using the fully coupled heart torso monodomain equations including a detailed description of human ventricular cellular electrophysiology. Myocardial and torso conductivities were based on the literature, as presented in Table 1.

Boundary conditions on all interior boundaries in contact with the torso, lungs, and cardiac cavities are zero flux for Vm; therefore, $-\mathbf{n} \cdot \mathbf{\Gamma} = 0$ where $\mathbf{n}$ is the unit outward normal vector on the boundary and $\mathbf{\Gamma}$ is the flux vector through that boundary for the intracellular voltage, equal to $\mathbf{\Gamma} = -\sigma \cdot \partial Vm / \partial \mathbf{n}$. For the variable Vm, the inward flux on these boundaries is equal to the outward current density $\mathbf{J}$ from the torso/chamber volume conductor; therefore, $-\sigma\, \partial Vm / \partial \mathbf{n} = \mathbf{n} \cdot \mathbf{J}$.

In the second part, we implement the performance of classical approaches for solving the ECG inverse problem using the epicardial potential formulation. The studied methods are the family of Tikhonov methods and the L regularization-based methods (Van Oosterom, 1999, 2001, 2003; Wang & Rudy, 2006).

ECG measurement was performed on the healthy volunteer in the Clinical Center Kragujevac, University of Kragujevac.

3 Results

The whole heart activation simulations from the lead II ECG signal at various time points on the ECG signal for patients #1 and #2 have been presented in Figs. 9 and 10. A comparison of the simulated ECG on the surface body with real ECG measurement at V1 for patients #1 and #2 has been presented in Figs. 11 and 12.

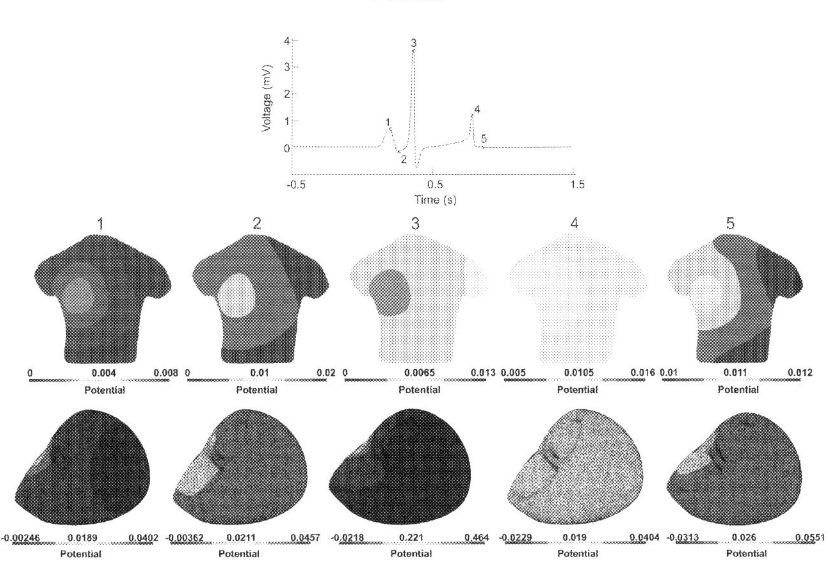

FIG. 9 Patient #1: Whole heart activation simulation from the lead II ECG signal at various time points on the ECG signal. There are 1–5 activation sequences corresponding to the ECG signal above. The color bar denotes the mV of the transmembrane potential.

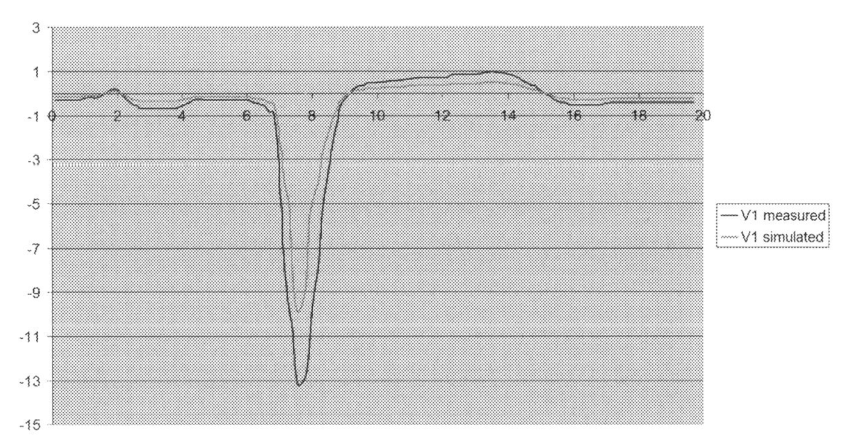

FIG. 10 Patient #2: Whole heart activation simulation from the lead II ECG signal at various time points on the ECG signal. There are 1–5 activation sequences corresponding to the ECG signal above. The color bar denotes the mV of the transmembrane potential.

The P—V diagram plots volume along the *X*-axis and pressure on the *Y*-axis. The area of the loop is equal to the stroke volume, which refers to the amount of blood pumped out of the left ventricle in one cardiac cycle. The effects of isolated changes in the preload are best demonstrated on the pressure-volume (P—V) diagram, which relates ventricular volume to the pressure inside the ventricle throughout the cardiac cycle. The maximum right point on the diagram is denoted as the end-diastolic volume (EDV), while the minimum left point is denoted as the end-systolic volume (ESV). Also, as EDV increases, the proportion of blood

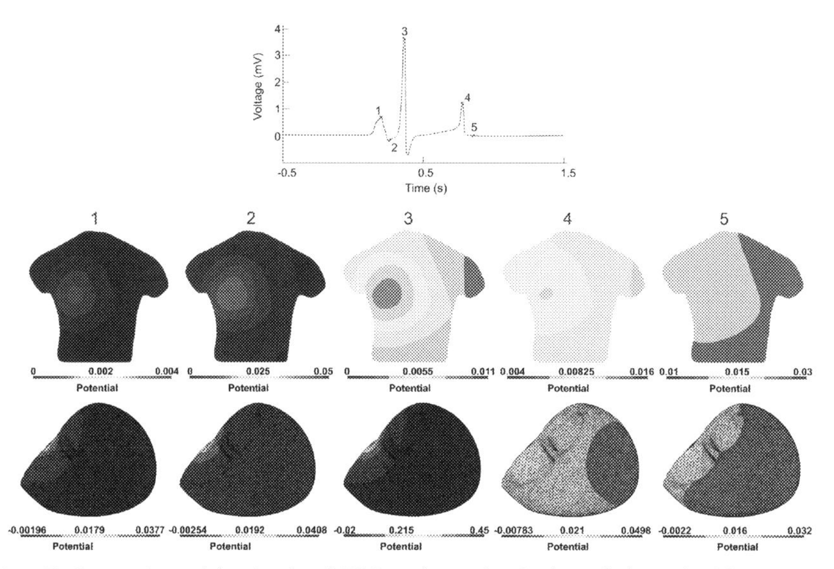

FIG. 11 Patient #1: Comparison of the simulated ECG on the surface body with the real ECG measurement at V1.

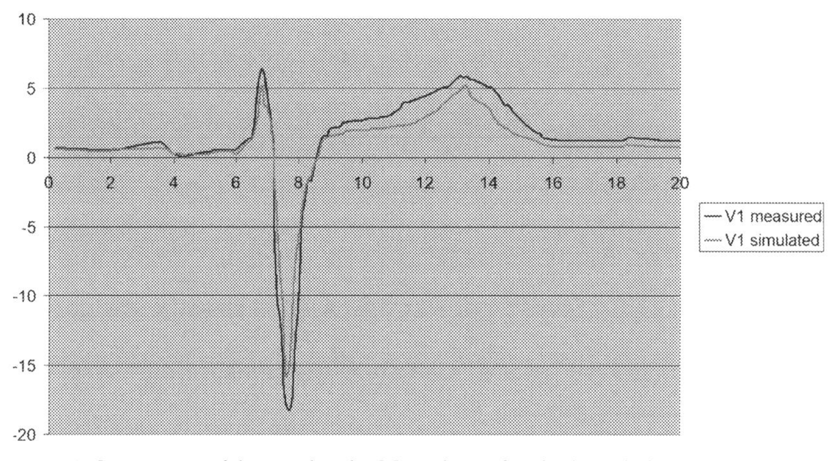

FIG. 12 Patient #2: Comparison of the simulated ECG on the surface body with the real ECG measurement at V1.

ejected by the heart increases slightly; this is the ejection fraction (EF) calculated by the equation: (EDV-ESV)/EDV. The reverse is also true. A decrease in the preload will result in a leftward shift down the end-diastolic P—V line, decreasing EDV, stroke volume, and causing a slight decrease in the ejection fraction (Villars et al., 2004).

A variety of commonly used medications affects the cardiac function. Some of the first-line treatments for heart failure, myocardial ischemia, and hypertension are described. Drugs that decrease the preload and have an influence on the cardiac PV diagrams are as follows (Sheth et al., 2015):

- Angiotensin-converting enzyme (ACE) inhibitors—interrupt the renin-angiotensin-aldosterone system (RAAS). RAAS is a complex system responsible for regulating the body's blood pressure. The kidneys release an enzyme called renin in response to the low blood volume, low salt (sodium) levels, or high potassium levels.
- Angiotensin receptor blockers (ARBs)—interrupt the RAAS.
- Nitrates—cause nitric oxide-induced vasodilation.
- Diuretics—promote the elimination of salt and water, resulting in a decreased overall intravascular volume.
- Calcium channel blockers—block calcium-induced vasoconstriction and decrease cardiac contractility.

The results obtained with a parametric model where PV diagrams depend on the change of Ca^{2+}, elasticity of the wall, and the inlet and outlet velocity profiles have been presented. It directly affects the ejection fraction.

The result of the PAK solver simulation (PAK, 2022) with different LV geometry and corresponding scenarios has been presented. The first part is related to the results obtained from the LV model with a 20% shorter base length. The second part is related to the results obtained from the LV model with a 50% longer base length and 50% thicker lateral wall. Both cases cover three scenarios: (i) influence of Ca^{2+} concentration, (ii) influence of the Holzapfel scale factor (elasticity), and (iii) influence of inlet and outlet velocities.

The presented approach with variation of LV geometry and simulations, which include the influence of different parameters on the PV diagrams, is directly interlinked with drug effects on heart function.

This work is in continuous progress and it includes the incorporation of different drugs that directly affect the cardiac PV diagrams and ejection fraction (e.g., angiotensin-converting enzyme inhibitors, angiotensin receptor blockers, nitrates, diuretics, calcium channel blockers).

3.1 Geometry of the left ventricle with a shorter base length

The results obtained from the left ventricle model with a 20% shorter base length and the geometry of this model are presented in Fig. 13.

3.1.1 Scenario1a: Influence of Ca^{2+} concentration

Triangular Ca^{2+} concentration function is shown in Fig. 14 along with the corresponding PV diagram. Parabolic Ca^{2+} concentration function and the corresponding PV diagram are shown in Fig. 15. Steep Ca^{2+} concentration function and the corresponding PV diagram are shown in Fig. 16. Shifted parabolic Ca^{2+} concentration function and the corresponding PV diagram are shown in Fig. 17. Parabolic wider Ca^{2+} concentration function and the corresponding PV diagram are shown in Fig.18.

The displacement field at 0.2, 0.5, and 0.6s is shown in Fig. 19. The velocity field at 0.2, 0.5, and 0.6s is shown in Fig. 20. The pressure field at 0.2, 0.5, and 0.6s is shown in Fig. 21.

The effect of different Ca^{2+} concentration functions on the ejection fraction is shown in Fig. 22.

FIG. 13 Geometry of the left ventricle model with a 20% shorter base.

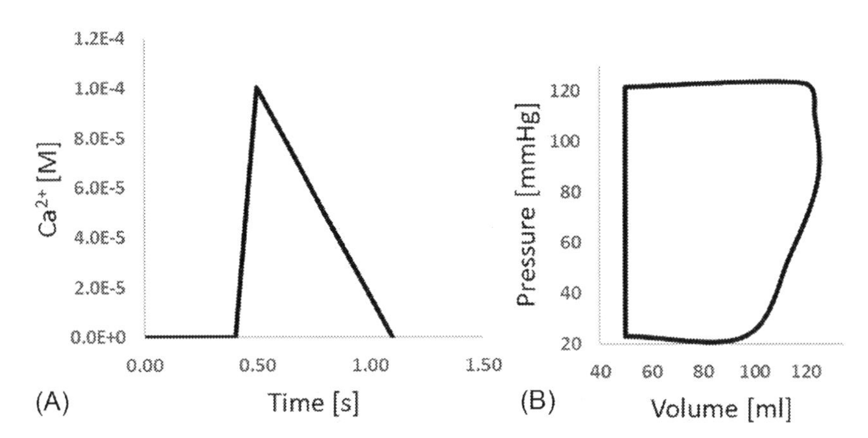

FIG. 14 Triangular Ca^{2+} concentration (A) and corresponding PV diagram (B).

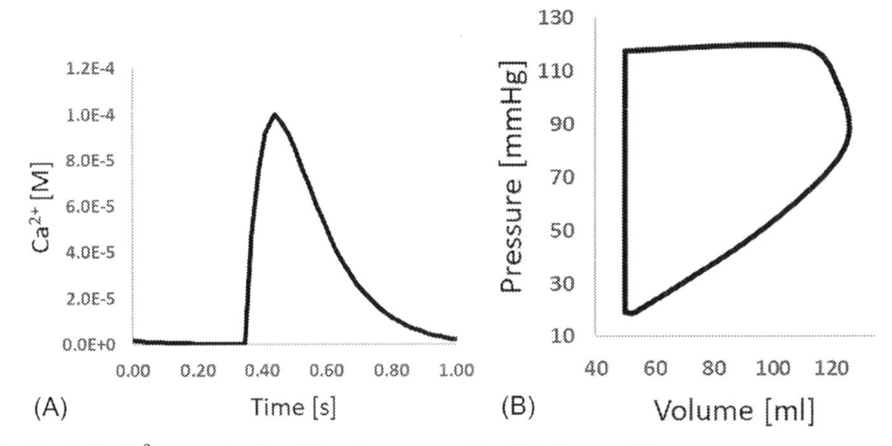

FIG. 15 Parabolic Ca^{2+} concentration (A) and corresponding PV diagram (B).

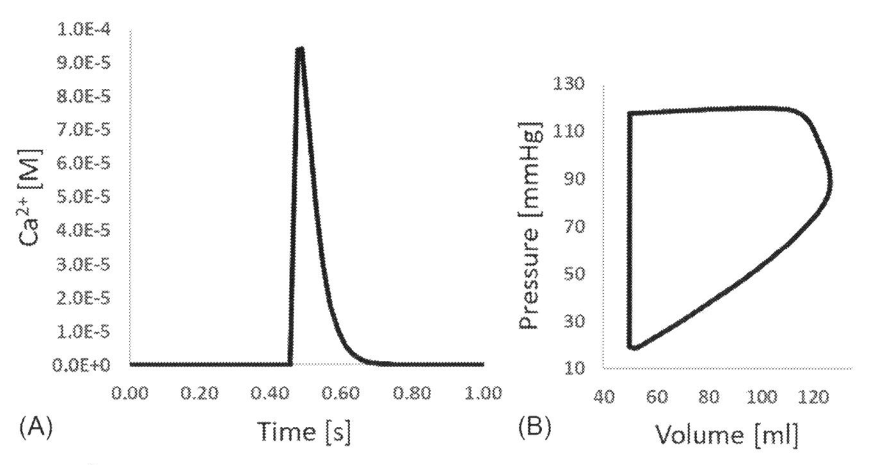

FIG. 16 Steep Ca^{2+} concentration (A) and corresponding PV diagram (B).

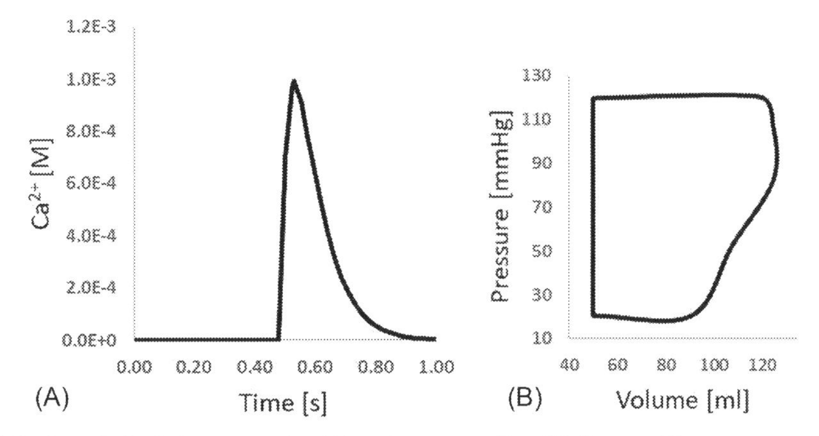

FIG. 17 Shifted parabolic Ca^{+} concentration (A) and corresponding PV diagram (B).

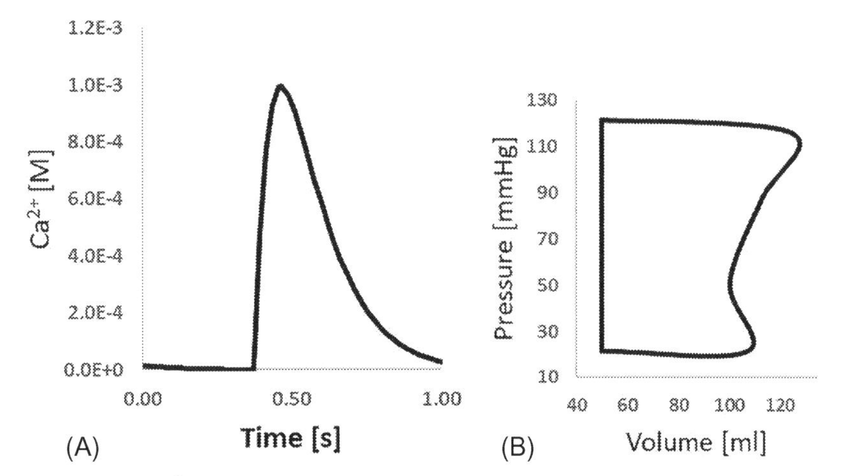

FIG. 18 Parabolic wider Ca^{2+} concentration (A) and corresponding PV diagram (B).

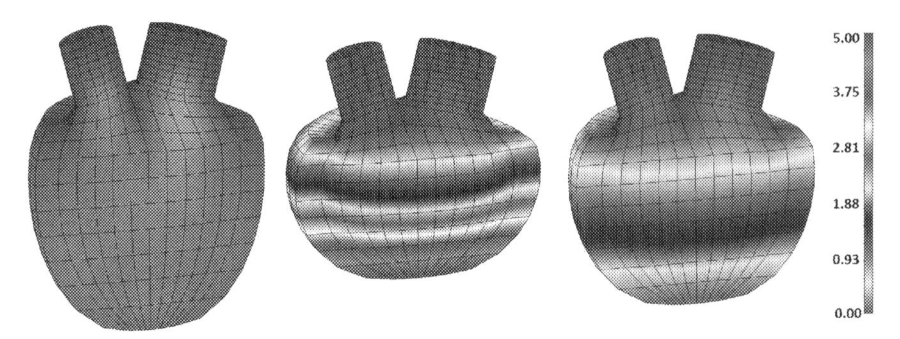

FIG. 19 Displacement field at 0.2, 0.5, and 0.6s for parabolic Ca^{2+} concentration function.

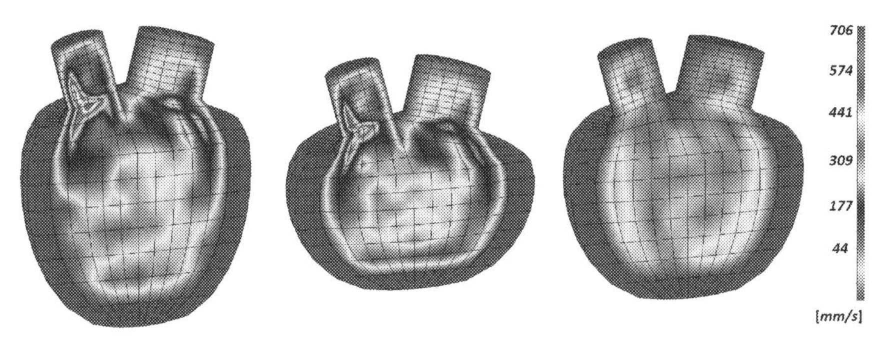

FIG. 20 Velocity field at 0.2, 0.5, and 0.6s for parabolic Ca^{2+} concentration function.

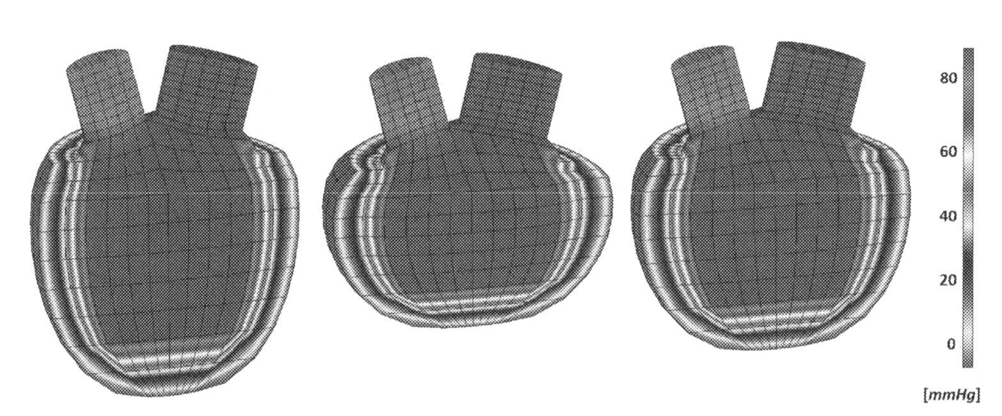

FIG. 21 Pressure field at 0.2, 0.5, and 0.6s for parabolic Ca^{2+} concentration function.

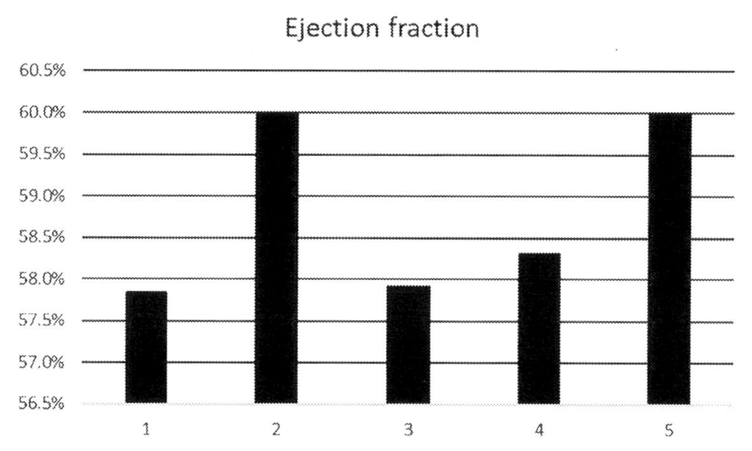

FIG. 22 Ejection fraction for (1) triangular, (2) parabolic, (3) steep, (4) shifted parabolic, and (5) parabolic wider Ca^{2+} concentration function.

3.1.2 *Scenario 2a: Influence of the Holzapfel scale factor (elasticity)*

In this section, we will examine the influence of elasticity on the cardiac cycle. PV diagrams for 20% higher and 20% lower elasticity are shown in Fig. 23. PV diagrams for 30% higher and 50% lower elasticity are shown in Fig. 24. The effect of elasticity on the ejection fraction is shown in Fig. 25.

3.1.3 *Scenario 3a: Influence of inlet and outlet velocities*

PV diagrams for 25% higher inlet velocities and 25% higher outlet velocities are shown in Fig. 26. V diagrams for 25% lower inlet velocity and 25% lower outlet velocity are shown in Fig. 27. In Fig. 28, we have shown the effect of inlet velocity on the ejection fraction. In Fig. 29, we have shown the effect of outlet velocity on the ejection fraction.

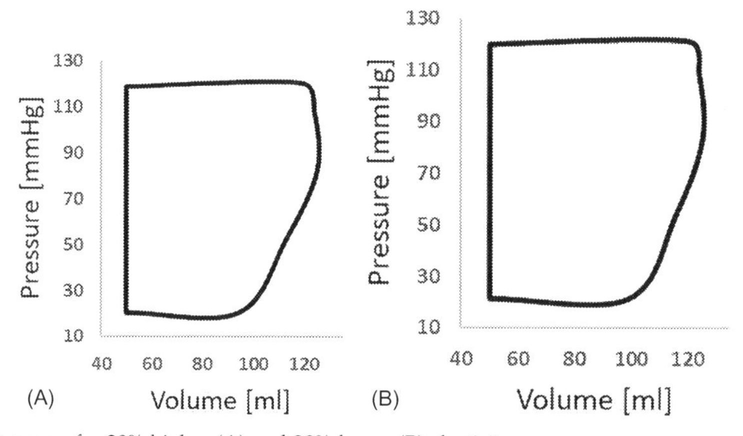

FIG. 23 PV diagrams for 20% higher (A) and 20% lower (B) elasticity.

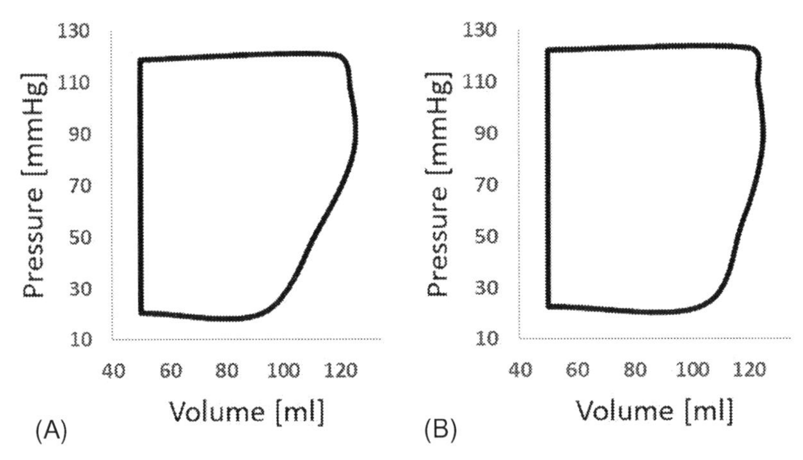

FIG. 24 PV diagrams for 30% higher (A) and 50% lower (B) elasticity.

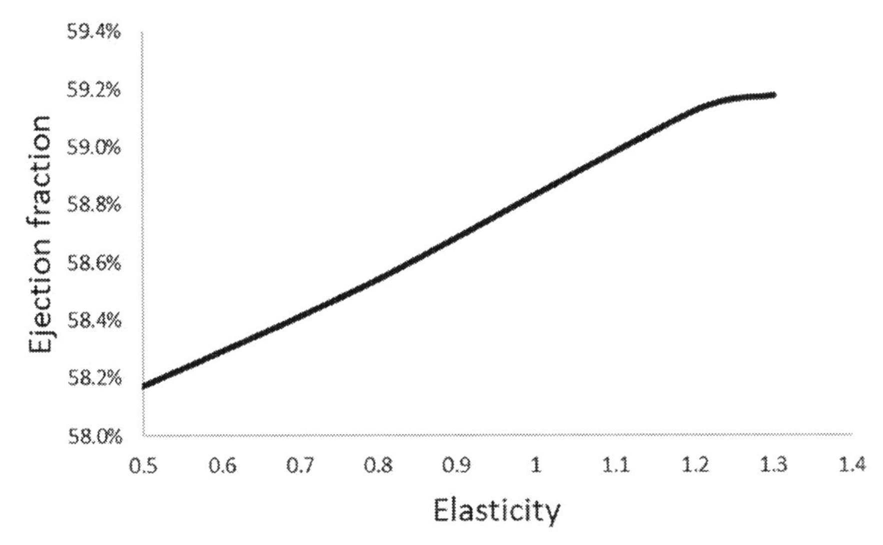

FIG. 25 Effect of elasticity on the ejection fraction.

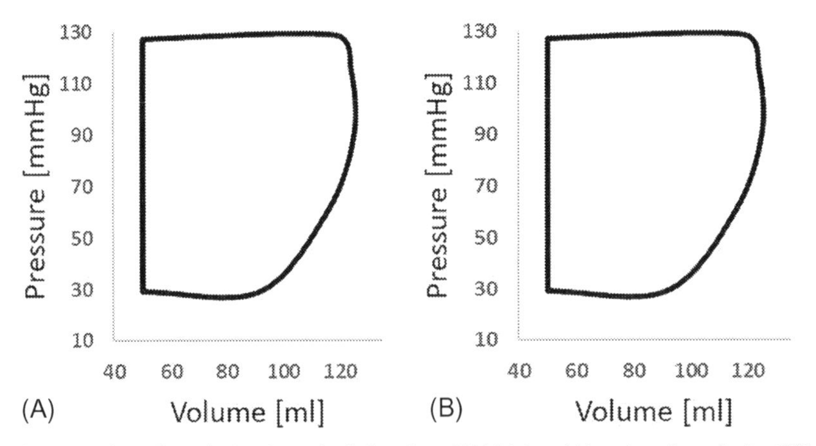

FIG. 26 PV diagrams for Inlet velocity from the left atrium 25% higher (A) and outlet velocity 25% higher (B).

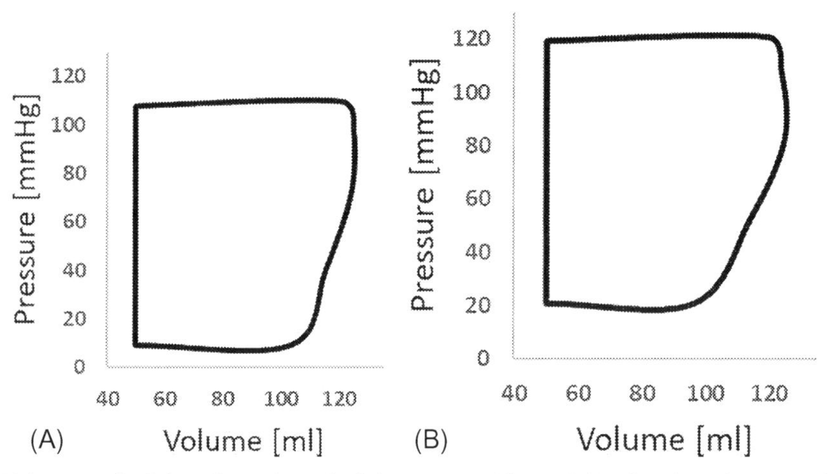

FIG. 27 PV diagrams for Inlet velocity from the left atrium 25% lower (A) and outlet velocity 25% lower (B).

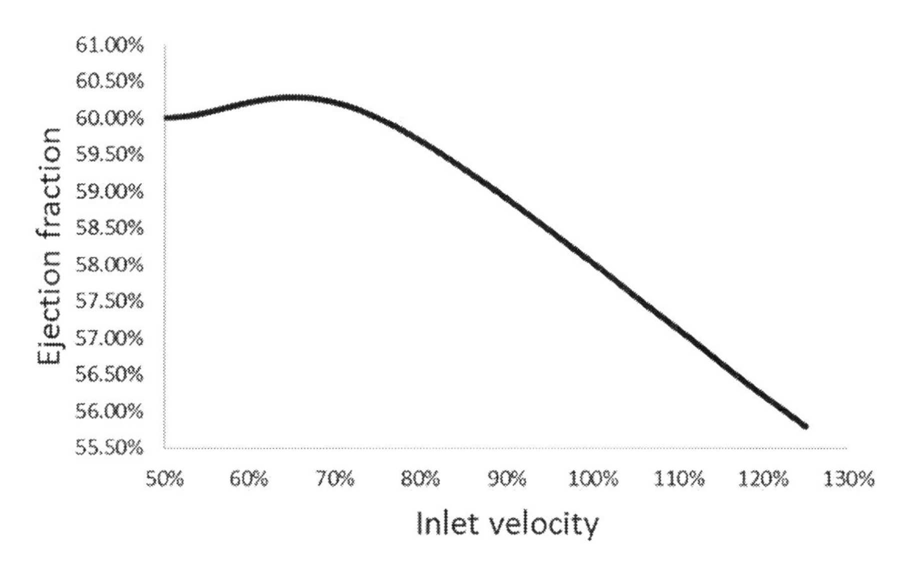

FIG. 28 Ejection fraction with respect to inlet velocity.

A realistic heart model has been produced using experimental data and DICOM files. The STL format with a left atrium (Fig. 30A, noted in blue) and chamber part (Fig. 30A, noted in yellow) with accompanying mitral valve cross section between (Fig. 20A, noted in green), and also aortic part (Fig. 20A, noted orange) of the model with aortic-cross section are included in the fluid part of the model, which is surrounded by a solid wall (Fig. 20A, wireframe). The finite element model consists of 139,896 hexahedral 3D elements, divided by 161,989 nodes. Model geometry is generated using STL files. Solid nodes are constrained around inlet/ outlet cross sections (Fig. 30A; red and magenta rings) and in the zone close to the mitral valve cross section. Other solid nodes are free. In Fig. 30C, 2 cross-section regions are marked to define the

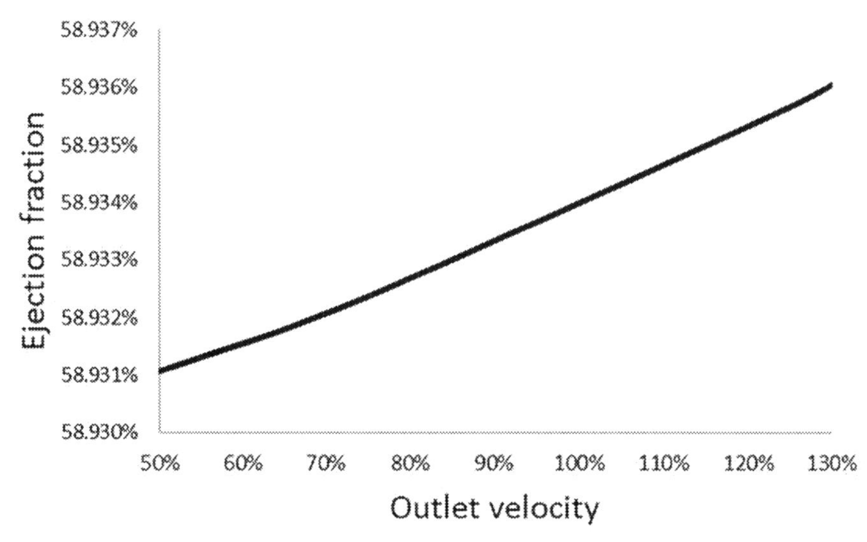

FIG. 29 Ejection fraction with respect to outlet velocity.

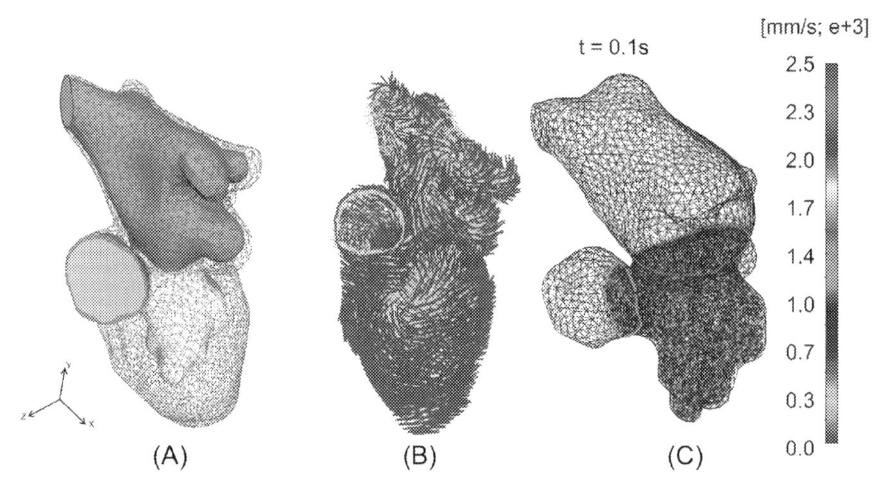

FIG. 30 (A) Realistic heart FE model with representative cross sections and fluid parts; (B) direction of fibers in the solid part of the realistic model; and (C) fluid velocity field at 0.1 s (mitral and aortic cross section noted).

prescribed inlet and outlet zones. Inside of the fluid domain, we have mitral valve cross section (part of the model between ventricle and atrium; Fig. 30C, red line) with inlet velocity function profile prescribed (Fig. 31A), and aortic valve cross section (part between ventricle and aortic branch, Fig. 30C, green line), with outlet velocity function profile prescribed (Fig. 31B). Fibers direction in the solid domain of realistic heart model are shown in Fig. 30B, and section C in the same figure shows the distribution of the velocity field in the realistic heart model at 0.1 s. It can be seen that velocity values are the highest at the inlet and outlet boundary cross-sections (red and green lines, Fig. 30C), which is logical due to the prescribed inlet

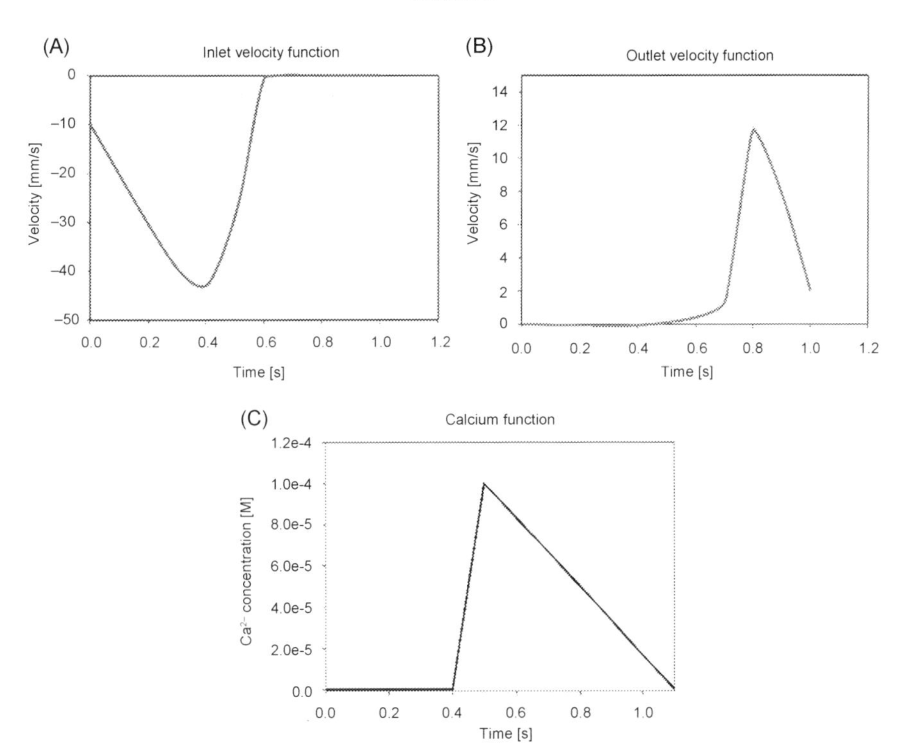

FIG. 31 (A) Inlet function of velocity at the mitral valve cross section; (B) outlet velocity function at the aortic valve cross section; and (C) Calcium concentration function used for activation of the muscle.

function and prescribed values at that cross section at the beginning of simulation. Regarding the material models used, we have selected the Holzapfel material model for obtaining passive stresses in the heart wall, and for muscle activation we used the Hunter material model for active stresses. Activation of the muscle is achieved using the calcium function, as displayed in Fig. 31C.

References

Baillargeon, B., Rebelo, N., Fox, D. D., Taylor, R. L., & Kuhl, E. (2014). The living heart project: A robust and integrative simulator for human heart function. *European Journal of Mechanics - A/Solids*, *48*, 38–47.

Bovendeerd, P. H. M., Arts, T., Huyghe, J. M., Van Campen, D. H., & Reneman, R. S. (1992). Dependence of local left ventricular wall mechanics on myocardial fiber orientation: a model study. *Journal of Biomechanics*, *2.5*(10), 1129–1140.

Fitzhugh, R. (1961). Impulses and physiological states in theoretical models of nerve membrane. *Biophysical Journal*, *1*, 445–466.

Gibbons Kroeker, C. A., Adeeb, S., Tyberg, J. V., & Shrive, N. G. (2006). A 2D FE model of the heart demonstrates the role of the pericardium in ventricular deformation. *American Journal of Physiology*, *291*(5), H2229–H2236.

Guccione, J. M., & McCulloch, A. D. (1993). Mechanics of active contraction in cardiac muscle: Part I—Constitutive relations for fiber stress that describe deactivation. *Journal of Biomechanical Engineering*, *115*, 72–81.

Guccione, J. M., Waldman, L. K., & McCulloch, A. D. (1993). Mechanics of active contraction in cardiac muscle: Part II—Cylindrical models of the systolic left ventricle. *Journal of Biomechanical Engineering*, *115*, 82–90.

H2020 project SILICOFCM. (2018–2022). *In silico trials for drug tracing the effects of sarcomeric protein mutations leading to familial cardiomyopathy* (p. 777204). www.silicofcm.eu.

Holzapfel, G. A., & Ogden, R. W. (2009). Constitutive modelling of passive myocardium: a structurally based framework for material characterization. *Philosophical Transactions of the Royal Society A, 367*, 3445–3475.

Kojic, M., Milosevic, M., Simic, V., Geroski, V., Ziemys, A., Filipovic, N., & Ferrari, M. (2019). Smeared multiscale finite element model for electrophysiology and ionic transport in biological tissue. *Computers in Biology and Medicine, 108*, 288–304. ISSN 0010-4825. https://doi.org/10.1016/j.compbiomed.2019.03.023.

Kojic, M., Milosevic, M., Simic, V., Milicevic, B., Geroski, V., Nizzero, S., Ziemys, A., Filipovic, N., & Ferrari, M. (2019). Smeared multiscale finite element models for mass transport and electrophysiology coupled to muscle mechanics. *Frontiers in Bioengineering and Biotechnology, 7*, 381. ISSN 2296–4185, pp. 1–16, 2296–4185.

McEvoy, E., Holzapfel, G. A., & McGarry, P. (2018). Compressibility and anisotropy of the ventricular myocardium: Experimental analysis and microstructural modeling. *Journal of Biomechanical Engineering, 140*(8), 140–148.

Nagumo, J., Arimoto, S., & Yoshizawa, S. (1962). An active pulse transmission line simulating nerve axon. *Proceedings of the IRE, 50*, 2061–2070.

PAK, 2022, Finite element program, Kragujevac, Serbia, http://www.bioirc.ac.rs/index.php/software/5-pak.

Pullan, A. J., Buist, M. L., & Cheng, L. K. (2005). *Mathematically modelling the electrical activity of the heart—From cell to body surface and back again.* World Scientific.

Sheth, P. J., Danton, G. H., Siegel, Y., Kardon, R. E., Infante, J. C., Jr., Ghersin, E., & Fishman, J. E. (2015). Cardiac physiology for radiologists: Review of relevant physiology for interpretation of cardiac MR imaging and CT. *Radiographics, 35*(5), 1335–1351.

Sommer, G., Schriefl, A. J., Andrä, M., Sacherer, M., Viertler, C., Wolinski, H., & Holzapfel, G. A. (2015a). Biomechanical properties and microstructure of human ventricular myocardium. *Acta Biomaterialia, 24*, 172–192.

Sommer, G., Schriefl, A. J., Andrä, M., Sacherer, M., Viertler, C., Wolinski, H., & Holzapfel, G. A. (2015b). Biomechanical properties and microstructure of human ventricular myocardium. *Acta Biomaterialia, 24*, 172–192.

Sovilj, S., Magjarević, R., Lovell, N., & Dokos, S. (2013). A simplified 3D model of whole heart electrical activity and 12-lead ECG generation. *Computational and Mathematical Methods in Medicine.* https://doi.org/10.1155/2013/134208.

Stevens, C., Remme, E., LeGrice, I., & Hunter, P. (2003). Ventricular mechanics in diastole: Material parameter sensitivity. *Journal of Biomechanics, 36*, 737–748.

Trudel, M.-C., Dub´e, B., Potse, M., Gulrajani, R. M., & Leon, L. J. (2004). Simulation of QRST integral maps with a membrane based computer heart model employing parallel processing. *IEEE Transactions on Biomedical Engineering, 51*(8), 1319–1329.

Van Oosterom, A. (1999). The use of the spatial covariance in computing pericardial potentials. *IEEE Transactions on Biomedical Engineering, 46*(7), 778–787.

Van Oosterom, A. (2001). The spatial covariance used in computing the pericardial potential distribution. In P. R. Johnston (Ed.), *Computational inverse problems in electrocardiography* (pp. 1–50). Southampton: WIT.

Van Oosterom, A. (2003). Source models in inverse electrocardiography. *Journal of Bioelectromagnetism, 5*, 211–214.

Vikhorev, P. G., & Vikhoreva, N. N. (2018). Cardiomyopathies and related changes in contractility of human heart muscle. *International Journal of Molecular Sciences, 19*(8), 2234.

Villars, P. S., Hamlin, S. K., Shaw, A. D., & Kanusky, J. T. (2004). Role of diastole in left ventricular function, I: Biochemical and biomechanical events. *American Journal of Critical Care, 13*(5), 394–403.

Wang, Y., & Rudy, Y. (2006). Application of the method of fundamental solutions to potential-based inverse electrocardiography. *Annals of Biomedical Engineering, 34*(8), 1272–1288.

CHAPTER

2

Deep learning approach in ultrasound image segmentation for patients with carotid artery disease

Branko Arsić[a,b]

[a]Bioengineering Research and Development Center, BioIRC, Kragujevac, Serbia [b]Faculty of Science, University of Kragujevac, Kragujevac, Serbia

1 Introduction

Precision medicine, also called personalized medicine, proposes the customization of health care where medical decisions, treatments, practices, or products are tailored to a subgroup of patients taking into account individual variability in genes, environment, and lifestyle for each person. A major unmet challenge in medicine is how to accurately predict the onset and course of a disease in an individual due to the complex and dynamic nature of disease progression, and specific individual comorbidities and lifestyles. To face the challenge the scientists are using advanced imaging technologies and omics datasets combined with various digital biosensors to capture physiological and behavioral data at a large scale thus forming tremendous repositories. The analysis of these increasingly huge datasets is not be possible without the help of statistics, artificial intelligence (AI), and machine learning algorithms (Lanza et al., 2020; Tyler et al., 2020).

Precision medicine is a global trend and a subject of study of many research groups and initiatives that deal with human brain (e.g., Alzheimer's disease), cancer research, drug discovery, etc. One such project is TAXINOMISIS (https://taxinomisis-project.eu/), which has received funding from the European Union's Horizon 2020 Research and Innovation Programme under grant agreement No 755320. The project is the first of its kind in cardiovascular and other chronic diseases in general, and it is strongly focused on the "personalized" element. The overall purpose of the project is to provide novel disease mechanism-based stratification for carotid artery disease patients to address the need for stratified and

https://doi.org/10.1016/B978-0-12-823956-8.00001-8

personalized therapeutic interventions nowadays. The model integrates clinical and personalized data, plaque and cerebral image processing and hemodynamics, with computer models and simulations for plaque growth and many more. This chapter contains a detailed description of the deep learning module which is used to provide the segmented ultrasound images of the human carotid artery.

Carotid artery stenosis (CAS) is one of the most common diseases in the human cardiovascular system. There are several diagnostic techniques that provide insight into the state of the carotid artery, including ultrasound (US), computed tomography angiography (CTA), and magnetic resonance angiography (MRA). Ultrasound is one of the noninvasive imaging techniques used in clinical diagnosing of carotid artery disease and it is usually the initial recommended CAS diagnostic examination because it is a very fast technique, readily available and relatively inexpensive. In the era of personalized medicine, automatic segmentation of the carotid artery lumen and wall from two-dimensional (2D) ultrasound images and three-dimensional (3D) carotid artery reconstruction is a crucial task toward fully automating the diagnosis procedure and patient's risk stratification. In order to determine the correct patient-specific diagnosis that also considers the individual anatomy of the particular patient, it is necessary to perform simulations using a patient-specific geometry. This step is impossible to do without knowing the location, shapes, and boundaries of the regions of interest. When the required boundaries are located by using some of the image processing techniques (Loizou, 2014) and their coordinates extracted, a patient-specific geometry can be created and used further for the blood flow simulations. However, segmenting the region of the lumen and wall is not a trivial task because the ultrasound images of the carotid bifurcation have typically low image quality (Hashimoto, 2011; Suri, 2008), incorporating significant noise, artifacts, shadowing, and reverberation (see Fig. 1). Also, the plaque itself remains a particularly challenging task, mainly due to the complex local geometry in combination with the fuzzy US appearance.

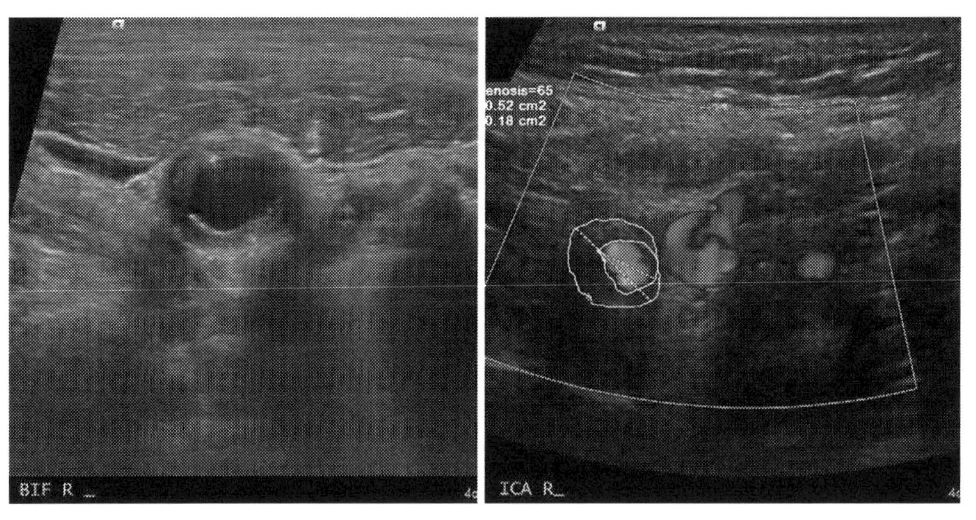

FIG. 1 US image of carotid artery, B-Mode and color-Doppler mode.

In this chapter, we are focused on a methodology for the image segmentation task of the carotid artery images obtained using the US examinations. An explanation of how to extract lumen and wall boundaries from the transversal US images of the CA, as well as from the longitudinal images, is given in more detail in the following sections. Besides a detection and localization of the stenosis is important, identification of atherosclerotic plaque components is essential to preestimate the risk of cardiovascular disease and stratify them as a high/low risk. A methodology for effective identification of the plaque types such as lipid, fibrous, and calcified tissue, by applying the deep learning methods on US images, is also presented in this chapter. This task is also very important because effectively identify the plaque types such as lipid, fibrous, and calcified tissue, by applying the deep learning methods to these images. All these data are necessary for the complete and valid 3D reconstruction of both human carotid arteries.

2 Carotid artery and personalized medicine

The carotid arteries are major blood vessels in the neck that supply blood to the brain, neck, and face. There are two carotid arteries, one on the right and one on the left. In the neck, each carotid artery branches into two divisions: the internal carotid artery supplies blood to the brain, and the external carotid artery supplies blood to the face and neck. Like all arteries, the carotid arteries are made of three layers of tissue: intima, the smooth innermost layer; media, the muscular middle layer; and adventitia, the outer layer. For more detail see Fig. 2.

Carotid artery disease (carotid artery stenosis) constitutes the primary cause of ischemic cerebrovascular events. In its early stages, carotid artery disease often does not produce any signs or symptoms and that is why regular checkups are important to help find potential health issues before they become a serious problem. The condition may go unnoticed until it is serious enough to deprive the brain of blood, causing a stroke or transient ischemic attack (TIA). A stroke is a medical emergency that can leave a patient with permanent brain damage

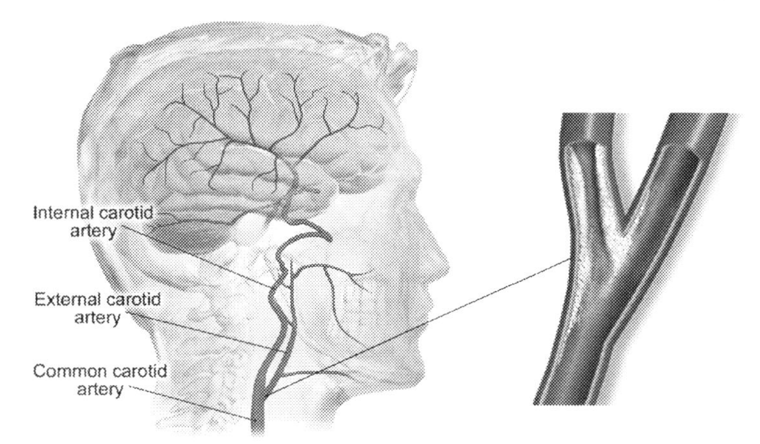

FIG. 2 Human carotid artery. *Adapted from "Common carotid artery," Wikipedia. n.d.[Online]. Available: https://en.wikipedia.org/wiki/Common_carotid_artery.*

and muscle weakness, but it can be also fatal in some cases. There are various factors that increase the risk of carotid artery disease such as high blood pressure, diabetes, family history, age, sleep apnea, tobacco use and many others. Following the goals of personalized medicine in general, a multidisciplinary, real-time approach for the stratification of patients with carotid artery disease is desired.

Accurately predicting the onset and subsequent course of a disease in an individual patient is a major unmet challenge in medicine, but a significant progress has been noticed in recent years (Ho et al., 2020; Pauli et al., 2017). The same goes for carotid artery disease. However, there exists a significant barrier to realizing the full potential of personalized medicine in our case as well, due to the lower image quality of US and the very abundance of heterogeneous and inconsistent data sources (Arsić et al., 2019). This complex approach, which is a subject of research of TAXINOMISIS project, should develop a new concept for the carotid artery disease stratification by analyzing the pathobiology of symptomatic plaques, identifying disease mechanisms, and developing a multiscale risk stratification model, which integrates clinical and personalized data, plaque, and cerebral image processing and computational modeling and novel biomarkers for high- vs low-risk states, in order to address the needs for stratified and personalized therapeutic interventions in the current era.

A patient-specific geometry of the carotid arteries could be created by using the methodology that consists of three main modules. US images obtained for a group of patients should be used in the deep learning (image segmentation) module to detect arterial structures, lumen and wall of the carotid artery, and determine the plaque constituents of atherosclerotic plaques from image data. This data will be sent as an input to the second module whose purpose is to reconstruct the three-dimensional geometry of the carotid bifurcation for a patient being examined by a doctor. Finally, this geometry should be used in the third module to model a complex process of blood flow, including the transport of low-density lipoprotein, macrophages, and cytokines through the arterial wall. Some examples of the simulations could be found in Djukic, Arsic, Djorovic, et al. (2020), Djukic, Arsic, Koncar, and Filipovic (2020) and Parodi et al. (2012). The overall concept is presented in Fig. 3.

The methodology described in this chapter could be a useful tool for the clinicians during ultrasound examination since it can provide valuable insight into the quantitative and qualitative state of the patient's CA in a short period of time and thus can help to achieve a more specialized treatment planning. In this chapter, we have mainly focused on the first module regarding the image segmentation techniques. This module is very complex and consists of many submodules and challenges which have to be solved before this module can be deployed into production. From the medical institutions and members of the project

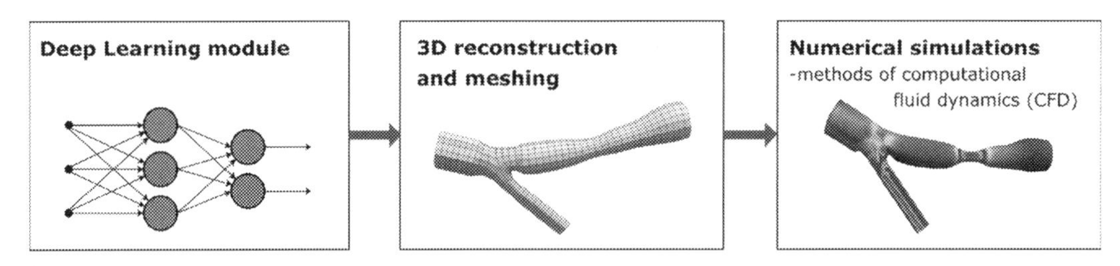

FIG. 3 The workflow of the US processing module.

consortium, we obtained the images in longitudinal and transversal modes taken from different CA positions, such as the place where a bifurcation is located and separate sets of images from the common and internal carotid artery.

After obtaining the US images for a patient, additional analysis can be performed. The US examination can be used to measure wall thickness and blood velocities in patients. However, there are some additional parameters that cannot be directly measured in this way but can be obtained using numerical simulations. For example, the finite element method (FEM) can be used to analyze blood velocity, pressure, and wall-shear stress distribution (Filipovic et al., 2012; Filipovic et al., 2013; Parodi et al., 2012).

3 Previous approaches in medical image segmentation

The ultrasound computer-aided diagnosis (CAD) system can be divided into four phases: image preprocessing, image segmentation, feature extraction, and lesion classification. The general flowchart of the ultrasound CAD system can be found in Huang et al. (2018). Taking into account that the US images have lower image quality, image preprocessing is almost always required. However, in the approach described here, no image filters were used, except image cropping for unifying input in a deep neural network.

Computer techniques have been intensively applied to analyze the US images. These techniques can be roughly separated into two general groups: traditional and deep learning (convolutional neural networks—CNNs) approaches.

3.1 The traditional image processing approach

The feature selection and extraction are the important steps in the traditional ultrasound CAD systems (Ravindraiah & Tejaswini, 2013), as the traditional CAD system focuses on the feature and the ultrasound image classification. The effective features can improve the accuracy and decrease the computational complexity of the system. The features adopted by traditional ultrasound CAD can be divided into four categories: texture, morphologic, model-based, and descriptor features.

One of the most common features in the ultrasound CAD systems is texture features that can reflect the character of the lesion surface. Here, we mention laws texture energy (LTE) (Laws, 1980), contrast of gray level values, local binary pattern (LBP) for the local texture of US image, graph-based techniques (Arsić et al., 2011), and the wavelet features derived from the wavelet transform of the US image. The wavelet features are used in Virmani et al. (2013) for the liver disease classification. The group of morphology features consists of Spiculation (Joo et al., 2004), elliptic-normalized circumference (ENC), elliptic-normalized skeleton (ENS), long axis-to-short axis ratio (L:S) (Chen et al., 2003). The model-based feature is one of the unique features of ultrasound images. It reflects the character of the backscattered echo from tissues (Shankar et al., 2001; Takemura et al., 2010).

Using ultrasound imaging of carotid plaques, several features that have been proposed to predict stroke in patients with asymptomatic carotid stenosis and thus contribute to risk stratification, have been confirmed in prospective studies. These texture features are used earlier in the asymptomatic carotid stenosis and risk of stroke (ACSRS) study. They include the severity of stenosis (Nicolaides et al., 2005), low grayscale median (GSM) (Madani et al., 2011; Nicolaides et al., 2010), plaque area $\geq$80 mm^2, a history of contralateral TIAs or stroke, the presence of a juxtaluminal black plaque area in the plaque image without a visible echogenic cap (JBA) $\geq$8 mm^2 (Griffin et al., 2010; Hashimoto et al., 2009; Kakkos et al., 2013) and the presence of discrete white areas (DWA) without acoustic shadowing indicating neovascularization (Nicolaides et al., 2010) and ulceration on 3D ultrasound (Kakkos et al., 2013).

After the selection and extraction of features, many classifiers can be easily adopted to classify the ultrasound images. The major classifiers employed by the ultrasound CAD system are Bayesian classifier, support vector machine, decision tree, artificial neural network, AdaBoost, etc. Nowadays the deep learning techniques are applied in the carotid US CAD system, together with other approaches for image segmentation. A completely user-independent algorithm for the segmentation of the wall of the common carotid artery (CCA) was developed in Nicolaides et al. (2010). Traditional feature extractors can be replaced by a convolutional neural network (CNN), since CNN's have a strong ability to extract complex features that express the image in much more detail, and they can be easily adopted to learn the task-specific features.

3.2 Deep convolution neural networks

In recent years, deep learning-based algorithms (Lan & Yoshua, 2016) show great promise in extracting features and learning patterns from complex data. Deep learning techniques have become the de facto standard for a wide variety of computer vision problems such as the automatic image segmentation, classification, and interpretation of medical image data in general. These developments have huge potential across a variety of areas, from image analysis to natural language processing, especially for medical imaging technology, medical data analysis, medical diagnostics, and health care in general. One of the main benefits of the deep learning approach is that it enables more optimal use of available information and improved prediction of the desired information. This is obtained by using processing layers (convolution, pooling, fully connected layers, etc.) that are more complex and more adapted to the data, in comparison with the generic imaging features (e.g., wavelets, spatial textures, statistical moments, etc.) that are analyzed in traditional US CAD systems (Alzubaidi et al., 2021). Feature extraction in this case is achieved automatically throughout the deep learning algorithms that have a multilayer data representation architecture. The first layers in an architecture extract the low-level features while the last layers extract the high-level features. This encourages researchers to extract discriminative features using the smallest possible amount of human effort and field knowledge (Lecun et al., 2015). However, the CNNs internal working mechanism is hard to understand because these learned features are hard to interpret from a human visual perspective.

Ultrasound images are affected by the multiplicative speckle noise, which tends to reduce the image quality, obscuring and blurring diagnostically important details. That is why CNNs offer a good alternative when compared to the traditional image processing techniques, where CNN is capable to extract new discriminative features using a combination of both global and local imaging information. The deep learning approach has been applied extensively in several medical imaging applications, such as for brain, lung, and breast imaging (Ravi et al., 2017; Shen et al., 2017). Some novel approaches regarding carotid intima-media thickness (IMT) and plaque detection are presented in Savaş et al. (2019) and Vila et al. (2020).

Besides carotid stenosis, identification of atherosclerotic plaque components is essential to preestimate the risk of cardiovascular disease and stratify them as a high or low risk (Boi et al., 2018). In the era of personalized medicine, plaque image analysis (Vancraeynest et al., 2011) has the potential to extract valuable information about the plaques and thus to identify more accurately the patients at risk of plaque rupture. It is now well established that the tissue composition plays a central role in the stability or vulnerability of atherosclerotic plaques (Alsheikh-Ali et al., 2010). Thus, it is important to merge the knowledge of medical and computer sciences and develop computational techniques that can automatically and objectively determine the atherosclerotic plaque constituents from imaging data (Boi et al., 2018), and in addition, compute the plaque development (Lekadir et al., 2017).

It is worth mentioning that the deep learning approach can be also used for more complex tasks such as multitask learning (MTL). Although MTL is not the subject of interest in this chapter, it is mentioned as the approach which has a good potential to face the problem of personalized medicine. The key idea of MTL is to model the relationship between relevant tasks (Liu et al., 2020) by learning both task common and task-specific features. In this way, each task can benefit from transferring related knowledge between tasks by multitask learning. For example, the characters of patients' tongues and face can be utilized to help doctors in diagnosing coronary heart disease (indicating disease location) (Wang et al., 2014).

The focus here is on the deep learning approach used to detect CA arterial structures. Data obtained after the image segmentation would be further used in a reconstruction task of the three-dimensional (3D) geometry of the carotid bifurcation thus providing real-time insight into the patient's condition to the doctors.

4 Dataset acquisition and description

In the study of carotid arteries, several diagnostic imaging techniques can help doctors to analyze various diseases such as stenoses, aneurysms, thromboses, dissections, and diseases caused by atherosclerotic plaques or congenital abnormalities. There are several noninvasive imaging techniques that provide insight into the state of the carotid artery, including digital subtraction angiography (DSA), ultrasound (US), computed tomography angiography (CTA), and magnetic resonance angiography (MRA). The radiologists recommend using US as an initial screening study because it is a safe and painless way to produce pictures of the inside of the body using sound waves. All image datasets (MRA, US, and CTA) are collected, validated, and labeled by the medical institutions, the participants of TAXINOMISIS project. At least two observers have participated in the image annotation. In addition, the segments of the carotid artery have been textually annotated on each image, also including if the examined artery was left or right.

Before the image data collection process started, the unique protocol for US examination had been created and shared among the clinical partners. It was a necessary step to enable greater interoperability and homogeneity of the US data. Moreover, this led to the harmonization of US data from all clinical centers, as far as possible, considering the diversity of ultrasound machines. It is still expected a significant enlargement of imaging dataset for further development and improvement of US processing module.

For the development of the deep learning module (the first of the three mentioned above), the US imaging data from the partners have been processed. The dataset consisted of 214 patients who underwent the US examination (baseline time point). Each patient had captured the common carotid artery, the branches and carotid bifurcation in transversal and longitudinal projections in B-mode. The average number of images per patient was 8.7 which leads to the estimated total number of 1861 images.

The preprocessing of US dataset included longitudinal and transversal images in B-mode and transversal images for the plaque characterization task. In this way, five different datasets for the deep learning models were created. The separate datasets for lumen and wall boundaries, for longitudinal and transversal positions, and the fifth dataset consists of images for plaque characterization. The following actions were performed for the first four datasets, while the actions for the plaque dataset are given in Section 5.2:

- Annotation of carotid lumen and wall area
- Resizing/Cropping of US images to 512×512 pixels
- Classification of longitudinal and transversal US images

Some examples of the annotated images can be seen in Fig. 4.

In this case, the first step of the preprocessing is the automated isolation of the image region which contains the arterial tree under reconstruction. This is performed by selecting a static 512×512 pixels window for both arterial models, left and right. Special attention is paid to the window coordinates in order that the whole arterial tree is visible in the region.

It should be noted that not all images were used for the image processing, such as images showing blood velocities within the artery which had lower zoom and were inappropriate for scaling (see Fig. 5). Also, all imaging data were anonymized respecting data protection and safety.

5 Methodology

5.1 Lumen and wall detection

In most cases, semantic image segmentation is performed in order to extract important features which can be further effectively used for the image classification task. Here, we went a step ahead, and want to provide a real-time engine that can detect all required boundaries and used them later for a 3D reconstruction task and numerical simulations.

The automatic carotid artery (lumen and wall) segmentation has been done using FCN-8s (Long et al., 2015; Simonyan & Zisserman, 2015), SegNet (Badrinarayanan et al., 2017), and U-Net (Ronneberger et al., 2015) based on deep convolutional networks. Besides the original versions of these architectures, the modified variants of the U-Net and SegNet networks, from the aspect of depth, were developed and tested. Although the SegNet CNN architecture has almost twice less trainable parameters compared to the used U-Net architecture, the results

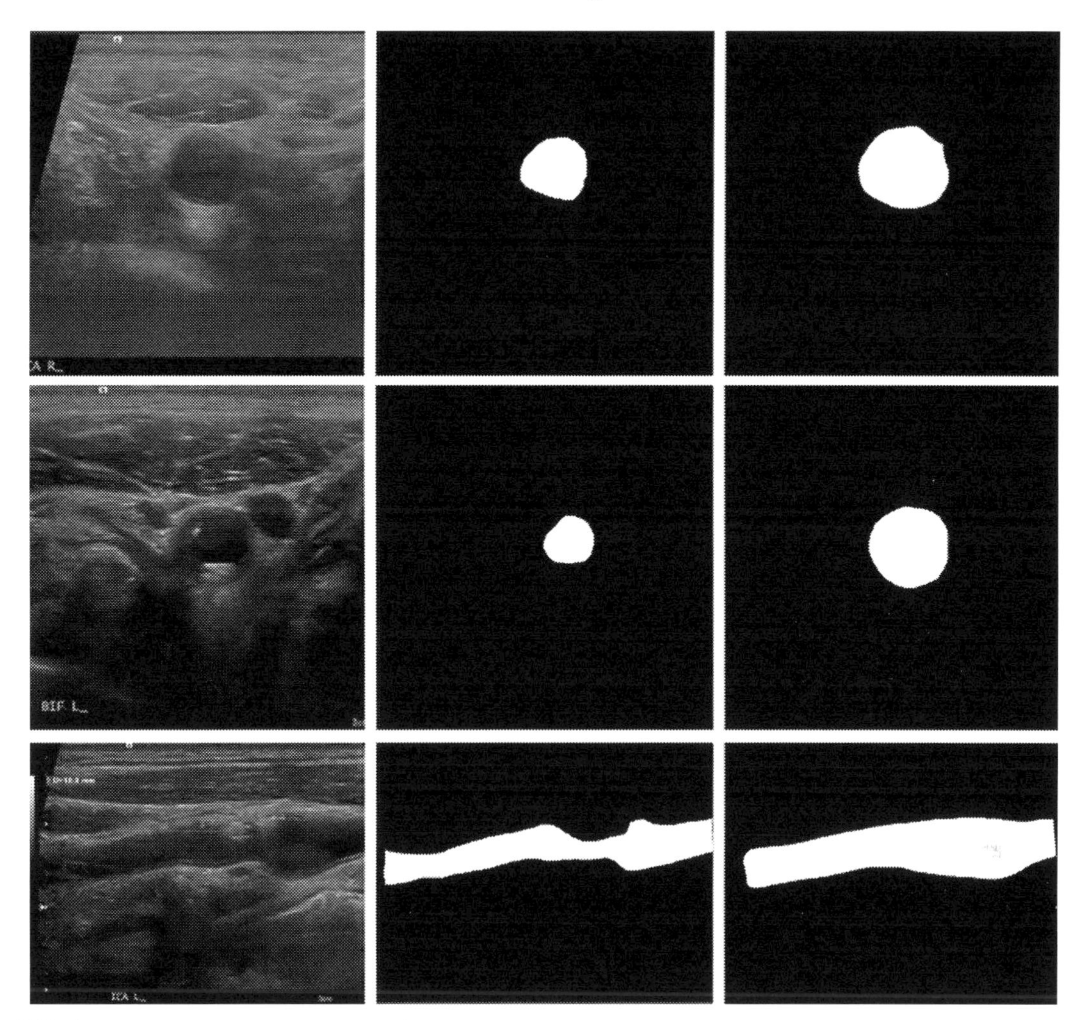

FIG. 4 Carotid ultrasound images. The first column represents the original images, the second column represents the lumen masks, and the third column represents the wall masks.

obtained using U-Net were much better and this segment is very important whether the most reliable 3D models are preferred.

U-Net is a convolutional neural network for image segmentation with the most important application being in the segmentation of medical images. It is based on encoder-decoder model. The encoder consists of convolutional and max-pooling layers which gradually decrease the spatial size of the image and increase the number of channels. After the encoder extracts the features, decoder part symmetrically performs upconvolution and convolution operations. Upconvolution doubles the spatial dimensions of the features and reduces the number of channels, which corresponds to the way encoder blocks is decreasing the image resolution and increasing the depth. In addition, skip connections are used in order to

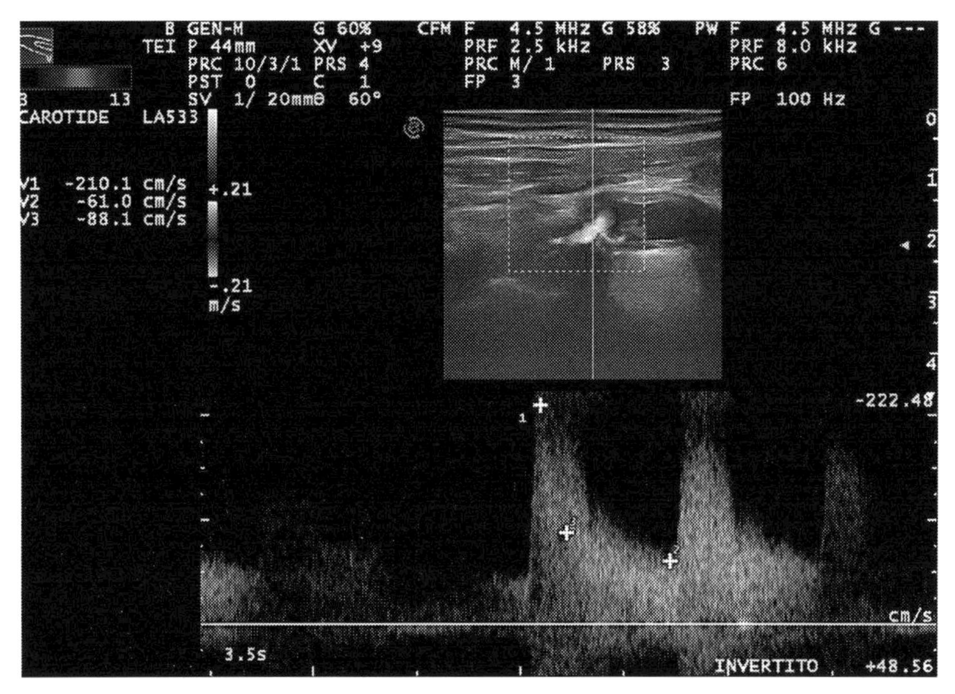

FIG. 5 Unused US image of the carotid artery.

improve the quality of decoder features. As it was mentioned before, a deeper variant of U-Net was tested too. This version is slightly modified from the original paper and it has two additional blocks in both encoder and decoder. The whole architecture could be found in Arsic et al. (2019). Some of the important parameters used in the U-Net model are discussed below.

This variant of U-Net uses batch normalization after each convolutional layer, which proves to work a lot better on our data than the original U-Net model (Zhou & Yang, 2019). All batch normalization layers are followed by a ReLU activation. The model is trained with a combination of binary cross entropy and soft dice coefficient as a loss function, which is expressed as

$$Loss = binary\_crossentropy\left(y_{true}, y_{pred}\right) + 1 - dice\_coeff\left(y_{true}, y_{pred}\right)$$

where y_{pred} and y_{true} denote the flattened predicted probabilities and the flattened ground truths of the image. Soft dice coefficient loss is described in Anbeek et al. (2005) and Chang et al. (2009).

A dataset used for the training phase consists of the US images of 214 patients. All subfolders corresponding to the patients are randomly divided into training, validation, and testing sets in the ratio of 8:1:1 at the carotid artery level (either for the left or for the right arterial model). A total of 1500 images have been taken out for training purposes. Moreover, the age, weight, and height of the patient are important factors that influence the final US images look, and they were also taken into consideration when the dataset was split. The examples of original and labeled images for the lumen and wall are shown in Fig. 4. Note that only

US images taken in B-mode (grayscale) are used here, because segmentation of the color-Doppler ultrasound is a trivial task.

5.2 Carotid plaque characterization

One of the main aims of TAXINOMISIS project is to identify and classify the atherosclerotic plaque components such as lipid core, fibrous and calcified tissue, by applying the deep learning methods on ultrasound imaging data. These types of computational methods can preestimate the risk of cardiovascular disease and stratify patients as a high/low risk (Vancraeynest et al., 2011), detecting the stabile and vulnerable atherosclerotic plaques. Considering that tissue composition plays a central role in the stability or vulnerability of atherosclerotic carotid plaques (Boi et al., 2018), it is important to merge the knowledge of medical and computer sciences and develop computational techniques that can automatically and objectively determine the atherosclerotic plaque constituents from imaging data (Lekadir et al., 2017) and in addition compute plaque development. Detection and classification of atherosclerotic plaque constituents is challenging due to the small size of the plaques and the complexity of the tissue appearance. On the other hand, US images of the carotid bifurcation have typically lower image quality, including different artifacts, which makes inconsistent detection and characterization of the plaques, even for an expert clinician.

When the lumen and wall of the carotid artery are detected in US images, the process of characterization of atherosclerotic carotid plaque can start. This prestep is important because the images used to train a new model are annotated by using the results obtained in the previous phase. Image annotation has included labeling of atherosclerotic plaque components such as lipid core, fibrous, and calcified tissue. The preprocessing of US images for this task is slightly different than that for the lumen and wall datasets and it included three steps:

- Overlapping the previously detected and segmented carotid lumen and wall area in transversal cross sections and extraction of plaque area.
- Creation of three different masks corresponding to plaque components (fibrous, lipid, calcified), and one mask for background color.
- Resizing/cropping of US images to 512 × 512 pixels.

Pixel map for the model was defined as follows (Fig. 6):

- background (0) is annotated with *black* color,

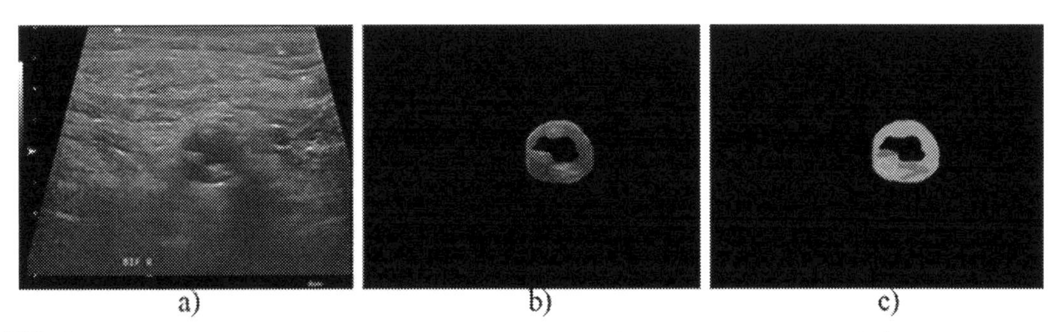

FIG. 6 Original data provided by clinical partners (A), extracted "ring" (B), and annotated plaque (C).

- fibrous plaque (1) is annotated with *yellow* color,
- lipid plaque (2) is annotated with *blue* color,
- calcified plaque (3) is annotated with *green* color.

In this task, the problem of atherosclerotic plaque components segmentation is defined as a multiclass segmentation model (or pixel labeling), where the aim is to label each pixel in an image with one of a number of classes. Four classes should be detected in images in our case: background (area outside the ring), fibrous, calcified, and lipid atherosclerotic plaque components. For the multiclass image segmentation task, three architectures were tested: *U-net*, *SegNet*, and PSPNet (pyramid scene parsing network) (Zhao et al., 2017). For trained and tested models image input size was 512×512 pixels.

First, the models were trained by using the categorical cross-entropy function as a loss function. However, the results were not satisfied due to the highly imbalanced classes in images. More than 90% of pixels belong to the background class, while the most prevalent component in the plaque zone is always a fibrous component. In order to handle highly imbalanced classes, we tested different loss functions instead of categorical cross-entropy loss, but the highest accuracy we achieved by using weighted categorical cross-entropy loss function with manually selected weights for each class. In order to increase the importance of nonbackground classes, various combinations of class weights were tested. In the end, a combination of class weights of [0.2, 2, 3, 3] was used for the model training. Metrics that were recorded during the training phase are accuracy (not recommended for highly imbalanced classes) and *mean IoU* (Intersection over Union). Note that a much better model could be obtained if the images are with reduced background, but all images used for lumen, wall, and plaque detection should have the same size so the pixels can be overlapped (pixel on pixel).

6 Results

In the case of lumen and wall detection, a binary classification task for image segmentation was considered. We compare our method against thresholding technique, SegNet (Badrinarayanan et al., 2017) and FCN-8s model with VGG16 as a backbone classifier (Long et al., 2015). However, FCN has almost twice as many trainable parameters than the U-Net network, so the U-Net can be trained faster, and it is more memory efficient. On the other side, U-Net architecture has almost twice as many trainable parameters as SegNet network. For the reliable 3D reconstruction task, the modified U-Net architecture (Arsic et al., 2019) was selected to perform image segmentation as a trade-off choice between accuracy and memory efficiency. The results for the test sets are shown in Table 1.

In Fig. 7, the second and third rows show the examples of the predicted images for the lumen and wall regions for one patient.

The results for the atherosclerotic plaque components segmentation are shown in Table 2 for the U-Net architecture because it always outperforms the SegNet and PSPNet models.

The different atherosclerotic plaque zones for three different patients are presented in Fig. 8. The patients are selected to depict different plaque compositions. The first row represents the original images, the second row represents manually labeled images, and the third row represents multiclass prediction results.

TABLE 1 U-Net results on test dataset for lumen and wall, in both projections, transversal and longitudinal.

Segmentation task	Projection	F1-score
LUMEN	Transversal	0.92
	Longitudinal	0.96
WALL	Transversal	0.84
	Longitudinal	0.81

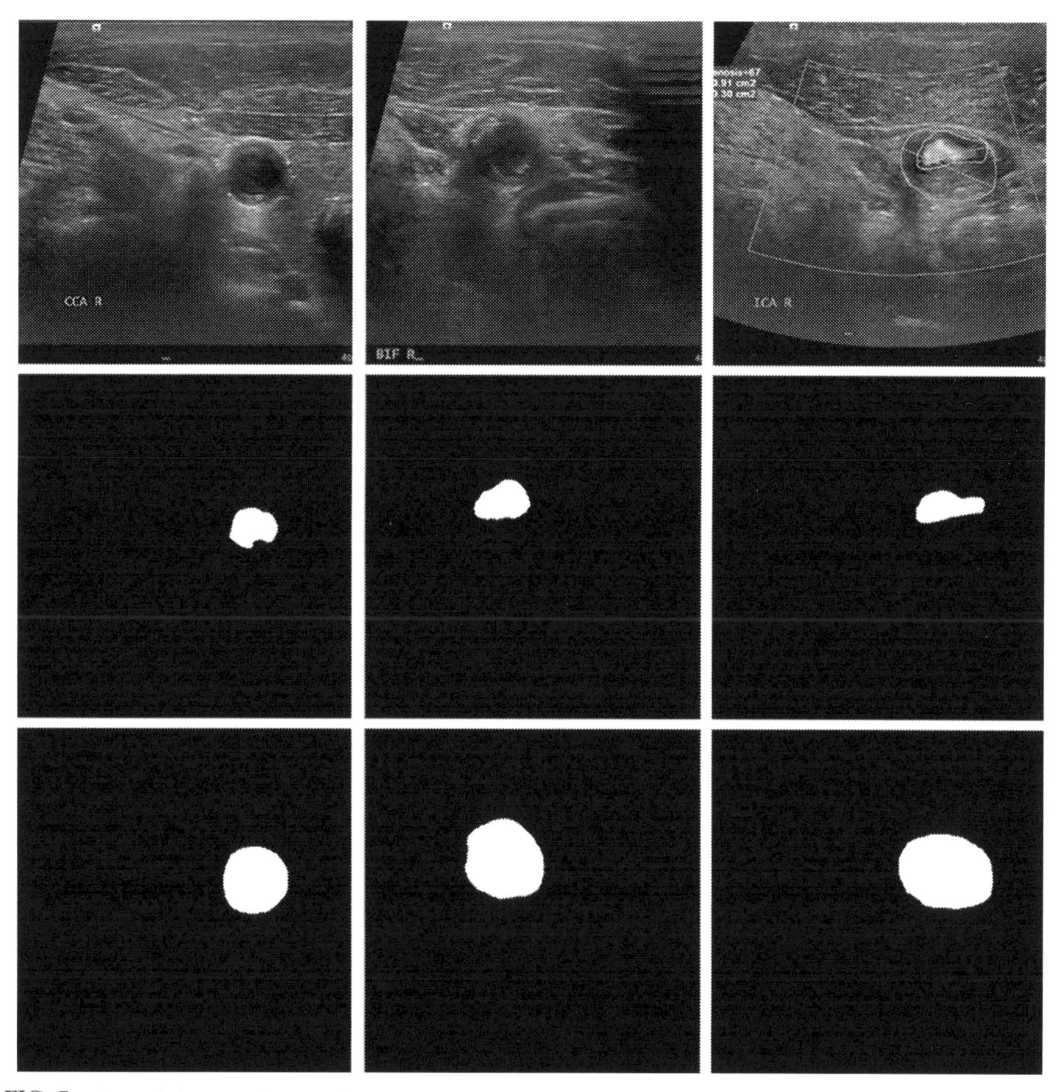

FIG. 7 Carotid ultrasound images for one patient. The first row represents the original images, the second and the third rows show the predicted lumen and wall regions, respectively.

TABLE 2 U-Net results on test dataset for the atherosclerotic plaque components segmentation.

Classification metric	TEST (Entire test set)
Mean IoU	0.606
Class-wise IoU	[0.998, 0.743, 0.220, 0.465]

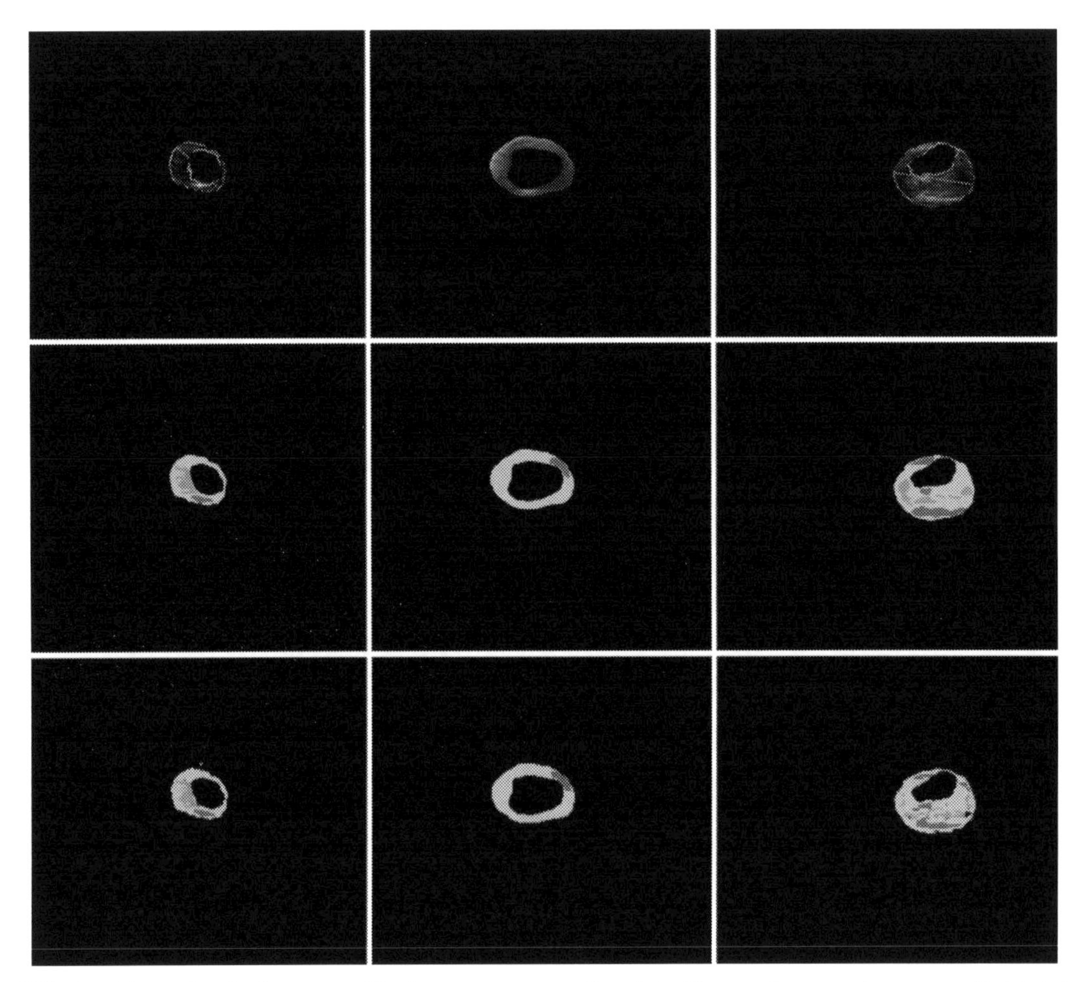

FIG. 8 Atherosclerotic plaque zone. The first row represents the original images, the second row manually labeled images, and the third row represents multiclass prediction results.

7 Conclusions

To pave the way for the advancement of precision medicine in cardiovascular surgery, TAXINOMISIS project aims to develop a rational new approach for the stratification of carotid artery disease patients by unwinding the pathobiology that underlies symptomatic plaques, discriminating distinct disease mechanism-driven states (endotypes) and biomarkers, and develop a multiscale risk stratification model. Nowadays, for delineation of vasculatures in the carotid artery, US examinations have been processed because of a noninvasive procedure without ionizing radiation. However, 2D ultrasound imaging has limitations and it cannot be a reliable assistant to the doctors during the patient examination. That is why the creation of the patient-specific geometry is important. Here, the deep learning techniques for the image segmentation task are used for obtaining the coordinates of lumen, wall, and plaque boundaries. These coordinates are required for the 3D reconstruction of the entire carotid artery, and they serve as a basis for further steps. The next level will include the blood flow modeling, as well as the lipoproteins transport into the reconstructed carotid arteries.

Acknowledgments

The research presented in this study was part of the project that has received funding from the European Union's Horizon 2020 research and innovation programme under grant agreement No. 755320-2—TAXINOMISIS. This article reflects only the author's view. The Commission is not responsible for any use that may be made of the information it contains. This research is also funded by the Serbian Ministry of Education, Science, and Technological Development [451-03-68/2020-14/200378 (Institute for Information Technologies, University of Kragujevac)]. All listed funding sources had no involvement in the study design; in the collection, analysis, and interpretation of data; in the writing of the manuscript; and in the decision to submit the manuscript for publication.

References

Alsheikh-Ali, A. A., Kitsios, G. D., Balk, E. M., Lau, J., & Ip, S. (2010). The vulnerable atherosclerotic plaque: Scope of the literature. *Annals of Internal Medicine*. https://doi.org/10.7326/0003-4819-153-6-201009210-00272.

Alzubaidi, L., et al. (2021). Review of deep learning: Concepts, CNN architectures, challenges, applications, future directions. *Journal of Big Data*, *8*(1), 53. https://doi.org/10.1186/s40537-021-00444-8.

Anbeek, P., Vincken, K. L., Van Bochove, G. S., Van Osch, M. J. P., & Van Der Grond, J. (2005). Probabilistic segmentation of brain tissue in MR imaging. *NeuroImage*. https://doi.org/10.1016/j.neuroimage.2005.05.046.

Arsić, B., Cvetković, D., Simić, S., & Škarić, M. (2011). Graph spectral techniques in computer sciences. *Applicable Analysis and Discrete Mathematics*, *6*(1), 1–30. https://doi.org/10.2298/aadm111223025a.

Arsić, B., Đokić-Petrović, M., Spalević, P., Milentijević, I., Rančić, D., & Živanović, M. (2019). SpecINT: A framework for data integration over cheminformatics and bioinformatics RDF repositories. *Semantic Web*, *10*, 795–813. https://doi.org/10.3233/SW-180327.

Arsic, B., Obrenovic, M., Anic, M., Tsuda, A., & Filipovic, N. (2019). Image segmentation of the pulmonary acinus imaged by synchrotron x-ray tomography. In *Proceedings—2019 IEEE 19th international conference on bioinformatics and bioengineering, BIBE 2019*. https://doi.org/10.1109/BIBE.2019.00101.

Badrinarayanan, V., Kendall, A., & Cipolla, R. (2017). SegNet: A deep convolutional encoder-decoder architecture for image segmentation. *IEEE Transactions on Pattern Analysis and Machine Intelligence*. https://doi.org/10.1109/TPAMI.2016.2644615.

Boi, A., et al. (2018). A survey on coronary atherosclerotic plaque tissue characterization in intravascular optical coherence tomography. *Current Atherosclerosis Reports*. https://doi.org/10.1007/s11883-018-0736-8.

Chang, H. H., Zhuang, A. H., Valentino, D. J., & Chu, W. C. (2009). Performance measure characterization for evaluating neuroimage segmentation algorithms. *NeuroImage*. https://doi.org/10.1016/j.neuroimage.2009.03.068.

Chen, C. M., et al. (2003). Breast lesions on sonograms: Computer-aided diagnosis with nearly setting-independent features and artificial neural networks. *Radiology*. https://doi.org/10.1148/radiol.2262011843.

Djukic, T., Arsic, B., Djorovic, S., Filipovic, N., & Koncar, I. (2020). Validation of the machine learning approach for 3D reconstruction of carotid artery from ultrasound imaging. In *2020 IEEE 20th international conference on bioinformatics and bioengineering (BIBE)* (pp. 789–794). https://doi.org/10.1109/BIBE50027.2020.00134.

Djukic, T., Arsic, B., Koncar, I., & Filipovic, N. (2020). 3D reconstruction of patient-specific carotid artery geometry using clinical ultrasound imaging. In *Workshop computational biomechanics for medicine XV, 23rd international conference on medical image computing & computer assisted intervention*. https://doi.org/10.5281/zenodo.4563917.

Filipovic, N., Teng, Z., Radovic, M., Saveljic, I., Fotiadis, D., & Parodi, O. (2013). Computer simulation of three-dimensional plaque formation and progression in the carotid artery. *Medical & Biological Engineering & Computing*. https://doi.org/10.1007/s11517-012-1031-4.

Filipovic, N., et al. (2012). ARTreat project: Three-dimensional numerical simulation of plaque formation and development in the arteries. *IEEE Transactions on Information Technology in Biomedicine*. https://doi.org/10.1109/TITB.2011.2168418.

Griffin, M. B., et al. (2010). Juxtaluminal hypoechoic area in ultrasonic images of carotid plaques and hemispheric symptoms. *Journal of Vascular Surgery*. https://doi.org/10.1016/j.jvs.2010.02.265.

Hashimoto, B. E. (2011). Pitfalls in carotid ultrasound diagnosis. *Ultrasound Clinics*. https://doi.org/10.1016/j.cult.2011.08.006.

Hashimoto, H., Tagaya, M., Niki, H., & Etani, H. (2009). Computer-assisted analysis of heterogeneity on B-mode imaging predicts instability of asymptomatic carotid plaque. *Cerebrovascular Diseases*. https://doi.org/10.1159/000229554.

Ho, D., et al. (2020). Enabling technologies for personalized and precision medicine. *Trends in Biotechnology, 38*(5), 497–518. https://doi.org/10.1016/j.tibtech.2019.12.021.

Huang, Q., Zhang, F., & Li, X. (2018). Machine learning in ultrasound computer-aided diagnostic systems: A survey. *BioMed Research International*. https://doi.org/10.1155/2018/5137904.

Joo, S., Yang, Y. S., Moon, W. K., & Kim, H. C. (2004). Computer-aided diagnosis of solid breast nodules: Use of an artificial neural network based on multiple sonographic features. *IEEE Transactions on Medical Imaging*. https://doi.org/10.1109/TMI.2004.834617.

Kakkos, S. K., et al. (2013). The size of juxtaluminal hypoechoic area in ultrasound images of asymptomatic carotid plaques predicts the occurrence of stroke. *Journal of Vascular Surgery*. https://doi.org/10.1016/j.jvs.2012.09.045.

Lan, C. A. G., & Yoshua, B. (2016). *Deep Learning—Ian Goodfellow, Yoshua Bengio, Aaron Courville—Google Books*. MIT Press.

Lanza, G., Giannandrea, D., Lanza, J., Ricci, S., & Gensini, G. F. (2020). Personalized-medicine on carotid endarterectomy and stenting. *Annals of Translational Medicine, 8*(19), 1274. https://doi.org/10.21037/atm-20-1126.

Laws, K. I. (1980). *Textured image segmentation*. Univ. South. California, IPI Rep.

Lecun, Y., Bengio, Y., & Hinton, G. (2015). Deep learning. *Nature*. https://doi.org/10.1038/nature14539.

Lekadir, K., et al. (2017). A convolutional neural network for automatic characterization of plaque composition in carotid ultrasound. *IEEE Journal of Biomedical and Health Informatics*. https://doi.org/10.1109/JBHI.2016.2631401.

Liu, B., Li, Y., Ghosh, S., Sun, Z., Ng, K., & Hu, J. (2020). Complication risk profiling in diabetes care: A Bayesian multi-task and feature relationship learning approach. *IEEE Transactions on Knowledge and Data Engineering, 32*(7), 1276–1289. https://doi.org/10.1109/TKDE.2019.2904060.

Loizou, C. P. (2014). A review of ultrasound common carotid artery image and video segmentation techniques. *Medical & Biological Engineering & Computing*. https://doi.org/10.1007/s11517-014-1203-5.

Long, J., Shelhamer, E., & Darrell, T. (2015). Fully convolutional networks for semantic segmentation. In *Proceedings of the IEEE Computer Society Conference on Computer Vision and Pattern Recognition*. https://doi.org/10.1109/CVPR.2015.7298965.

Madani, A., Beletsky, V., Tamayo, A., Munoz, C., & Spence, J. D. (2011). High-risk asymptomatic carotid stenosis ulceration on 3D ultrasound vs TCD microemboli. *Neurology*. https://doi.org/10.1212/WNL.0b013e31822b0090.

Nicolaides, A. N., et al. (2005). Severity of asymptomatic carotid stenosis and risk of ipsilateral hemispheric ischaemic events: Results from the ACSRS study. *European Journal of Vascular and Endovascular Surgery*. https://doi.org/10.1016/j.ejvs.2005.04.031.

Nicolaides, A. N., et al. (2010). Asymptomatic internal carotid artery stenosis and cerebrovascular risk stratification. *Journal of Vascular Surgery, 52*(6), 1486–1496. https://doi.org/10.1016/j.jvs.2010.07.021.

Parodi, O., et al. (2012). Patient-specific prediction of coronary plaque growth from CTA angiography: A multiscale model for plaque formation and progression. *IEEE Transactions on Information Technology in Biomedicine*. https://doi.org/10.1109/TITB.2012.2201732.

Pauli, C., et al. (2017). Personalized in vitro and in vivo cancer models to guide precision medicine. *Cancer Discovery, 7*(5), 462–477. https://doi.org/10.1158/2159-8290.CD-16-1154.

Ravi, D., et al. (2017). Deep learning for health informatics. *IEEE Journal of Biomedical and Health Informatics*. https://doi.org/10.1109/JBHI.2016.2636665.

Ravindraiah, R., & Tejaswini, K. (2013). A survey of image segmentation algorithms based on fuzzy clustering. *International Journal of Computer Science and Mobile Computing*, *2*(7), 200–206.

Ronneberger, O., Fischer, P., & Brox, T. (2015). U-net: Convolutional networks for biomedical image segmentation. In *Lecture Notes in Computer Science (including subseries Lecture Notes in Artificial Intelligence and Lecture Notes in Bioinformatics)*. https://doi.org/10.1007/978-3-319-24574-4_28.

Savaş, S., Topaloğlu, N., Kazcı, Ö., & Koşar, P. N. (2019). Classification of carotid artery intima media thickness ultrasound images with deep learning. *Journal of Medical Systems*, *43*(8), 273. https://doi.org/10.1007/s10916-019-1406-2.

Shankar, P. M., et al. (2001). Classification of ultrasonic B-mode images of breast masses using Nakagami distribution. *IEEE Transactions on Ultrasonics, Ferroelectrics, and Frequency Control*. https://doi.org/10.1109/58.911740.

Shen, D., Wu, G., & Il Suk, H. (2017). Deep learning in medical image analysis. *Annual Review of Biomedical Engineering*. https://doi.org/10.1146/annurev-bioeng-071516-044442.

Simonyan, K., & Zisserman, A. (2015). Very deep convolutional networks for large-scale image recognition. In *3rd International Conference on Learning Representations, ICLR 2015—Conference Track Proceedings*.

Suri, J. S. (2008). *Advances in diagnostic and therapeutic ultrasound imaging*. Artech House.

Takemura, A., Shimizu, A., & Hamamoto, K. (2010). Discrimination of breast tumors in ultrasonic images using an ensemble classifier based on the adaboost algorithm with feature selection. *IEEE Transactions on Medical Imaging*. https://doi.org/10.1109/TMI.2009.2022630.

Tyler, J., Choi, S. W., & Tewari, M. (2020). Real-time, personalized medicine through wearable sensors and dynamic predictive modeling: A new paradigm for clinical medicine. *Current Opinion in Systems Biology*, *20*, 17–25. https://doi.org/10.1016/j.coisb.2020.07.001.

Vancraeynest, D., Pasquet, A., Roelants, V., Gerber, B. L., & Vanoverschelde, J. L. J. (2011). Imaging the vulnerable plaque. *Journal of the American College of Cardiology*. https://doi.org/10.1016/j.jacc.2011.02.018.

Vila, D. M. M., et al. (2020). Semantic segmentation with DenseNets for carotid artery ultrasound plaque segmentation and CIMT estimation. *Artificial Intelligence in Medicine*. https://doi.org/10.1016/j.artmed.2019.101784.

Virmani, J., Kumar, V., Kalra, N., & Khandelwal, N. (2013). SVM-based characterization of liver ultrasound images using wavelet packet texture descriptors. *Journal of Digital Imaging*. https://doi.org/10.1007/s10278-012-9537-8.

Wang, Y., et al. (Feb. 2014). Therapeutic effect in patients with coronary heart disease based on information analysis from traditional Chinese medicine four diagnostic methods. *Journal of Traditional Chinese Medicine = Chung i tsa chih ying wen pan*, *34*(1), 34–41. https://doi.org/10.1016/s0254-6272(14)60051-0.

Zhao, H., Shi, J., Qi, X., Wang, X., & Jia, J. (2017). Pyramid scene parsing network. In *vol. 2017-Janua. Proceedings—30th IEEE conference on computer vision and pattern recognition, CVPR 2017* (pp. 6230–6239). https://doi.org/10.1109/CVPR.2017.660.

Zhou, X. Y., & Yang, G. Z. (2019). Normalization in training U-net for 2-D biomedical semantic segmentation. *IEEE Robotics and Automation Letters*. https://doi.org/10.1109/LRA.2019.2896518.

CHAPTER

3

Simulation of stent mechanical testing

Dalibor D. Nikolic and Nenad Filipovic
Bioengineering Research and Development Center (BioIRC), Kragujevac, Serbia

1 Introduction

Coronary artery disease (CAD) is one of the leading causes of death in the world. This CAD occurs as a consequence of the growth of atherosclerotic plaques in the coronary arteries. There are numerous causes of plaque growth and development; however, mechanical factors such as shear stress in arteries caused by the artery geometry have a significant effect on the location and rate of plaque development. This topic is explained in detail in the following papers (Filipovic et al., 2013, 2012; Parodi et al., 2012), while the development of plaques in a certain disease of the myocardial bridge is explained in more detail in the paper by Nikolić et al. (2014). Depending on the stages of the disease, there are several treatments. In the initial stages, treatment is based on **pharmacological therapy**—giving appropriate drugs can increase blood flow through the coronary arteries. While in severe diseases **bypass surgery is approached**—surgical intervention under general anesthesia based on the creation of bridges that bypass narrowing and clogging of blood vessels—a more modern approach is applied **percutaneous transluminal coronary angioplasty (PTCA)**—coronary artery narrowing to restore normal blood flow. Today, about 500,000 such procedures are performed in Europe every year. Percutaneous transluminal coronary angioplasty (PTCA) is one of the newer methods of treatment and has been introduced as a minimally invasive treatment for CAD. A stent with an angioplasty balloon is inserted through the peripheral arteries at the site of the lesion through a catheter. The catheter balloon is inflated to nominal pressure, compressing the atherosclerotic plaque along the arterial wall and placing the stent. In this way, the blood vessel at the site of the lesion expands to its normal diameter. This restores vascular patency and improves blood flow to the myocardium. PTCA generally results in a high degree of clinical success; however, 1% of patients redevelop initial symptoms within 6–12 months. This renarrowing of the treated artery is caused by restenosis, which is the main limitation and disadvantage of this procedure (Martin & Boyle, 2011). The basic

Cardiovascular and Respiratory Bioengineering
https://doi.org/10.1016/B978-0-12-823956-8.00003-1

implant necessary to perform a PTCA procedure is a "stent" (Roguin, 2011). A stent is a mostly tubular endoprosthesis made of biocompatible metal alloys that is inserted through a catheter and placed at the site of narrowing with the aim of expanding or opening the lumen of hollow tubular structures, most often blood vessels but also other tubular canals (urinary, bile, etc.).

The development of stents in terms of new designs and materials is necessary for medicine. Advances in stent technology and especially advances in the technology of stent delivery and placement systems (catheters), improving their flexibility, have expanded the clinical application of stents; so, today, it is possible to install stents even in very complex lesions (Kandzari et al., 2002).

However, 20%–50% of restenoses occurred in patients whose primary stenoses were treated with bare-metal stents (BMC) alone. This phenomenon is partly related not only to the design of the stent (Kastrati et al., 2001; Morton et al., 2004) but also to the interaction of arteries and balloons during stent implantation (Morton et al., 2004). In part, restenosis is responsible for the inflammatory response of the blood vessel wall to a foreign body—the stent, as well as local hemodynamic factors on the arterial wall itself (e.g., shear stress) (Frank et al., 2002).

In order to reduce the occurrence of restenosis, drug-eluting stents (DES) have been developed. This type of stent is coated with antiinflammatory and antiproliferative chemicals—drugs, in order to prevent local inflammatory processes in the tissue and slow down the rapid growth of cells. In this way, the rate of restenosis is reduced to below 10% (Morice et al., 2002; Serruys et al., 2006; Sousa et al., 2005) but it is important to emphasize that in this way the mechanical injuries that occur during stent implantation are not reduced. A new generation of implant devices and bioresorbable vascular scaffolds (BVS), have been developed from biodegradable materials and these devices degrade over time; so, in case of restenosis, there is a possibility of reintervention in the same lesion. Also, restenosis in DES stents occurs due to the uneven distribution of the stent struts with which it acts on the wall of the blood vessel (Takebayashi et al., 2004). Great efforts are currently being made to obtain a simpler development that would reduce the cost of DES stents, as there is currently a very large difference compared to new generations of BMC stents (Kaiser et al., 2005), and in addition, DES stents have proven to be high risk in interventions where it is necessary to stop the circulation, i.e., to make an artificial blood clot in an artery (in-stent thrombosis, i.e., blood clot formation) (Virmani et al., 2004). The process of treating the diseased part of the artery should be temporary, but the installation of metal stents is not such an ideal solution permanently; so, nowadays, a lot of work is being done on the development of biodegradable stents that should replace all metal endoprostheses in the future.

Although the stent is primarily designed for blood vessels, this device has a very high potential to be used to treat other diseases that cause narrowing or closing of cavities in the body. In tracheobronchial obstruction (e.g., patients with severe dyspnea, often on the verge of suffocation), symptoms can be alleviated by the use of respiratory stents (Walser et al., 2004). Stents are also used for problems with urethral canal narrowing and/or prostate obstruction (Hussain et al., 2004). Stents are used to support and open duodenal canals, etc. Certainly, even with these interventions, no matter how minimally invasive they are, attention must be paid to the design of the stent in order to minimize possible complications (e.g., restenosis, stent migration, artery straightening, etc.) (Borisch et al., 2005; Saad et al., 2003).

All future improvements of current stent generations are reduced to optimizing existing designs and redesigning individual parts of stents to meet the specific behavior and geometry of the target lesion while overcoming any shortcomings.

Compared to experimental research (Narracott et al., 2007; Rieu et al., 2003; Poerner et al., 2004; Ormiston et al., 2004), computer models are an excellent tool for research, development, and optimization of mechanical characteristics of stents. Some test methods are very difficult or even impossible to implement, so the greater is the benefit of computer models and simulations, which are the only alternative in these situations. The advantages of computational simulations are also in the great freedom for testing: it is possible to create several different test scenarios, use several different geometries and materials for the model and compare them with each other before making prototypes (Perry et al., 2002).

Because of all this, computer models and simulations are a very valuable and unavoidable step in the process of stent design and development. But, as the numerical models are close to real (they are not the same) certain checks need to be done through experimental testing.

Migliavacca et al. (2002) investigated the effect of the geometrical parameters of stents (artery surface ratio, strut thickness, cell length) on deployment characteristics. Results for the Palmaz-Schatz stent were compared with the Carbostent and Multilink Tetra stent (Migliavacca et al., 2002). The expansion of each stent was obtained through a pressure load applied to the inner surface. The results of their research showed that geometrical parameters had a significant influence on the deployment characteristics. In the case when the metal to artery surface ratio is lower, this is associated with higher rates of radial and longitudinal stent recoil and also lower rates of dog-boning effects. This study demonstrated the huge potential of the finite element method in the design and optimization of stent designs. The second generation of stents—drug-eluting balloon expandable stents—are commonly used to treat coronary lesions. In some cases, biodegradable parts are used in the stent design: the stent backbone can be made of metal (stainless steel, cobalt-chromium, platinum-chromium) or of a biodegradable polymer (e.g., PLLA, Tyrocore), but in both cases they are covered by biodegradable drug-eluting polymeric coatings. In the following chapter, simulation on DES stent designs will be presented. The SYNERGY BP Everolimus-Eluting Platinum Chromium Coronary Stent is a partially Bioresorbable Vascular Scaffold (BVS) produced by Boston Scientific Limited (Galway, Ireland).

In order to improve the design of biodegradable stents, it is necessary to predict how a given stent structure degrades over time in response to cyclic loading. This complex and related problem of contact mechanics and radial loading, although difficult to monitor and understand in typical physiological conditions, can be solved with many FEM packages. Soares et al. (Soares et al., 2007, 2008; Soares, Jr, & Rajagopal, 2010; Soares, Rajagopal, & Moore Jr, 2010) proposed a description of material properties that include the effects of mechanical deformation under conditions of accelerating biodegradation over time. This is the basic model according to which the material is characterized by a neo-Hookean, purely elastic model, at fixed degradation.

The energy density function for neo-Hookean material depends on the shear modulus and the first principal invariant of the right Cauchy-Green deformation tensor.

The shear modulus is usually a material constant estimated from load-deformation experiments; but, in the case of a biodegradable material, this material property changes over time as the material degrades.

The degradation parameter is a 3D spatially and time-dependent degradation field, and for completely degraded material it has the value 1 while for undegraded material it has the value 0. Soares et al. (2008) proposed a deformation rate function the case of uniaxial extension. As mentioned above, using biodegradable materials for implantable stents brings numerous advantages. Unfortunately, their performance is not sufficiently well characterized, either in silico or in vivo. In this chapter, simulations on the DES type of stent that finds the greatest application in the treatment of coronary heart disease are presented as well as experimental testing and comparison of results.

The finite element model for a DES stent [SYNERGY BP (bioabsorbable polymer) Everolimus-Eluting Platinum Chromium Coronary Stent—Boston Scientific Limited] and material model is based on experimental curves implemented in FEA. In the Results section inflation test, radial force test, crush test with two plates, local compression, longitudinal tensile strength, and three-point bending are presented. A comparison of the results between simulation and real experiments is given in the form of graphs and coefficient of determination and correlation coefficient.

2 Methodology

The finite element method (FEM), sometimes referred to as finite element analysis (FEA), is a computational technique used to obtain approximate solutions of boundary value problems in engineering. Simply stated, a boundary value problem is a mathematical problem in which one or more dependent variables must satisfy a differential equation everywhere within a known domain of independent variables and satisfy specific conditions on the boundary of the domain. Boundary value problems are also sometimes called field problems. The field is the domain of interest and most often represents a physical structure. The field variables are the dependent variables of interest governed by the differential equation. The boundary conditions are the specified values of the field variables (or related variables such as derivatives) on the boundaries of the field. Depending on the type of physical problem being analyzed, the field variables may include physical displacement, temperature, heat flux, and fluid velocity, to name only a few (Hutton, 2003). For solving these simulations, we used the in-house FEM solver PAK (Kojic et al., 2001).

2.1 PAK software

The PAK solver performs solid analysis for material and geometrical nonlinear problems as well as contact problems with preprocessing and postprocessing of stent geometries, and returns structured 3D data grids containing the strain, stress, and pressure. In the general PAK analysis module, three components are implemented:

Preprocessing unit: The preprocessing unit uses the reconstructed geometry from the CAD file for generating the 3D finite element mesh, optimizing and adapting it in the appropriate format for complex structural flow simulations. The user sets different parameters from the graphical user interface (GUI) and generates the model in accordance with those parameters. The whole model is exported in a textual DAT file.

Finite element solver: The finite element solver is the main processing unit created using the PAK finite element solver software that uses a DAT file generated by the preprocessing

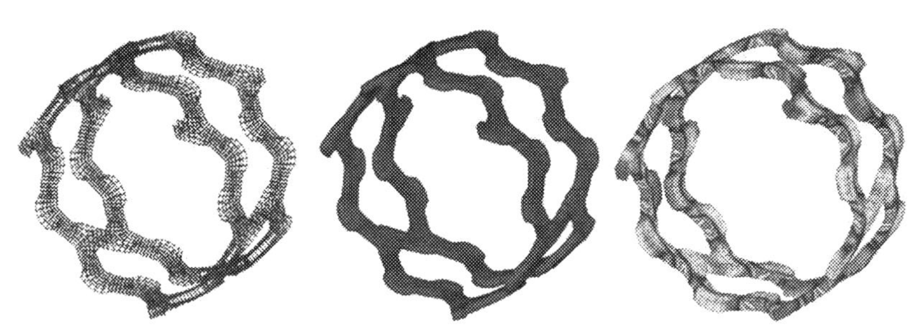

FIG. 1 Generating finite element model and simulation results.

unit for performing mechanical structure simulations. All simulation results are stored in a UNV textual file for every mesh node and element.

Postprocessing unit: The postprocessing unit uses the UNV file generated in the previous step to import all data, extract the parameters' results in a structured 3D grid and, finally, save the data to the repository for further visualization. BIOIRC created a special visualization component in Windows Forms and WPF utilizing OpenGL-based technologies. The component is integrated to the main window and the final user can inspect the parameters' results in 3D by rotating, zooming, and/or cutting the 3D grid. The same process is used for pre- and poststenting use cases and models as well. Also, all results were generated in the VTK file format for visualization on a cloud platform (Fig. 1).

2.2 Finite element model for SYNERGY BP

The SYNERGY BP Everolimus-Eluting Platinum Chromium Coronary Stent is a partially Bioresorbable Vascular Scaffold (BVS) produced by Boston Scientific Limited (Galway, Ireland). The geometric model of the SINERGY BP stent was modeled in a precrimping configuration. Detailed geometry parameters are presented in Fig. 2. The main part of the stent—strut has 89 μm with 79 μm thickness, while after expansion the stent outer diameter OD is

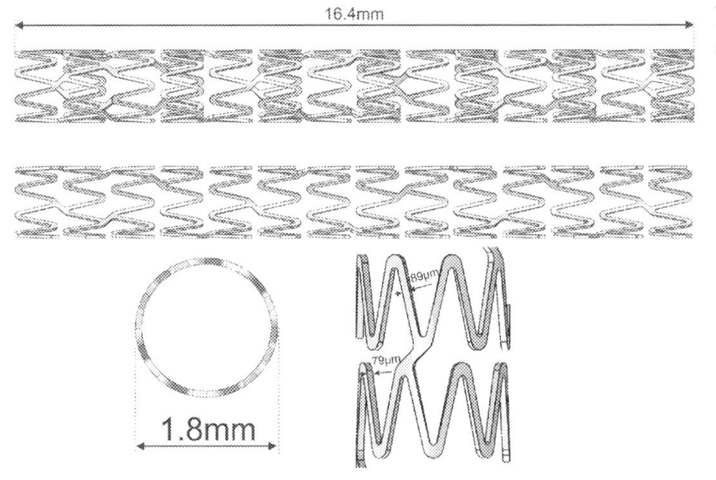

FIG. 2 SYNERGY BP stent detailed geometry parameters.

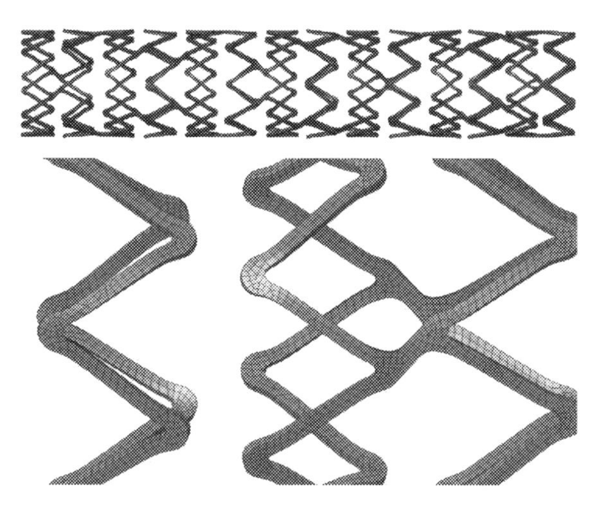

FIG. 3 SYNERGY BP discretized model in hexahedral elements.

3 mm and nominal length of 16.17 mm. The aim of the simulations study is to perform structural analyses of the stent device; it is important to build a representative discretized model of the stent with 36,786 hexahedral elements (Fig. 3).

2.3 Material model

The SYNERGY BP stent is produced from a platinum-chromium alloy (Pt—Cr), which exhibits bilinear elastoplastic behavior. For this device, the following material parameter values were used from a paper by O'Brien et al. (2010). The values for the material density, Young's modulus, Poisson's ratio, and yield stress were provided by the manufacturer (Table 1). For solving model in-house a PAK (Kojic, Filipovic, Slavkovic, Zivkovic, & Grujovic, 1996) solver was used.

TABLE 1 Material property SYNERGY BP.

Material property SYNERGY BP material model	
Density	**9.9e − 09 ton/mm^3**
Elastic	
Young's modulus	203,000 MPa
Poisson's ratio	0.3
Plastic	
Yield stress (MPa)	Plastic strain
480.96	0.
1207.63	0.37

3 Results

According to the ISO 25539 standard, the following standard tests have been performed: Simulated use—pushability, torquability, trackability, recoil, crush resistance, flex/kink, longitudinal tensile strength, crush resistance with parallel plates, local compression, radial force, foreshortening, dog boning, three-point bending, inflation, and radial fatigue test. Some of those tests request very complex boundary conditions and usually not enough parameters for defining a good input FE model. But, the information necessary for engineers for developing a stent faze are: radial and axial conformability. This information can be obtained from only a few of the listed tests (inflation test, radial force test, crush test with two plates, local compression, longitudinal tensile strength, three-point bending), it is not necessary to simulate all the tests from the ISO 25539 standard. During the stent development phase engineers need to understand where our stent design is in relation to a competing model, and this can be very easily concluded based on just a few tests which allow us to compare axial and radial conformability.

3.1 Inflation test

The purpose of an inflation test is to determine the diameter required to inflate the balloon to the nominal recommended pressure by analyzing the outer diameter (measured in three positions: proximal, middle, distal) and length. From the inflation test, results for recoil, foreshortening, and dog boning can be obtained. Foreshortening in very important test and determines the length to diameter relationship of the stent. Foreshortening represents the difference between the length of the stent in the crimped and expanded states. The dog-boning test evaluates the difference between the diameter of the inflated implant on the proximal and distal ends and on the middle of the implant when the implant is expanded under the maximum recommended inflation pressure. All of this information might be useful for planning clinical treatment. The results from the test are presented in the form of diameter/pressure curves. In order to simulate physiological conditions, testing occurs at 37°C temperature in pH7.4 phosphate-buffered saline. That environment, pH, and temperature represent the clinically relevant environment. For the in vitro experiment, a special equipment was used. For inflation, a computer-controlled Nexus 5000 with 99.96% accuracy syringe pump was used. A very precise pressure sensor with 0.1% measurement error and a Keyence laser optical micrometer was used for the measurement of diameter (Fig. 4). Each sample is fixed in a specially designed fixator.

After the start of the inflation process, on every increase of 50 kPa the pump temporarily stops and performs the stent outer diameter measurement with the laser micrometer, and then the process continues. Stent manufacturers use a variety of methods to determine the optimal stent inflation rate. They usually make a compromise between the inflation rate and internal stress in the stent material.

The aim of manufacturers is to perform the surgical procedure as quickly as possible without compromising the integrity of the implant. Achieving this goal is a very demanding process and it very much depends on the experience of manufacturer engineers.

Under laboratory conditions to obtain consistently test results a low inflation speed is used (around 3 kPa/s) which is much less than the manufacturer's suggestion; in this way, we

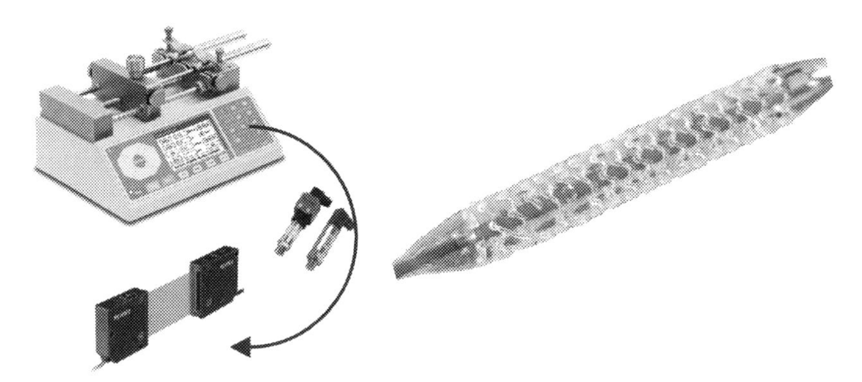

FIG. 4 Inflation test—Experimental set up.

provide the balloon material more time to adapt to new pressures and avoid possible inflation speed impact on the balloon material deformation. This inflation speed does not influence the numerical results. In silico simulations, the stent is positioned in the way that the central stent axis is collinear with the Z axis. All nodes at one end were fixed node movement in the Z direction. Small inflation speed in simulations provides numerical stability. To simulate inflation, the pressure is set on the internal surface of the balloon. The contact boundary condition between the stent internal surfaces and the balloon outer surfaces is set. All the boundary conditions for mimicking the real test are presented in the Fig. 5.

The stress distribution from the simulation of the inflation test of the SYNERGY BP device is presented in Fig. 6. Comparison of diameter-pressure curve results, between real experimental data and simulations for the SYNERGY BP devices, is presented in Fig. 7. In the real inflation test 14 SYNERGY BP units were tested, but only two representative diameter-pressure curves from actual tests are presented in Fig. 7. However, the data from all

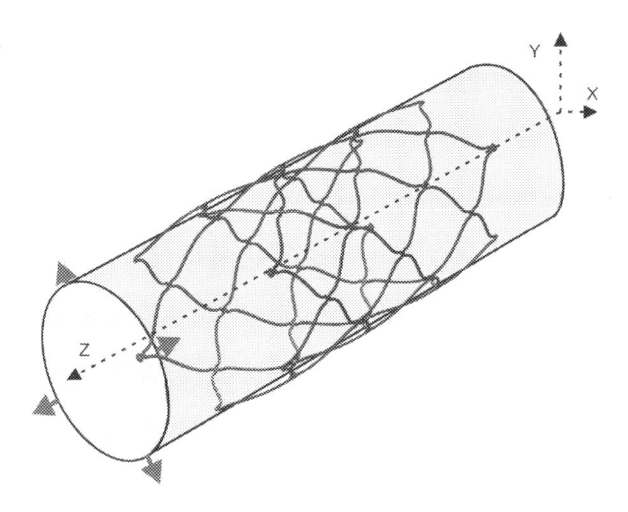

FIG. 5 Inflation test—Boundary condition set up.

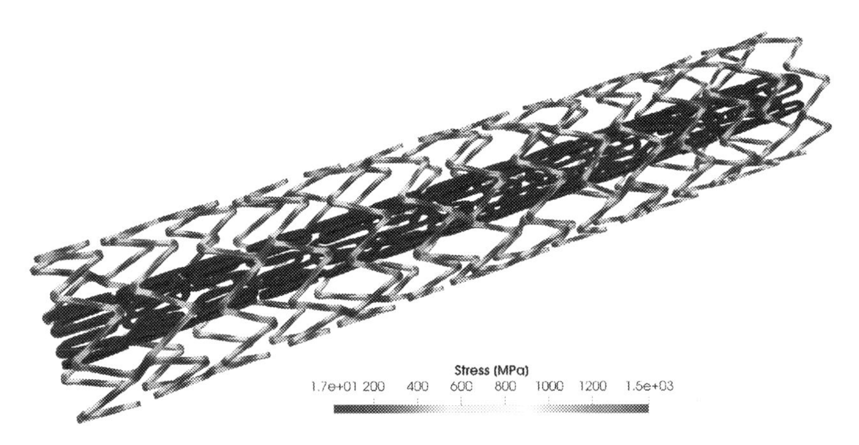

FIG. 6 Inflation test—Stress distribution field.

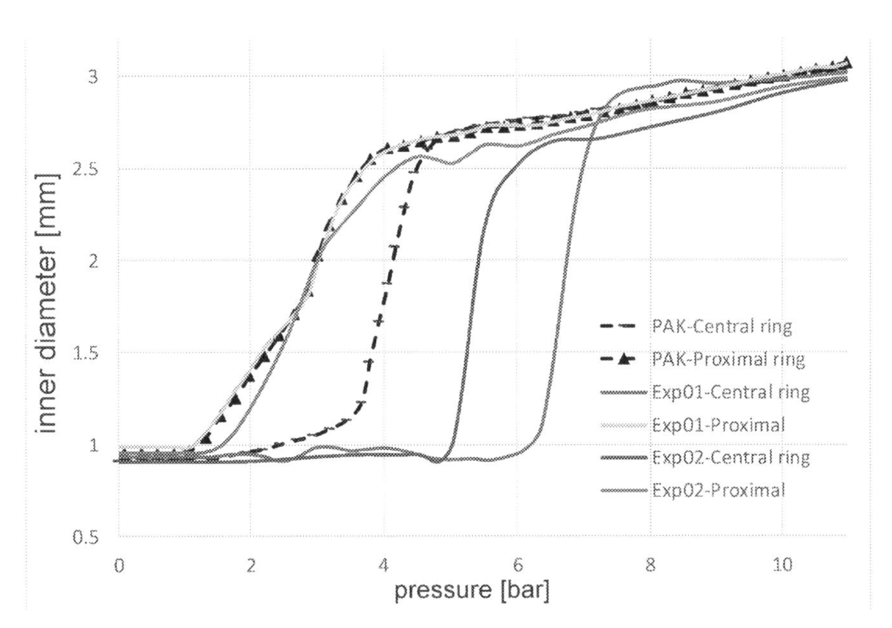

FIG. 7 Inflation test—Diameter-pressure curves.

SYNERGY BP, $n = 14$ curves were used in the calculation of the R^2 and the results are reported in Table 2.

From the simulation of the inflation test are also obtained results for foreshortening and dog-boning (Table 3).

3.2 Radial force test

The main purpose of this test is to determine the load/deformation characteristics of the stent, while a circumferentially uniform radial load is applied. The implant is compressed

TABLE 2 Inflation test results.

Inflation test results	
Position of measurements	
Proximal/distal—Coefficient of determination R^2	0.9990
Proximal/distal—Correlation coefficient R	0.9995
Proximal/distal—Significance level	$P < 0.0001$
Center position—Coefficient of determination R^2	0.7770
Center position—Correlation coefficient R	0.8815
Center position—Significance level	$P < 0.0001$

TABLE 3 Foreshortening and dog-boning test results.

Foreshortening results	
Length change	**mm (%)**
Experiment data	0.23 (0.071)
In-silico data	0.21 (0.065)
Dog-boning	
Experiment data	3.25%
In-silico data	3.1%

using a uniform compression, starting with an outer diameter equal to the maximum indicated vessel diameter. The "Mylar" loop crimping device is used to apply radial force on the outer surface of the stent. The load and the associated diameter are recorded while compressing the stent until an appreciable reduction on force occurs or a diameter reduction of at least 50% is reached. The sample is fixed inside of the "Mylar" loop (Fig. 8). The axial movement on the machine pulls one end of "mylar" material and "closing" the loop. Decreasing loop diameter perform radial compression device. During the whole process, the machine measures load and displacement.

The stent is positioned in such a way that the central stent axis is collinear with the Z axis. Node movement at one end of the stent is fixed in the Z direction. The radial force is set on the outer surface of the cylinder placed outside of the stent with the purpose of simulating the "Mylar" loop crimping device. The contact boundary condition is set between the stent outer surfaces and the cylinder internal surfaces. Boundary conditions for mimicking the real test are presented in Fig. 9.

The stress distribution from the simulation of crimping the SYNERGY BP device has been presented in Fig. 10.

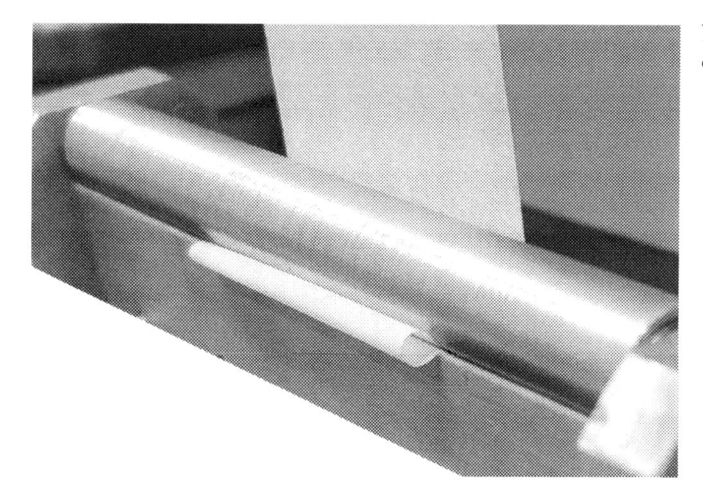

FIG. 8 Radial force test—"Mylar" loop crimping device.

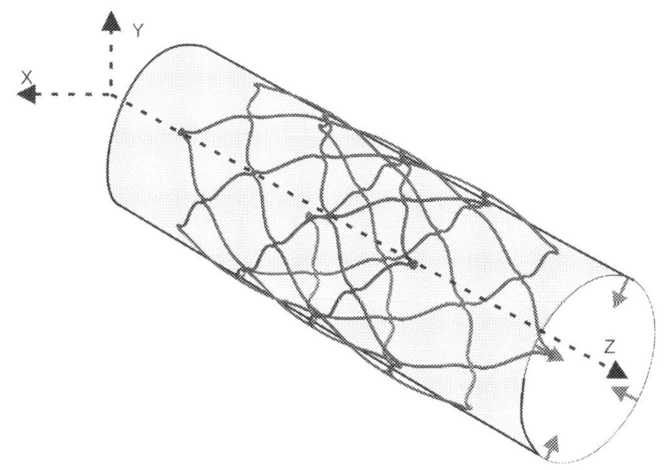

FIG. 9 Radial force test—Boundary condition set up.

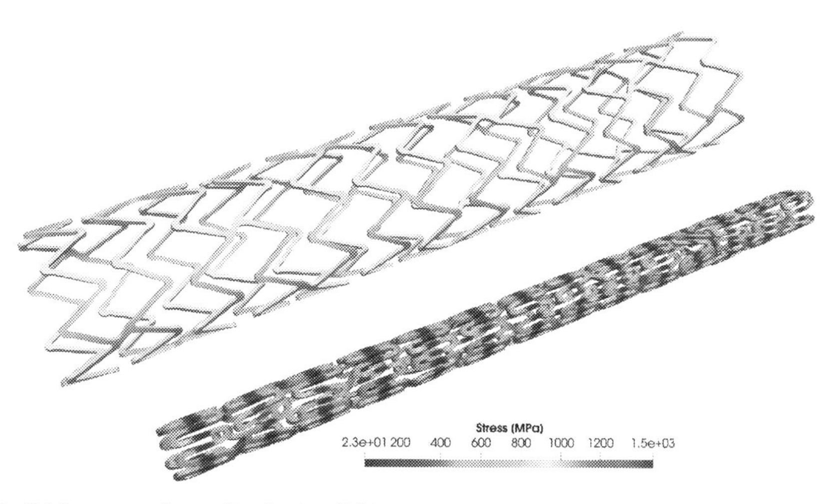

FIG. 10 Radial force test—Stress distribution field.

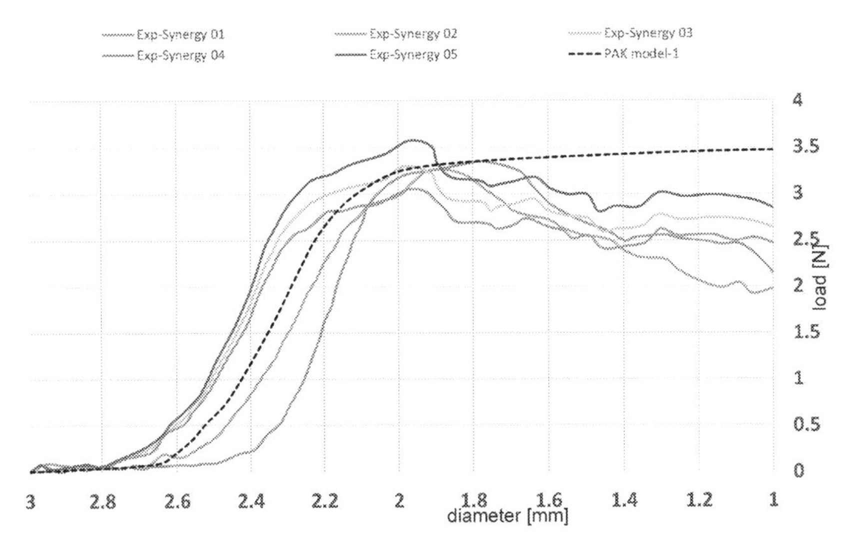

FIG. 11 Radial force test—Diameter–load curve results.

Comparison of the diameter-load curve results, between real experimental data and the simulation, is presented in Fig. 11. In the experimental compression test 5 SYNERGY BP device units were tested. The data from all SYNERGY BP: $n=5$ curves were used in the calculation of the R^2 and the results are reported in Table 4.

Comparison of the real experimental data for stent recoil and the results of the simulation required simplification in order to allow interpretation to take place. A comparison of the mean values between the real test and the simulation is presented in Table 5.

TABLE 4 Radial force test results.

Radial force test	SYNERGY BP $N=5$
Coefficient of determination R^2	0.8099
Correlation coefficient R	0.9
Significance level	0.0001

TABLE 5 Recoiling results.

Model	SYNERGY BP
Experiment data	3.0%
In-silico data	2.79%

3.3 Crush test with two plates

This testing is defined by ISO 25339 as well as part of ASTM F2081. This testing is applicable to the clinical application of the 3.0 mm × 16 mm SYNERGY stent from BSL; the same dimensions were deployed into all IFU-specified blood vessel sizes.

According to the standard, the purpose of the crush test with two planes is to determine the load required to cause clinically relevant buckling or a deflection equivalent to a diameter reduction of at least 50%. In this test method, it is also necessary to determine the load/deformation characteristics of the stent while a uniform axial load is applied.

The testing occurs at room temperature but before the start of the test all samples spend about 30 min in pH7.4 phosphate-buffered saline at 37°C. For the test environment, pH and temperature are selected to simulate a clinically relevant environment. Plate crush tests were conducted on 3.0 mm × 16 mm SYNERGY stents though the AMETEK Brookfield tensile machine equipped with a load cell of maximum 50 N, resolution of 0.00490 N, and displacement accuracy of 0.05 mm. The sample is fixed between two plates (Fig. 12). Start the axial load machine compressing the device. During the whole process, the machine measures load and displacement to build a force-displacement curve. The protocol consists of firstly moving the upper plate in the displacement-controlled mode toward the stent until a permanent deformation is caused and, secondly, by slowly raising the upper plate to unload the stent structure. The test was performed on five partially SYNERGY stents. Comparing the force-displacement results from Fig. 15 in all tests samples the elastic and plastic phases are deemed repeatable.

In order to simulate a real boundary condition, the stent is positioned in such a way that the central stent axis is collinear with the Z axis, between the two rigid plates (top and bottom). All nodes at one end had fixed node movement in the Z direction. All stent nodes located in the XZ coordinate plane are bounded in the Y direction and all stent nodes located in the YZ coordinate plane are bounded in the X direction to achieve a more stable simulation. The bottom plate is fixed and axial force is set on the top plate. The contact boundary condition is set

FIG. 12 Crush test with two plates experimental set up.

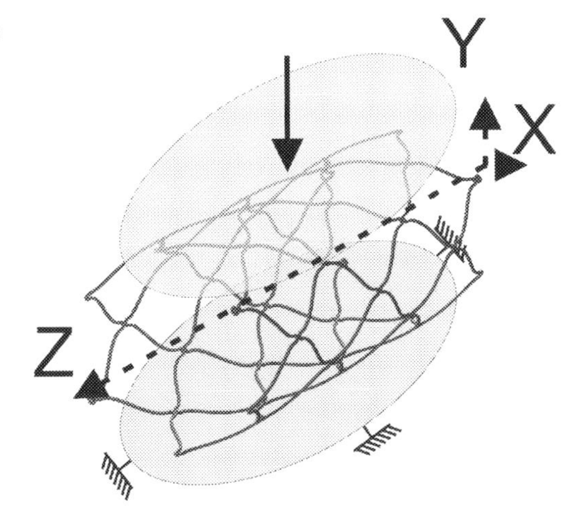

FIG. 13 Crush test with two plates boundary condition set up.

between the stent outer surfaces and the surfaces of the plates. All the boundary conditions for mimicking the test are presented in Fig. 13. The stress distribution from the crush test of SYNERGY BP device is presented in Fig. 14.

In the form of graph comparison of the diameter-load curve results, between real experimental data and the simulation, is presented in Fig. 15 SYNERGY BP devices.

In the real crush test 5 SYNERGY BP units were tested. The data from all SYNERGY BP: $n=5$ curves were used in the calculation of the R^2 and the results are reported in Table 6.

3.4 Local compression

According to the standard, the local compression test determines the deformation of the device in response to a localized compressive force to a diameter reduction of at least 50%.

The test was performed on an axial load testing device AMETEK Brookfield tensile machine equipped with a very precise load cell of 50 N, resolution of 0.00490, and displacement accuracy of 0.05 mm. The testing occurs at room temperature but before the start of the test all samples spend about 30 min in pH7.4 phosphate-buffered saline at 37°C. For the test

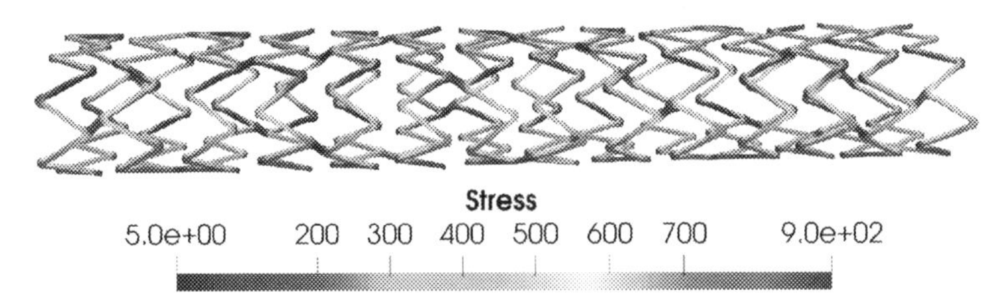

FIG. 14 Crush test with two plates—Stress distribution field.

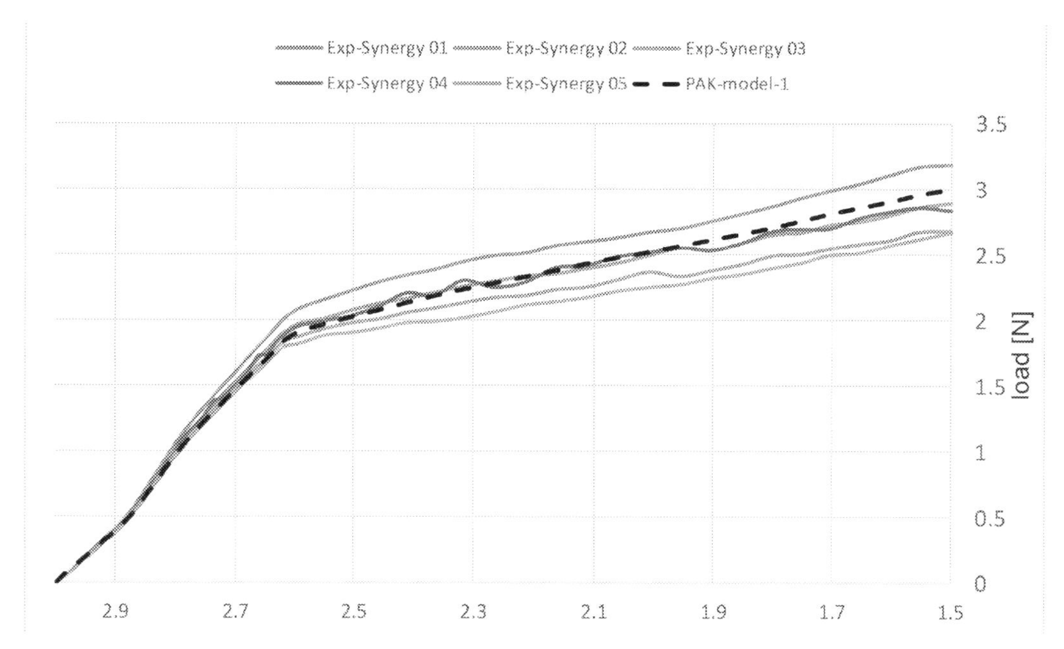

FIG. 15 Crush test with two plates—Diameter–load curve results.

TABLE 6 Crush test, two plates test results.

Crush test, two plates test results	SYNERGY BP $N=5$
Coefficient of determination R^2	0.9957
Correlation coefficient R	0.9979
Significance level	0.0001

environment, pH and temperature are selected to simulate a clinically relevant environment. The sample is fixed between the plate and trapped spike with a diameter of 1 mm (Fig. 16). The axial load machine compressing the device at a localized point is started. During the whole process, the machine measures force and displacement. Two 3.0 mm × 16 mm SYNERGY stents in their open (diameter 3 mm) configuration were tested.

In this simulation, the stent is positioned in such a way that the central stent axis is collinear with the Z axis, between two plates (small plate on the top cylindrical shape 1 mm in diameter and large plate on the bottom). Nodes movement at one end of the stent is fixed in the Z direction. All stent nodes located in the XZ coordinate plane are bounded in the Y direction and all stent nodes located in the YZ coordinate plane are bounded in the X direction. Axial force is set on the small top plate. The bottom plate is fixed. The contact boundary condition is set between the stent outer surfaces and plate surfaces. Boundary conditions for mimicking the test are presented in Fig. 17.

FIG. 16 Local compression set up.

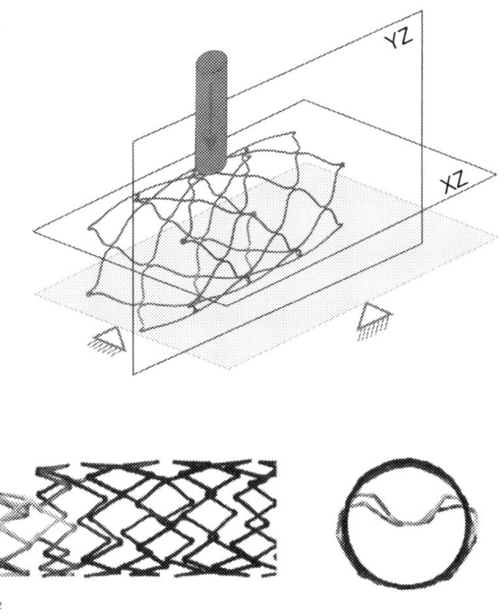

FIG. 17 Local compression—Boundary set up.

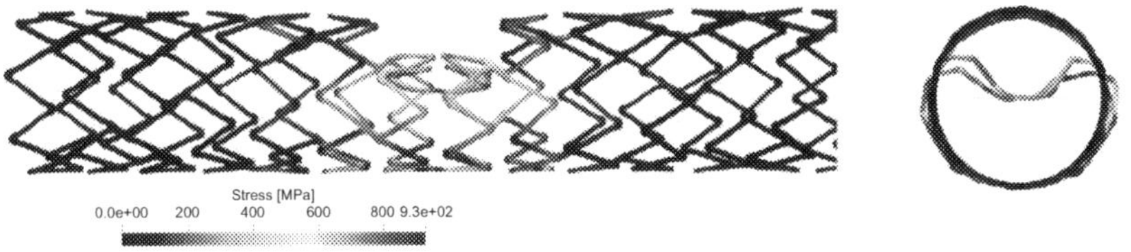

FIG. 18 Local compression—Stress distribution field.

The stress distribution from the crush test of the SYNERGY BP device is presented in Fig. 18.

In the form of a graph comparison of the displacement–load curve results, between real experimental data and the simulation is presented in Fig. 19.

In the real local compression 5 SYNERGY BP units were tested. The data from all SYNERGY BP: $n=5$ curves were used in the calculation of the R^2 and the results are reported in Table 7.

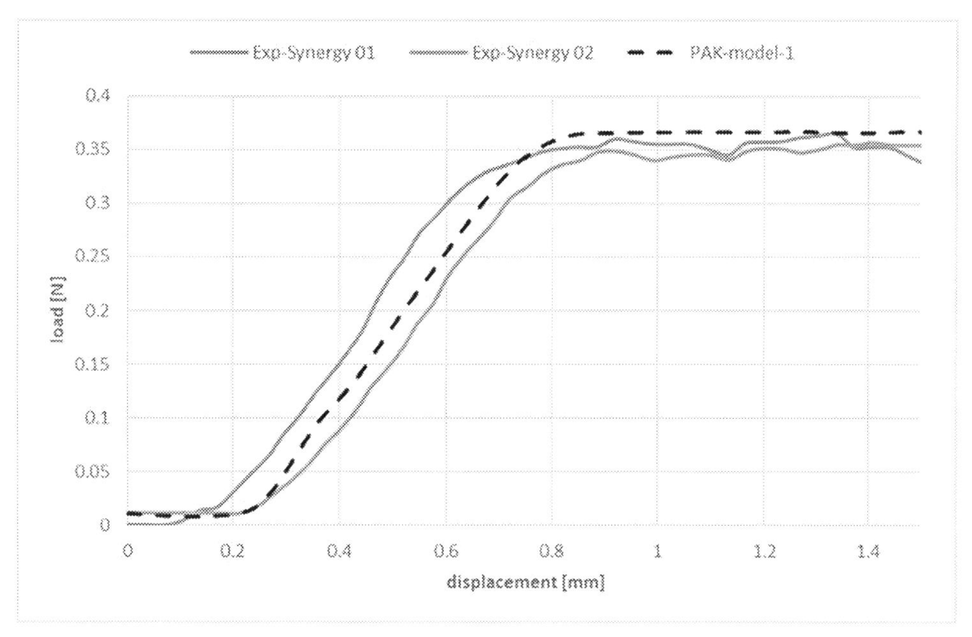

FIG. 19 Local compression test—Displacement–load curve results.

TABLE 7 Local compression test.

Local compression test results	SYNERGY BP $N=5$
Coefficient of determination R^2	0.9967
Correlation coefficient R	0.9984
Significance level	0.0001

3.4.1 Longitudinal tensile strength

The purpose of this test is to determine the longitudinal tensile strength of the struts, joints, and/or fixed connections of a stent device. A tensile load is applied in the longitudinal direction until the tested bond breaks or loses functional integrity. The test was performed on an axial load testing device with very precise low force load cell and a specially designed tool for small sized devices. For the first test a 9.8 N load cell was used, with a resolution of 0.00098 N. In the second test 50 N was used, with a resolution of 0.0049 N. In both cases, a 0.05 mm displacement resolution was used. The testing occurs at room temperature but before the start of the test all samples spend about 30 min in pH 7.4 phosphate-buffered saline at 37°C. For the test environment, pH and temperature are selected to simulate a clinically relevant environment. The test was performed on the unstressed SYNERGY stents. The sample is fixed between specially designed fixtures (Fig. 20). The axial tensile load is started. In the test, the material elastic response is analyzed. During the whole process, the machine measures load and displacement.

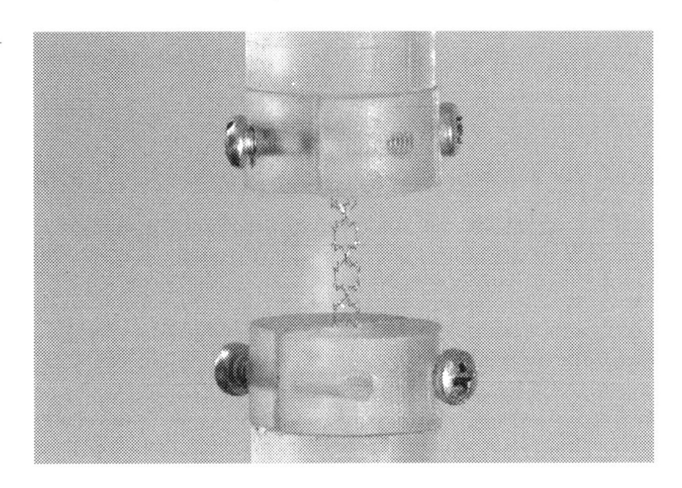

FIG. 20 Longitudinal tensile strength—Experimental set up.

Tensile tests were identified as the simplest experiments to characterize the mechanical response of coronary stents. During the traction test, imposing a displacement such as to bring the device to work in the plastic field, it is possible to carry out an assessment of the yield stress of the material. In this simulation, the stent is positioned in the way that the central stent axis is collinear with the Z axis. All stent nodes located in the XZ coordinate plane are bounded in the Y direction and all stent nodes located in the YZ coordinate plane are bounded in the X direction. Nodes on one side of the stent are fixed in the Z direction, and axial force is set on the nodes on the opposite side in the Z direction. Boundary conditions for mimicking the test are presented in Fig. 21.

Fig. 22 presents the stress distribution from the simulation of the longitudinal tensile strength testing of the SYNERGY device. Comparison of the displacement-force curve results, between real experimental data and the simulation are presented in Fig. 23.

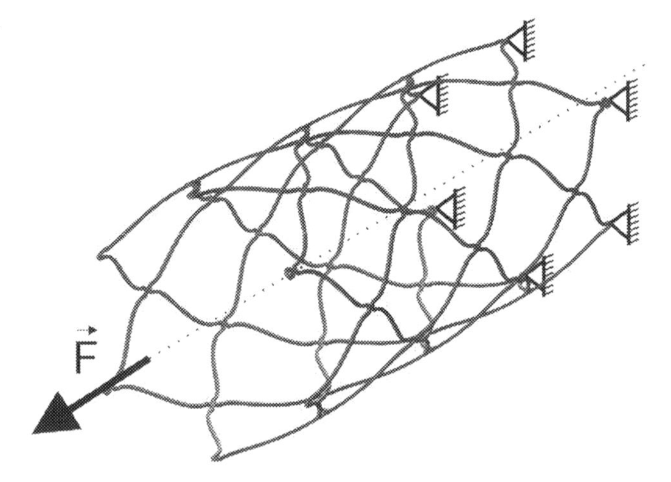

FIG. 21 Longitudinal tensile strength—Boundary condition set up.

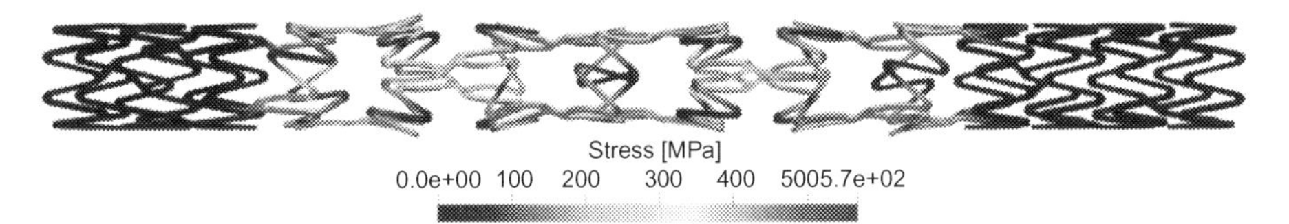

FIG. 22 Longitudinal tensile strength—Stress distribution field.

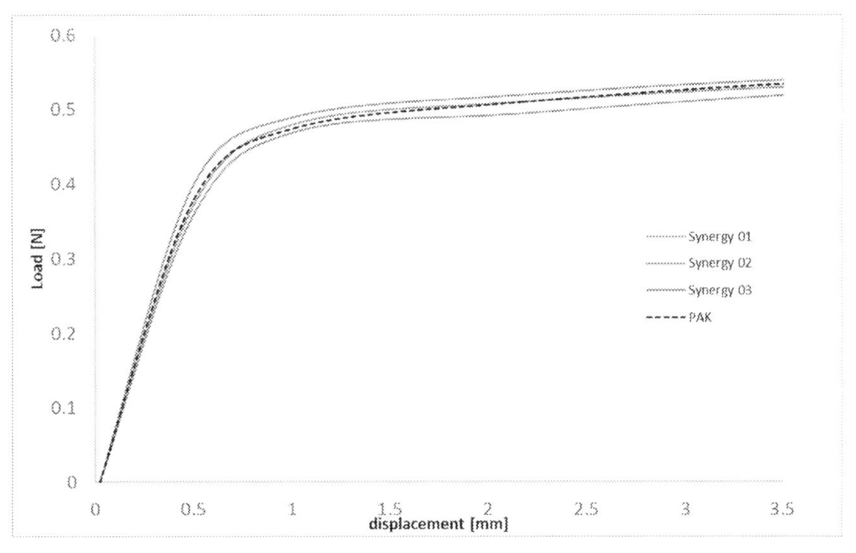

FIG. 23 Longitudinal tensile strength—Displacement–load curves result.

In the real inflation test, 5units were tested. Only two representative diameter-pressure curves from the actual test articles are shown in Fig. 23. However, the data from all $n=5$ curves were used in the calculation of the R^2 and the results are reported in Table 8.

By analyzing displacement-force curves created from simulation and from real experiment results, it can be concluded that the longitudinal tensile test simulation mimics the real longitudinal tensile test with very good precision.

TABLE 8 Statistical analysis of results—Longitudinal tensile test.

Model	SYNERGY $N=5$
Coefficient of determination R^2	0.9944
Correlation coefficient R	0.9984
Significance level	0.0001

3.5 Three-point bending

The ASTM F2606 test standard provides guidelines for quantitatively characterizing balloon-expandable stent flexibility using three-point bending procedures. It is provided for characterizing deployed stent flexibility. The test was performed on an axial load testing device AMETEK Brookfield machine with a displacement accuracy of 0.05 mm, equipped with very precise low force load cell of 9.8 N (resolution of 0.00098 N) and specially designed tool for small sized devices. The testing occurs at room temperature but before the start of the test all samples spend about 30 min in pH 7.4 phosphate-buffered saline at 37°C. For the test environment, pH and temperature are selected to simulate a clinically relevant environment. The sample is fixed between 3 point band fixtures (Fig. 24). The axial load machine compressing device at a localized point is started. During the whole process, the machine measures load and displacement.

In this simulation, the stent is positioned in such a way that the central stent axis is collinear with the Z axis. All stent nodes located in the XZ coordinate plane are bounded in the Y direction and all stent nodes located in the XY coordinate plane are bounded in the Z direction. Three rigid body cylinders are positioned in such a way that their central axis is collinear with the X axis. Axial force is applied via the top cylinder, and the bottom cylinders are fixed. The contact boundary condition is set between the stent outer surfaces and the outer surfaces on cylinders. Boundary conditions for mimicking the test are presented in Fig. 25.

Fig. 26 presents the stress distribution from the simulation of three-point bending testing off the SYNERGY device. The comparison of the displacement-force curve results, between real experimental data and the simulation is presented in Fig. 27. In the real inflation test 10 units were tested; only two representative diameter-pressure curves from actual test articles are shown in Fig. 27. However, the data from all $n=10$ curves were used in the calculation of the R^2 and the results are reported in Table 9.

By analyzing displacement-force curves created from simulation and from real experiment results, it can be concluded that the three-point test simulation mimics the real three-point test with very good precision. The precision of simulation is quantified by the coefficient of

FIG. 24 Three point bending test—Experimental set up.

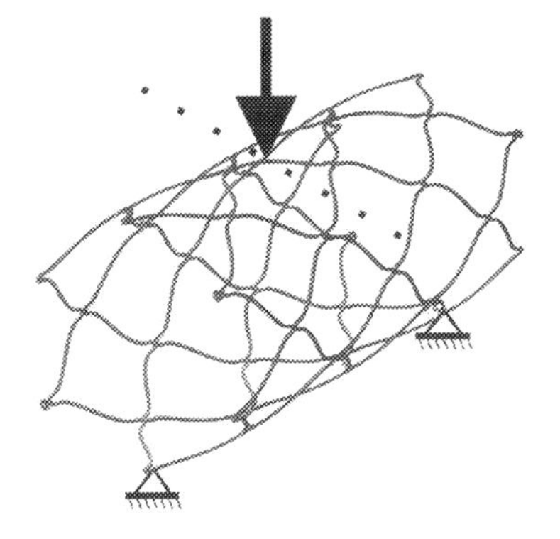

FIG. 25 Three-point bending test—Boundary condition set up.

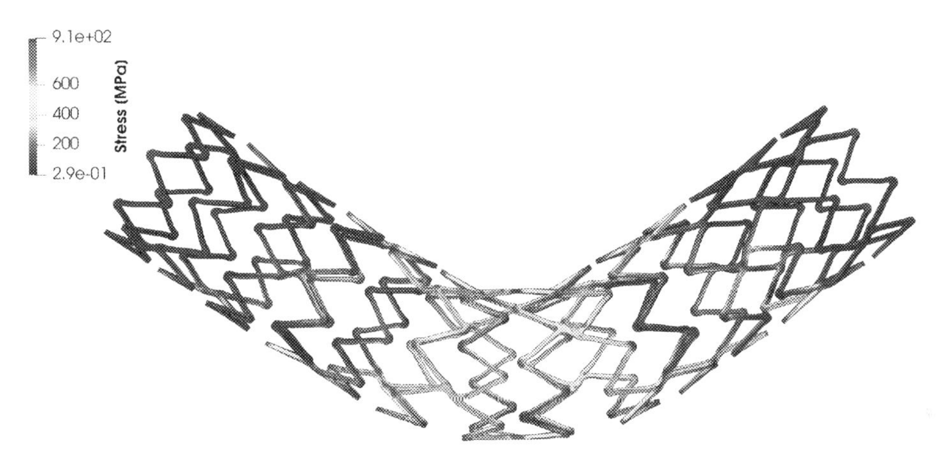

FIG. 26 Three point bending test—Stress distribution field.

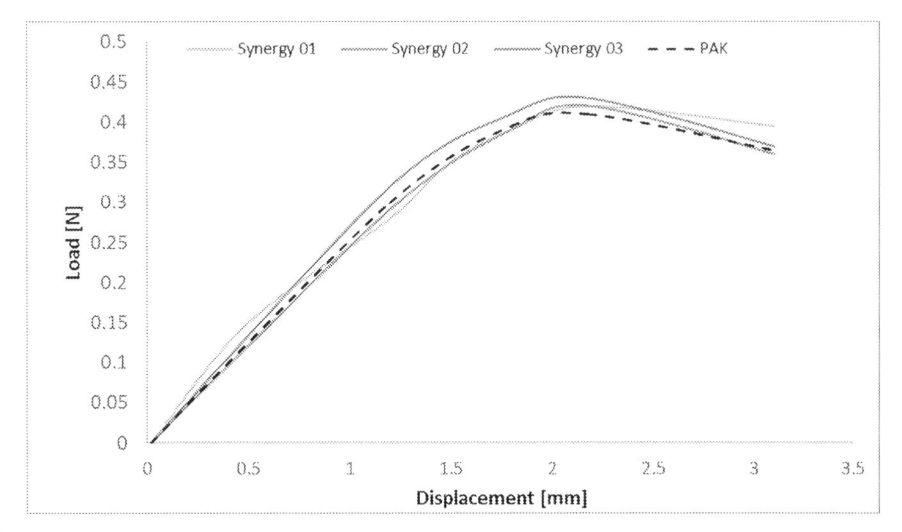

FIG. 27 Three-point bending test—Displacement–load curves result.

TABLE 9 Statistical analysis of results—Thee point bending.

Model	SYNERGY $N=10$
Coefficient of determination R^2	0.99
Correlation coefficient R	0.995
Significance level	0.0001

determination (R^2) and correlation coefficient (R) calculated between the simulation curve and all points from all experimental curves. There is a strong correlation between simulation and real experiments for the SYNERGY device quantified with $R^2=0.99$ and $R=0.995$, which showed that data points lie within mean $\pm$ 1.96 standard deviations.

The main purposes of simulations are to show that axial and radial conformability can be very precisely simulated and compared between two different devices. In this way, designing and optimization of stent devices can be faster and the whole process will cost less.

By comparing diameter-pressure curves (Inflation test: Fig. 7), displacement–load curves (Local compression, Fig. 19, Longitudinal tensile strength, Fig. 23, Three-point bending tests, Fig. 27) and diameter-load curves (Radial force test, Fig. 11, Crush test with two plates, Fig. 15) created from the simulation with data from real experiments, it can be concluded that the simulation mimics the real inflation test with very high precision. The precision of simulation is quantified by the coefficient of determination (R^2) and correlation coefficient (R) calculated between the simulation curve and all points from all experimental curves.

4 Conclusions

In this study, authors presented in vitro and in silico mechanical tests of stent devices—SYNERGY BP representing a partially BVS device from Boston Scientific Limited. A comparison between the results of the real test and the simulation are presented in the form of comparable curves. In this chapter inflation test, radial force test, crush test with two plates, local compression, longitudinal tensile strength, and three-point bending tests are presented. A comparison of the test results interpreted in the form of graphs (diameter-load curves or pressure-diameter curves) for the partially BVS SYNERGY BP (device already approved for commercialization) show very good correlation with results from the real mechanical tests. For most of the results, there are strong correlations between simulation and real experiments for the coefficient of determination ($R^2>0.99$) and the correlation coefficient ($R>0.99$).

In the inflation test for the SYNERGY BP stent, the coefficient of determination ($R^2=0.777$) and the correlation coefficient ($R=0.8815$) show a little "deviation" at the central stent position. This can be attributed to the speed of process transition between the real test and the simulation.

In the radial force test, the coefficient of determination ($R^2=0.8099$) and the correlation coefficient ($R=0.9$) are lower in the case of the SYNERGY BP device due to the set up of the radial mechanical test for a small stent diameter.

In the crush test with two plates, the coefficient of determination ($R^2 = 0.8099$) and the correlation coefficient ($R = 0.9$) are lower in the case of the SYNERGY BP device due to the set up of the radial mechanical test for a small stent diameter.

In local compression, the coefficient of determination ($R^2 = 0.9961$) and the correlation coefficient ($R = 0.9981$) represent the very good correlations between simulation and real experiments.

In the longitudinal tensile strength test, the coefficient of determination ($R^2 = 0.9944$) and the correlation coefficient ($R = 0.9984$) in the elastic phase of simulation represent the strong correlations between simulation and real experiments.

In the three-point bending, the coefficient of determination ($R^2 = 0.99$) and the correlation coefficient ($R = 0.995$) represent the strong correlations between simulation and real experiments.

All simulations need several hours of preparation for boundary conditions, prescribing material models, material properties, and nonlinear contact. The execution of the simulation takes a few hours, which is attributed to the complexity of most of the tests. The residual stress and initial geometry for different tests derive directly from the previous running tests such as crimping and expanding. Another huge advantage of computer simulations is in the additional results obtained from them, such as the field of stress distribution and deformation. In this way, it is possible to recognize critical places in the design at an early stage of development. It has been shown that in silico tests can mimic most of the appropriate ISO standards for mechanical in vitro stent devices testing. It opens a new avenue for in silico tests, which can partially or fully replace the real mechanical testing required for submission to a regulatory body.

References

Borisch, I., Hamer, O. W., Zorger, N., Feuerbach, S., & Link, J. (2005). In vivo evaluation of the carotid wallstent on three-dimensional contrast material-enhanced MR angiography: Influence of artifacts on the visibility of stent lumina. *Journal of Vascular and Interventional Radiology*, *16*(5), 669–677.

Filipovic, N., Nikolic, D., Saveljic, I., Milosevic, Z., Exarchos, T., Pelosi, G., & Parodi, O. (2013). Computer simulation of three-dimensional plaque formation and progression in the coronary artery. *Computers and Fluids*, *88*, 826–833.

Filipovic, N., Rosic, M., Tanaskovic, I., Milosevic, Z., Nikolic, D., Zdravkovic, N., Peulic, A., Kojic, M. R., Fotiadis, D. I., & Parodi, O. (2012). ARTreat project: Three-dimensional numerical simulation of plaque formation and development in the arteries. *IEEE Transactions on Information Technology in Biomedicine*, *16*(2), 272–278. https://doi.org/10.1109/TITB.2011.2168418.

Frank, A. O., Walsh, P. W., & Moore, J. E., Jr. (2002). Computational fluid dynamics and stent design. *Artificial Organs*, *26*(7), 614–621.

Hussain, M., Greenwell, T. J., Shah, J., & Mundy, A. (2004). Long-term results of a self-expanding wallstent in the treatment of urethral stricture. *BJU International*, *94*(7), 1037–1039.

Hutton, D. V. (2003). *Fundamentals of Finite Element Analysis*. ISBN 9780072922363.

Kaiser, C., Brunner-La, R. H. P., Buser, P. T., Bonetti, P. O., Osswald, S., Linka, A., Bernheim, A., Zutter, A., Zellweger, M., Grize, L., & Pfisterer, M. E. (2005). Incremental cost-effectiveness of drug-eluting stents compared with a third-generation bare-metal stent in a real-world setting: Randomised Basel stent Kosten Effektivitats trial (BASKET). *Lancet*, *366*(9489), 921–929.

Kandzari, D. E., Tcheng, J. E., & Zidar, J. P. (2002). Coronary artery stents: Evaluating new designs for contemporary percutaneous intervention. *Catherization and Cardiovascular Interventions*, *56*(4), 562–576.

Kastrati, A., Mehilli, J., Dirschinger, J., Pache, J., Ulm, K., Schuhlen, H., Seyfarth, M., Schmitt, C., Blasini, R., Neumann, F. J., & Schomig, A. (2001). Restenosis after coronary placement of various stent types. *The American Journal of Cardiology*, *87*(1), 34–39.

Kojic, M., Filipovic, N., Živkovic, M., Slavkovic, R., & Grujovic, N. (2001). *PAK-FS finite element program for fluid-structure interaction*. Serbia: Kragujevac.

Martin, D., & Boyle, F. J. (2011). Computational structural modelling of coronary stent deployment: A review. *Computer Methods in Biomechanics and Biomedical Engineering*, *14*(4), 331–348. https://doi.org/10.1080/10255841003766845.

Migliavacca, F., Petrini, L., Colombo, M., Auricchio, F., & Pietrabissa, R. (2002). Mechanical behavior of coronary stents investigated through the finite element method. *Journal of Biomechanics*, *35*(6), 803–811. https://doi.org/10.1016/s0021-9290(02)00033-7.

Morice, M. C., Serruys, P. W., Sousa, J. E., Fajadet, J., Ban Hayashi, E., Perin, M., Colombo, A., Schuler, G., Barragan, P., Guagliumi, G., Molnar, F., & Falotico, R. (2002). A randomized comparison of a sirolimus-eluting stent with a standard stent for coronary revascularization. *The New England Journal of Medicine*, *346*(23), 1773–1780.

Morton, A. C., Crossman, D., & Gunn, J. (2004). The influence of physical stent parameters upon restenosis. *Pathol Biol (Paris)*, *52*(4), 196–205.

Narracott, A. J., Lawford, P. V., Gunn, J. P., & Hose, D. R. (2007). Balloon folding affects the symmetry of stent deployment: Experimental and computational evidence. *Annual International Conference of the IEEE Engineering in Medicine and Biology Society*, *2007*, 3069–3073. https://doi.org/10.1109/IEMBS.2007.4352976 (PMID: 18002642).

Nikolić, D., Radović, M., Aleksandrić, S., Tomasević, M., & Filipović, N. (2014). Prediction of coronary plaque location on arteries having myocardial bridge, using finite element models. *Computer Methods and Programs in Biomedicine*, *117*(2), 137–144. https://doi.org/10.1016/j.cmpb.2014.07.012.

O'Brien, B. J., Stinson, J. S., Larsen, S. R., Eppihimer, M. J., & Carroll, W. M. (2010). A platinum-chromium steel for cardiovascular stents. *Biomaterials*, *31*(14), 3755–3761. https://doi.org/10.1016/j.biomaterials.2010.01.146.

Ormiston, J. A., Currie, E., Webster, M. W., Kay, P., Ruygrok, P. N., Stewart, J. T., Padgett, R. C., & Panther, M. J. (2004). Drug-eluting stents for coronary bifurcations: Insights into the crush technique. *Catheterization and Cardiovascular Interventions*, *63*, 332–336.

Parodi, O., Exarchos, T., Marraccini, P., Vozzi, F., Milosevic, Z., Nikolic, D., Sakellarios, A., Siogkas, P., Fotiadis, D., & Filipovic, N. (2012). Patient-specific prediction of coronary plaque growth from CTA angiography: A multiscale model for plaque formation and progression. *IEEE Transactions on Information Technology in Biomedicine: a publication of the IEEE Engineering in Medicine and Biology Society*, *16*(5), 952–965.

Perry, M., Oktay, S., & Muskivitch, J. C. (2002). Finite element analysis and fatigue of stents. *Minimally Invasive Therapy & Allied Technologies*, *11*(4), 165–171.

Poerner, T. C., Ludwig, B., Duda, S. H., Diesing, P., Kalmar, G., Suselbeck, T., Kaden, J. J., Borggrefe, M., & Haase, K. K. (2004). Determinants of stent expansion in curved stenotic lesions: An in vitro experimental study. *Journal of Vascular and Interventional Radiology*, *15*, 727–735.

Kojic, M., Filipovic, N., Slavkovic, R., Zivkovic, M., Grujovic, N. (1996). PAK software, http://www.bioirc.ac.rs/index.php/software/5-pak (last visited: May 1, 2021).

Rieu R., Barragan P., Garitey V., Roquebert P.O, Fuseri J., Commeau P., Sainsous J., (2003) Assessment of the trackability, flexibility, and conformability of coronary stents: A comparative analysis. Catheterization and Cardiovascular Interventions, 59: 496–503,.

Roguin, A. (2011). Stent: The man and word behind the coronary metal prosthesis. *Circulation. Cardiovascular Interventions*, *4*(2), 206–209.

Saad, C. P., Murthy, S., Krizmanich, G., & Mehta, A. C. (2003). Selfexpandable metallic airway stents and flexible bronchoscopy: Longterm outcomes analysis. *Chest*, *124*(5), 1993–1999.

Serruys, P. W., Kutryk, M. J. B., & Ong, A. T. L. (2006). Coronary-artery stents. *The New England Journal of Medicine*, *354*(5), 483–495.

Soares, J. S., Jr, M. J. E., & Rajagopal, K. R. (2008). Constitutive framework for biodegradable polymers with applications to biodegradable stents. *ASAIO Journal*, *54*(3), 295–301. https://doi.org/10.1097/MAT.0b013e31816ba55a.

Soares, J. S., Jr, M. J. E., & Rajagopal, K. R. (2010). Modeling of deformation accelerated breakdown of polylactic acid biodegradable stents. *Journal of Medical Devices*, *4*. https://doi.org/10.1115/1.4002759, 041007.

Soares, J. S., Moore, J. E., Jr., & Rajagopal, K. R. (2007). In F. Mollica, L. Preziosi, & K. R. Rajagopal (Eds.), *Modeling of biological materials* (pp. 125–177). Birkhauser Basel.

Soares, J. S., Rajagopal, K. R., & Moore, J. E., Jr. (2010). Deformation-induced hydrolysis of a degradable polymeric cylindrical annulus. *Biomechenics and Modeling in Mechanobiology*, *9*, 177–186. https://doi.org/10.1007/s10237-009-0168-z.

Sousa, J. E., Costa, M. A., Abizaid, A., Feres, F., Seixas, A. C., Tanajura, L. F., Mattos, L. A., Falotico, R., Jaeger, J., Popma, J. J., Serruys, P. W., & Sousa, A. G. M. R. (2005). Four-year angiographic and intravascular ultrasound follow-up of patients treated with sirolimus-eluting stents. *Circulation*, *111*(18), 2326–2329.

Takebayashi, H., Mintz, G. S., Carlier, S. G., Kobayashi, Y., Fujii, K., Yasuda, T., Costa, R. A., Moussa, I., Dangas, G. D., Mehran, R., Lansky, A. J., Kreps, E., Collins, M. B., Colombo, A., Stone, G. W., Leon, M. B., & Moses, J. W. (2004). Nonuniform strut distribution correlates with more neointimal hyperplasia after sirolimus-eluting stent implantation. *Circulation*, *110*(22), 3430–3434.

Virmani, R., Guagliumi, G., Farb, A., Musumeci, G., Grieco, N., Motta, T., Mihalcsik, L., Tespili, M., Valsecchi, O., & Kolodgie, F. D. (2004). Localized hypersensitivity and late coronary thrombosis secondary to a sirolimus-eluting stent: Should we be cautious? *Circulation*, *109*(6), 701–705.

Walser, E. M., Robinson, B., Raza, S. A., Ozkan, O. S., Ustuner, E., & Zwischenberger, J. (2004). Clinical outcomes with airway stents for proximal versus distal malignant tracheobronchial obstructions. *Journal of Vascular and Interventional Radiology*, *15*(5), 471–477.

4

ECG simulation of cardiac hypertrophic condition

Igor Saveljic and Nenad Filipovic
Bioengineering Research and Development Center (BioIRC), Kragujevac, Serbia

1 Introduction

The heart is the central organ of the cardiovascular system. It enables normal blood circulation and sends blood to every part of our body. Since ancient times, people considered the heart to be one of the most important organs in the body of animals and humans. The oldest scientific treatise on the heart comes from the ancient Egyptian manuscript of Ebers dated around 1550 BCE. The Egyptians considered the heart to be the center of mental life, primarily of thought. Their anatomical representations of the heart were vague and inaccurate, but at the same time they knew that the heart is a hollow organ filled with blood, that it is the starting point of large blood vessels, and that its beating creates the pulse of arteries (Jean, 2013). Aelius Galenus (129–199), a famous Greek physician and philosopher in the Roman Empire, concluded that the heart is the mover of blood, which moves through the blood vessels (Nutton, 1973). He was well acquainted with the heart chambers, pericardium, heart valves, coronary arteries, and foramen ovale. Galen's teaching on the heart was not applied until the 16th century, when William Harvey (1578–1657), an English physician, described the blood flow to the heart as a propulsion pump (Schalick, 2006). He proved experimentally that the heart is a pump that pushes blood into the great arteries according to the principles of thermodynamics, accepts blood that returns to it from the veins and thus maintains constant blood circulation in the body with its rhythmic contraction and stretching (Gregory, 2001; Harvey, 1889).

Cardiovascular and Respiratory Bioengineering
https://doi.org/10.1016/B978-0-12-823956-8.00006-7

2 Anatomy of the heart

The heart is a hollow muscular organ that pumps blood through blood vessels by constant rhythmic contractions. It consists of two separate pumps: the right heart, which pumps blood through the lungs, and the left heart, which pumps blood through peripheral organs. In fact, each of these two separate hearts is a pulsating pump made up of two cavities: the atrium and the ventricle. The atrium functions as a reservoir of blood and as an entrance to the chamber, but it also pumps to a lesser extent, thus contributing to the movement of blood into the chamber. However, most of the force that pushes blood through the lungs and through the peripheral circulatory system occurs in the ventricle.

2.1 The structure of the heart

The heart is an organ approximately the size of a compressed fist, with the average weight ranging from 200 to 450g. The heart beats about 100,000 times a day. During that time, it pumps out about 7500 L of blood. The heart is located asymmetrically in the middle and lower part of the mid-chest, almost always on the left side, between the inner sides of the left and right lungs. The heart has the shape of an irregular three-sided pyramid (Standring, Borley, & Gray, 2008).

2.1.1 The outer structure of the heart

The heart is made up of three layers: the epicardium, the myocardium, and the endocardium, and is lined with a heart tissue (Fig. 1).

The pericardium (heart tissue) envelops the heart. It consists of a double membrane, the inner part of which adheres to the myocardium, and the outer part is fixed to the fibrous

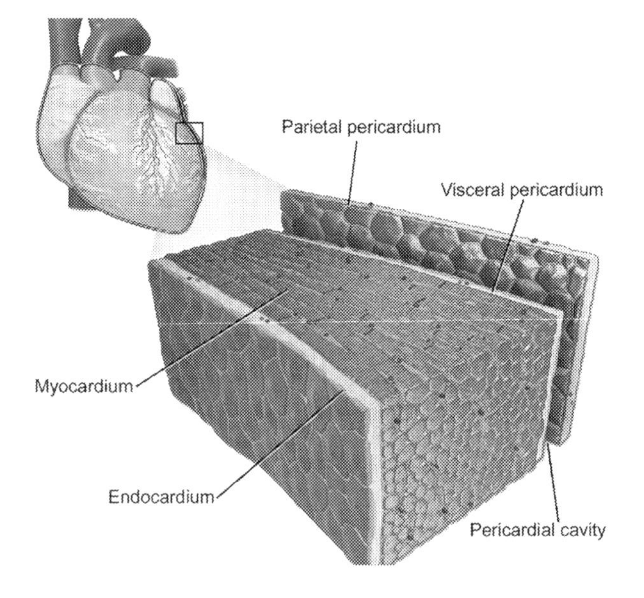

FIG. 1 Heart wall structure. The outer structure of the heart—the epicardium, the myocardium, and the endocardium. *From https://en.wikipedia.org/wiki/Pericardium.*

veins, two of which go to the sternum, three to the diaphragm and two to the spinal column. A thin layer of fluid separates these two layers, reduces friction during heart contractions, and allows the heart to move, even though it is attached to the body. The pericardium consists of fibrous connective tissue. The inner wall of the pericardium is the epicardium.

The endocardium is a thin sheath that covers the inside of the heart cavities (chambers, atria) and all the protrusions and formations that are located there. The endocardium consists of endothelium, subendothelial connective tissue, and a muscular-elastic layer (smooth muscle cells). The endocardium connects the subendocardial connective tissue with the myocardium of the heart, in which large, bright Purkinje cells can be seen, which belong to the conducting musculature of the heart. Below the endocardium is connective tissue with elastic fibers and smooth muscle threads of the cortex preventing the folding of the endocardium in systole.

The myocardium or heart muscle makes up the largest mass of the heart. It consists of special transverse-striped muscle fibers that are thicker in the ventricular area and thinner in the atrium area. The transverse-striated musculature of the heart is also called the cardiac skeleton. Cardiac muscle tissue is composed of two types of transverse striated fibers. One type makes the working musculature of the heart responsible for contractions, and the other the conducting musculature, which contains little contractile fibers and is responsible for creating and conducting impulses to the contractile fibers. Cardiac muscle fibers are composed of serially bound cells (cardiomyocytes) that contain actin and myosin filaments and a large number of mitochondria necessary for constant energy production.

2.1.2 *The inner structure of the heart*

The heart is a hollow transverse-striped muscle, which can be divided into the right venous and left arterial halves. Each half is composed of two cavities: the atrium and the ventricle (Guyton, 1991). The atria and ventricles on the left side of the heart form the left heart, and the atria and ventricles on the right side the right heart (Fig. 2). The left and right atria and the left and right ventricles are separated from each other by a ventricular and

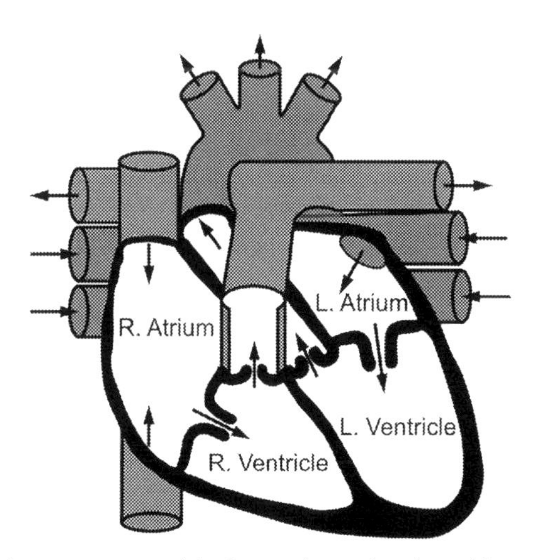

FIG. 2 Heart chambers. The inner structure of the heart—heart chambers. *No permission required.*

intercortical septum that prevents mixing of the blood of the left and right hearts. The walls of the atria are thinner than the walls of the chambers. The muscles of both chambers are well developed, but the muscles of the left ventricle are more developed and the thickest when viewed in cross section. The reason for this is that the left ventricle pumps blood into the systemic circulation where the pressure is much higher, unlike the right ventricle that pumps blood into the pulmonary circulation (Brandenburg et al., 1989).

The superior vena cava and the inferior vena cava flow into the right atrium. The superior vena cava brings blood from the head, neck, both arms, and chest cavity. The inferior vena cava brings blood from the abdominal and pelvic cavities and both legs. The coronary venous sinus flows in front of and below the mouth of the inferior vena cava. The coronary sinus brings blood from the heart veins. Thus, in the right atrium, all venous blood is collected from the body.

Four pulmonary veins flow into the left atrium. Two veins bring arterial blood from each lung. Blood from the left atrium is pushed through the left atrioventricular orifice into the left ventricle.

The right ventricle has the shape of a three-sided pyramid with the base facing the atrium and the apex facing the top of the heart. Blood from the right ventricle is pushed during systole through the right arterial orifice into the pulmonary artery. The left ventricle is the largest and strongest heart cavity, which creates blood pressure. The muscles of the left ventricle are quite thick, and are thicker than the wall of the right ventricle. The right wall forms the cardiac interventricular septum, which predominantly forms the musculature of the left ventricle and has a greater thickness. A small part of the interventricular septum is thin and does not contain muscle fibers.

The atrial-ventricular orifices are openings through which the atria communicate with the ventricles. At these ears, there is a valvular apparatus (heart valves), which prevent the return of blood from the chambers to the atria. There is a three-leaved (tricuspid) valve on the right atrial-ventricular orifice, and a two-leafed (mitral) valve on the left. The valves close or open the orifices at a certain stage of cardiac action. The phase of filling with blood is called diastole, and the phase of contraction of the heart muscle and emptying of the cavity is called systole. In the atrial diastole phase, the ventricle is in the systole phase. Then, the three-leafed valve and the two-leafed valve at the atrial-ventricular orifices are closed, and the pulmonary valves and the aortic valve at the arterial orifices are opened and the ventricle is emptied. In the atrial systole phase, the ventricle is in the diastole phase. At this point, the tricuspid valve and the bicuspid valve are open, and the pulmonary valve and aortic valve are closed and the ventricle fills with blood.

2.2 Cardiac circulation

The heart muscle is a pump that works constantly, and it is of great importance that it is constantly supplied with a sufficient amount of blood, nutrients, and oxygen. The blood that fills the heart cavities belongs to the functional bloodstream and cannot feed the walls of the heart. That is why the muscular system of the heart has a special blood flow or cardiac circulation, which consists of arteries, arterioles, capillaries, venules, and veins. About 5%–10% of the cardiac output goes to the heart circulation, which ensures the flow of 250 to 350 cm^3 of blood per minute through the heart vessels, during rest. This amount of blood is called coronary flow, which can be increased by 4–5 times during hard muscular work. The cardiac circulation consists of the arterial and venous system of blood vessels.

2.2.1 Arterial system of the heart

The arterial system of the heart consists of the right coronary and left coronary arteries (Fig. 3). The coronary arteries separate from the root of the aorta, the initial dilated part, called the bulbus, which with the corresponding left and right crescent valves forms the Walsalve sinus of the aorta. The coronary arteries are a network of blood vessels that extend along the surface of the heart, providing oxygenated blood to the heart muscle. These arteries branch into arterioles at their ends. The arterioles then branch into a large number of capillaries.

The initial segment of the left coronary artery is called the left main stem, and two branches are separated from it: the anterior descending branch and the circumflex branch. The left coronary artery supplies the left half of the heart. The anterior descending branch of the left coronary artery descends to the anterior side of the heart, providing blood to the anterior part of the septum with septal branches, and the anterior wall of the left ventricle with diagonal branches. The circumflex is located between the left atrium and the ventricle, extends along the left side toward the back of the heart, and supplies blood to the walls of the ventricles and the left atrium from the back. It continues into blunt marginal branches, which supply blood to the lateral side of the left ventricle.

The right coronary artery is located between the right atrium and the ventricle, descending along the posterior wall of the heart, bringing blood to the right half of the heart. It branches into an acute sharp marginal branch, which supplies the lateral walls of the right ventricle, the branch of the AV node, and the posterior descending branch.

All of these heart arteries branch into smaller blood vessels that cover the entire heart, and some go deeper into the heart muscle itself. The smallest branches are called capillaries, and in them erythrocytes deliver oxygen to the heart muscle tissue, and bind to carbon dioxide and other unnecessary metabolic products that drain from the heart. Coronary veins collect oxygen-poor blood from heart cells and carry it to the right atrium.

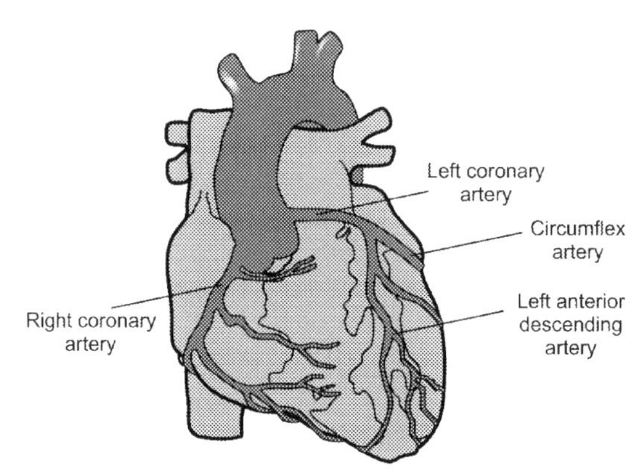

FIG. 3 Coronary arteries of the heart. The arterial system of the heart: the right coronary and left coronary arteries. *No permission required.*

2.3 Cardiac physiology

The heart is a muscle pump that, with its rhythmic contractile activity, enables a constant flow of blood through the circulatory system and all organs and tissues. Unlike other organs, the heart is in constant rhythmic activity and, except for the diastolic phases, there is no possibility of resting. The intensity of the work that the heart does is not constant, but changes depending on the needs of the organism. The heart, therefore, must have a highly developed ability to adapt. There is another characteristic of the heart that distinguishes it from other organs in a unique way. It is a high degree of automation, which enables its permanent activity even without extracardiac impulses, nervous or humoral. The impulse that originates in the center of automation spreads through the heart muscle thanks to its ability to conduct. Conductivity is the ability of the whole heart muscle, but it is specially developed in a specific conduction system of the heart. The conduction system of the heart consists of specialized cells that can generate and conduct impulses. These cells do not contract; they are less differentiated and more resistant. The conduction system of the heart is responsible for cardiac automatism. The conduction system of the heart has the following functions:

- Generation of impulses,
- Impulse generation control,
- Conduction of impulses to the working musculature.
- The conduction system of the heart consists of:
- SA node (sinoatrial node).
- AV node (atrioventricular node).
- His bundle.
- Left and right branches of the His bundle.
- Purkinje fibers.

The heart works automatically because it has cells that have a special role in the creation and transmission of impulses, which include cells of a specific autonomous system and atrial and ventricular myocardial working muscle cells.

Cells of a specific autonomous system can generate and transmit impulses. This system consists of the conductive musculature of the heart, which is similar in structure to the cells of the working musculature. The difference is in the color that is lighter than the cells of the working musculature, the size that is slightly smaller, and the smaller the number of myofibrils. In this system, histologically, there are three types of cells: P-cells, transitional and Purkinje cells. P-cells are located in the central part of the sinus node or atrioventricular node in a slightly smaller number. Nerve endings have been found around these cells and it is thought that P-cells could be the main creators of impulses, because the activity of the atrioventricular node can be initiated by impulses of the vegetative (sympathetic and parasympathetic) system. Transitional cells are located around P-cells. They most likely do not have much power in generating impulses and their role consists more in transmitting impulses from P-cells to Purkinje cells. They are located in the marginal parts of the sinus node and form a connection with the cells of the working muscles of the atria. There are Purkinje cells in the convex part of the atrioventricular node, where they receive impulses from the atria, then in the atrioventricular node, especially at the transition of the atrioventricular node in the His bundle, in the His bundle and its branches, all the way to the working musculature cells.

The cells of the working musculature of the atria and ventricles are significantly larger and are composed of numerous myofibrils, and those of sarcomeres in which the basic contractile structures are located. In each sarcomere, there is a hexagonal arrangement of myosin and actin filaments, which are separated while the cell is at rest. Activation of a muscle cell results in the fusion of actin and myosin into actinomyosin, which leads to shortening (contraction) of the sarcomere and development of mechanical force (pressure) within the muscle cells and muscle fibers. Upon completion of the contraction, actinomyosin separates into actin and myosin fibers.

Fig. 4 shows a specialized excitation and conduction system of the heart that controls cardiac contractions. The figure shows a sinus node (also called a sinoatrial or SA node) in which a rhythmic impulse is generated under normal conditions: first, an internodal pathway that conducts an impulse from the sinus node to the atrioventricular (AV) node; next, the AV node in which the impulse from the atrium is delayed before passing to the ventricle; finally, the AV bundle that conducts the impulse from the atrium to the ventricle and the left and right branches of the Purkinje fibers that conduct the impulse to all parts of the ventricles (Pullan et al., 2005).

The sinus node (often called the sinoatrial node) is a small, flat, ellipsoidal band of specialized heart muscle about 3 mm wide, about 15 mm long and about 1 mm thick. It is located on the upper posterolateral wall of the right atrium just below and slightly lateral to the opening of the superior vena cava. The muscle fibers of this node have almost no contractile muscle elements and are only 3–5 μm in diameter, unlike the atrial muscle fibers that surround them, which are 10–15 μm in diameter. Nevertheless, the muscle fibers of the sinus node are directly connected to the atrial fibers so that any action potential that begins in the sinus node instantly propagates along the atrial muscle wall. Some heart fibers have the ability to self-excite, a process that allows for automatic rhythmic discharge and contraction. This is especially true for the fibers of the specialized conduction system of the heart, including the fibers of the sinus node. For this reason, the sinus node controls the rhythm of the whole heartbeat.

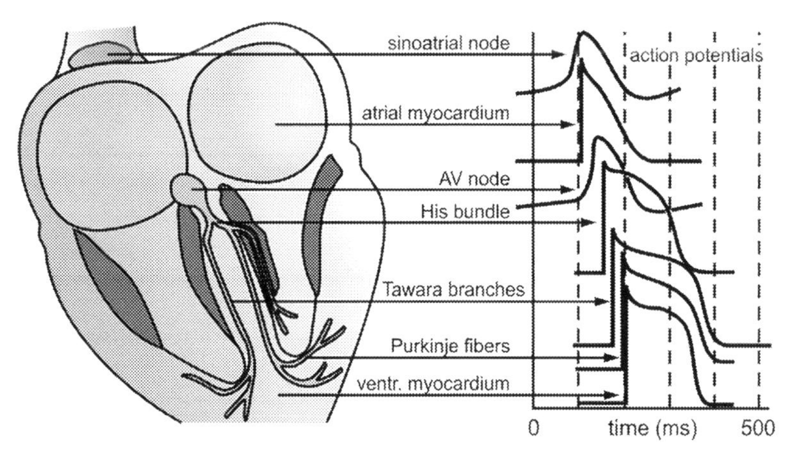

FIG. 4 The conduction system of the heart. A specialized excitation and conduction system of the heart that controls cardiac contractions. *No permission required.*

The atrial conduction system is organized so that the heart impulse does not travel from the atrium to the ventricle too quickly, and it is this delay that provides enough time for the atria to expel all the blood into the ventricles before the ventricular contraction begins. In fact, the AV node and its surrounding conductive fibers are responsible for this delay. The AV node is located on the posterior wall of the right atrium just behind the tricuspid valve. The cardiac impulse, after passing through the internodal pathway, reaches the AV node approximately 0.03 s after originating in the sinus node. Then, there is a delay of another 0.09 s in the AV node itself before the impulse enters the penetrating part of the AV beam where it enters the chamber. A last delay of 0.04 s occurs mainly in this penetrating His bundle that is composed of multiple small bundles passing through the fibrous tissue separating the atria from the ventricles. Therefore, the total delay in the AV node and the His bundle is 0.13 s, which, with an initial delay of 0.03 s for the signal to arrive from the sinus node and the AV node, is 0.16 s until the signal finally reaches the contractile muscles of the ventricles. Specialized Purkinje cells lead from the AV node through the AV beam to the chambers. Apart from the initial part of these fibers where they break through the AV fibrous barrier, they have a functional characteristic completely opposite to that of the AV node fibers. They are very large fibers, even larger than normal ventricular fibers, and conduct an action potential at a rate of 1.5–4 m/s, a velocity that is 6 times greater than in normal ventricular fibers and about 150 times greater than in the AV node. This allows for an almost instantaneous transmission of the heart pulse through the rest of the ventricular muscle. The rapid conduction of the action potential through Purkinje cells is believed to be caused by a high degree of permeability between the interstitial joints in the inserted discs, between the individual cells that make up the Purkinje fibers. Therefore, the ions easily pass from one cell to another, thus increasing the velocity of transmission. Purkinje fibers also have very few myofibrils, meaning that they contract either very little or not at all during the impulse transmission process.

Under normal conditions, the impulse originates in the sinoatrial node. In some abnormal conditions, this is not the case. Several other parts of the heart can cause internal rhythmic excitation in the same way as a sinus node. This is especially true for the AV node and Purkinje fibers. AV node fibers and Purkinje fibers have a frequency of 15–40 times/min. The sinus node has a frequency of 70–80 times/min. The sinus node controls the rhythm of the heart because the period of its rhythmic emptying is faster than in any other part of the heart that is capable of the same. Because of that, the sinus node is always the pacemaker of a healthy heart (Podrid & Kowey, 2010).

In rare cases, a place in the atrial or ventricular muscles develops an excessive possibility of excitation and becomes a pacemaker. A pacemaker located somewhere outside the sinoatrial node is called an ectopic pacemaker. An ectopic pacemaker causes an abnormal rhythm of contractions in various parts of the heart and can significantly impair pumping. If for some reason the sinoatrial node fails, its role is taken over by the AV node but with a lower frequency. If the AV node also fails, the heart will continue to work, but again at a lower frequency, because Purkinje fibers continue to take on the role of a pacemaker.

3 Electrocardiography

When a heartbeat passes through the heart, an electric current also propagates through the heart to the tissues that surround it. A small part of that current reaches the surface of the skin. If we place electrodes on the skin on opposite sides of the heart, the electrical potentials generated by the heart can be recorded. Such a recording is known as an electrocardiogram (ECG) (Noble et al., 1990).

The electrocardiogram is the oldest cardiac method that has not been suppressed by newer cardiac methods, but is enriched with each new achievement. The development of the ECG is attributed to the Nobel laureate (1924) and physicist, physician Willem Einthoven (1860–1927). The ECG registers the electrical activity of the heart muscle cells. This is achieved by recording on a treadmill, changing the potential via electrodes on the arms, legs, and precordially (on the chest). Before Einthoven invented this machine, it was known that the heartbeat produces electrical impulses but instruments from that time were not sensitive enough to detect these impulses unless they were connected directly to the heart, which was a problem. In early 1901, Einthoven completed a series of tests for a device called the String galvanometer. The device used a coil of very thin wire that passed between strong electromagnets (AlGhatrif & Lindsay, 2012). When an electric current passed through a coil, an electromagnetic field would move that coil. The original machine required water cooling for strong electromagnets and weighed about 280 kg. It took five people for the machine to work. This device increased the sensitivity of galvanometers of that time so that the electrical activity of the heart could be measured on the skin (Fig. 5).

With the advancement of science and technology, electrocardiographs have changed a lot. No more water cooling is needed as well as five people for the machine to work. Today's electrocardiographs can be divided into two large groups according to the criterion of mobility, i.e., portable and nonportable. Portable devices are much more expensive and this is primarily influenced by the degree of autonomy they have, i.e., the possibility of operation without connection to the electrical network. By registering changes in the electrical potential at the level of the heart muscle cell membrane, information is obtained on heart rate, rhythm (arrhythmias, tachycardia, bradycardia, extrasystoles), quality of the conduction system (possible AV blocks, His bundle branch blocks), as well as signs of ischemia, acute and subacute infarction, heart load, hypertrophy and dilatation (increase in the volume of a certain heart cavity), etc. (Carley, 2003).

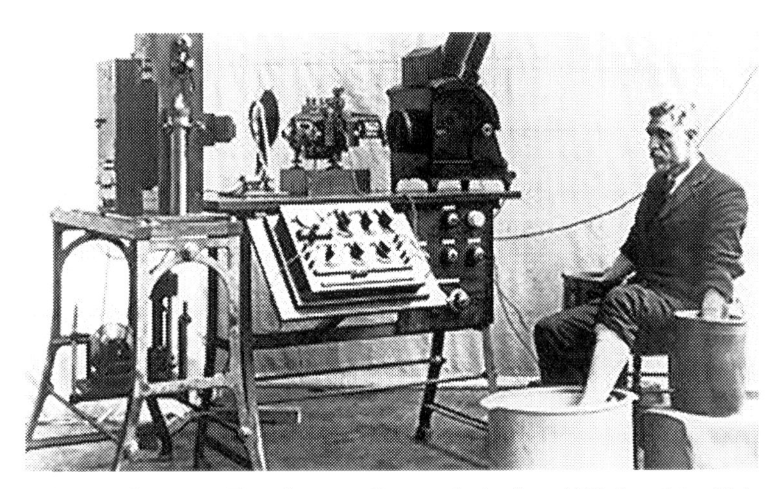

FIG. 5 Electrocardiogram from 1903. First electrocardiogram device from 1903. *From https://doi.org/10.3402/jchimp.v2i1.14383.*

The quality interpretation of ECG recordings is crucial in ECG diagnostics. There are no side effects or contraindications for performing ECG diagnostics. The waves on the ECG primarily reflect the electrical activity of the myocardial cells, which make up the largest part of the heart. ECG waves have three basic characteristics:

1. duration measured in fractions of a second (milliseconds),
2. amplitude measured in millivolts (mV), and
3. configuration, i.e., the shape and appearance of the wave, which is a subjective criterion.

The ECG strip is a roll of graphic paper, with vertical and horizontal division. The horizontal distribution represents the time where one square represents 0.04 s. Voltage is measured on the vertical axis, where the range of one square corresponds to one-tenth of a millivolt and the height of one large square is 0.5 mV.

The characteristic parts of the ECG signal are P, Q, R, S, and T waves and are shown in Fig. 6. The QRS complex consists of Q, R, and S waves. We have already explained what happens to the heart muscle when it is stimulated by electrical impulses.

The P wave represents the electrical activity of the contractions of both atria. The QRS complex represents an electrical impulse on the way from the AV node through Purkinje fibers to myocardial cells. The QRS complex represents the electrical activity of irritated chambers. The Q wave is the first descending part of the QRS complex and it is important to know that the Q wave is often not present on the ECG. The first ascending wave following the Q wave is the R wave. The ascending R wave is followed by the descending S wave. The difference between Q and S is that in front of the Q wave there is no ascending wave and in front of the S wave there is an ascending wave. The T wave represents the repolarization of the chambers so that they can be re-stimulated by an electrical impulse. We can understand this wave as a "reset" of heart cells. One cardiac cycle consists of the P wave, the QRS complex, and the T wave. This cycle is constantly repeated.

3.1 Twelve-channel ECG

Impulse detection is achieved in such a way that part of the electrical impulses that propagate through the heart propagate to the surrounding tissue. A small portion of these impulses reaches the skin. Electrodes located on the patient's skin can detect these electrical impulses. The signals detected by the electrodes are called leads or more precisely the signal detected by the pair of electrodes makes one lead. The standard modern ECG has 12 leads

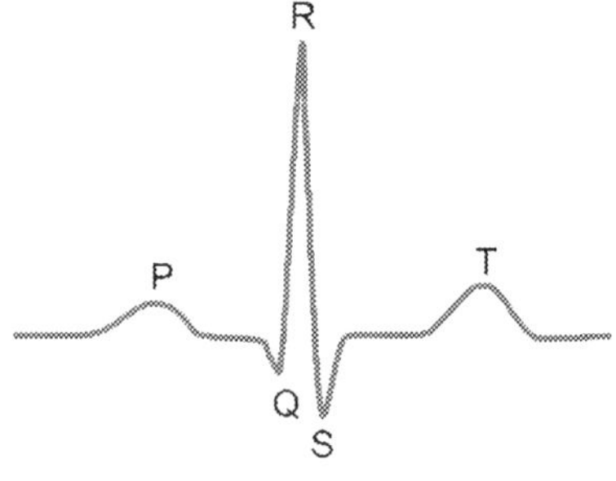

FIG. 6 P wave, QRS complex and T wave. The characteristic parts of the ECG signal: P wave, QRS complex and T wave. *No permission required*

(channels), which are defined based on the position and orientation of the electrodes on the body (Trudel et al., 2004). Each of these twelve leads observes the heart from its specific angle, which increases its sensitivity for certain parts of the heart, and decreases for others. Two electrodes are placed on the patient's arms and two on the legs. This forms the basis for six limb drains comprising three standard bipolar (DI, DII, DIII) and three unipolar drains (aVR, aVL, aVF). Six electrodes are placed on the chest and they represent six precordial leads (V1, V2, V3, V4, V5, V6). Precordial leads are leads that are obtained from electrodes that are placed on the patient's chest, in front of the heart (precordial). The position of the precordial positive leads is shown in Fig. 7. To obtain six precordial leads, six positive electrodes were placed in six different places around the chest. Precordial drains are projected through the AV node toward the patient's back because there are negative poles of these drains on the back. These drains are marked with markings V1 to V6. Leads V1 and V2 are placed above the right side of the heart and they are called right precordial leads. Leads V5 and V6 are placed above the left side of the heart and they are called left precordial leads. Leads V3 and V4 are placed above the interventricular septum. The interventricular septum is the joint wall of the left and right ventricles. In this part, the AV beam is divided into left and right branches.

Fig. 8 illustrates an electrocardiogram recorded with six standard chest leads. Because the surface of the heart is close to the chest wall, each lead records the electrical potentials of the

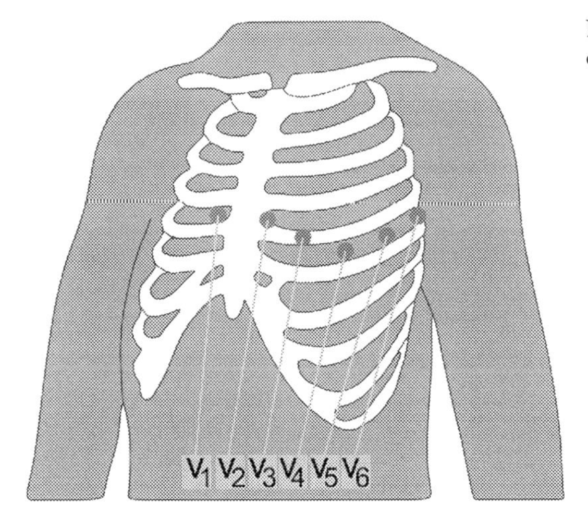

FIG. 7 Position of precordial electrodes. The position of the six precordial positive leads. *No permission required.*

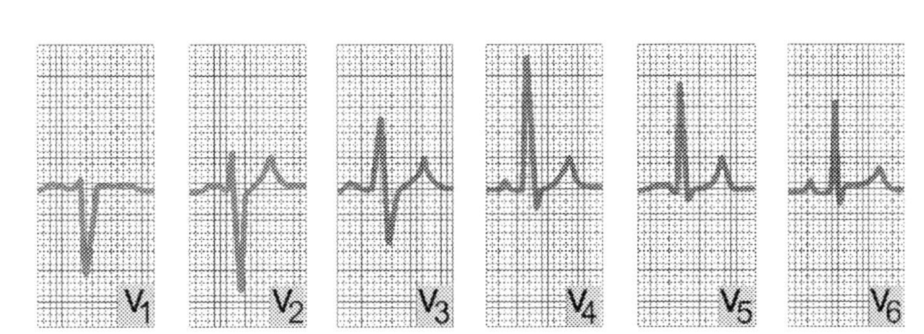

FIG. 8 Electrocardiogram recorded with precordial leads. An electrocardiogram recorded with six standard chest leads. *No permission required.*

heart muscle directly below the electrode. Therefore, relatively small abnormalities in the ventricles, especially on the anterior ventricular wall, can cause changes in electrocardiograms recorded in individual chest leads. In leads V1 and V2, the images of the QRS complex are mainly negative because the electrode is closer to the base of the heart than to the vertices in these derivatives, and the base is in the direction of electronegativity by most of the ventricular depolarization process. Also for the opposite reason, other excerpts are mostly positive (Ebrard et al., 2009).

The overall goal of performing electrocardiography is to obtain information about the structure and function of the heart. The medical benefits of this information are numerous and mainly relate to the need to master knowledge of the structure and/or function. The ECG has a specific diagnostic value in the following circumstances:

- dysfunction of the rhythm generator (SA node) in situations in which the role of the rhythm generator is taken over by another (ectopic) center which may be in the atria, in the AV node, or in the chambers,
- cardiac arrhythmias (arrhythmia or variable rhythm, tachycardia or rapid heartbeat, bradycardia or slow heart rate, extrasystoles and jumps, heart blocks, ventricular fibrillation, atrial fibrillation),
- cardiomyopathy (increase in wall thickness),
- slowed down conduction of impulse action potentials through atria and ventricles.
- ischemia and myocardial infarction,
- pericarditis,
- systemic heart disease,
- determining the effect of drugs,
- electrolyte imbalance, especially potassium.

4 Cardiomyopathy

Cardiomyopathy is most easily defined as a disease of the heart muscle. It is a disease that has many causes, signs, and symptoms and different ways of diagnosis and therapy. The main feature of cardiomyopathy is that the heart muscle changes, in a way that it becomes larger, thicker or harder, and in rare cases the muscle tissue is replaced by fibrotic tissue and scarring. As the myocardium changes, the heart becomes weaker and is not able to pump blood so efficiently or maintains a normal electrical rhythm. As a result, heart rhythm disorders, arrhythmias, complications with valves and the worst outcome—heart failure can occur. There are several main types of cardiomyopathy: ischemic, hypertensive, dilated or congestive, hypertrophic, infiltrative, and restrictive. The three most common types are restrictive, dilated, and hypertrophic cardiomyopathy.

4.1 Dilated or congestive cardiomyopathy

Dilated or congestive cardiomyopathy is a disease of the heart muscle that leads to heart failure and is dominated by ventricular dilatation and systolic dysfunction. The most common form is cardiomyopathy. Symptoms include shortness of breath, general weakness and peripheral edema. In the progress of the disease, three stages can be distinguished: asymptomatic stage, moderately severe, and severe clinical picture. The cause may be genetic or related to infection or environmental conditions. The hereditary form covers 30%–50% of all cases of the disease, and is inherited in an autosomal dominant manner. The myocardium dilates, thins, and compensates hypertrophied, resulting in functional mitral or tricuspid regurgitation and atrial dilation. Thus, the main pathological characteristics of dilated cardiomyopathy are: dilatation of all four heart cavities, especially the left ventricle, increased left ventricular mass, hypertrophic, elongated myocytes with reduced myofibril count and interstitial fibrosis. In most patients, the disorder affects both ventricles. Treatment: Symptomatic, therapy of congestive heart failure (reduce salt intake, diuretics, digitalis, vasodilators, ACE inhibitors, and sometimes beta-blockers). Drug treatment has not proved to be very successful; so lately, heart transplantation has come into consideration as an option to increase survival rates.

4.2 Restrictive cardiomyopathy

Restrictive cardiomyopathy is the rarest form of cardiomyopathy, and the main pathological feature is the restriction of ventricular filling and the existence of inextensible ventricular walls that resist diastolic filling and thus cause general weakness and shortness of breath on exertion. The contractile function itself and the thickness of the ventricular wall are normal, but it is an abnormal phase of relaxation or filling. Due to the inability to properly fill the ventricles, blood returns to the atria, and consequently to the lungs and the rest of the body. The right ventricle is most often affected, leading to peripheral edema, accumulation of ascites and hepatomegaly, but with unchanged heart size. The cause is unknown in most cases. The goal of treatment is to treat the cause of the disease, if it is known, to reduce symptoms while reducing the increased pressure filling the ventricles without significantly reducing the cardiac output. Drugs are used to reduce venous blood flow in cardiac diuretics, and nitrates in small initial doses to avoid a drop in blood pressure. Sinus rhythm maintenance is required because arrhythmias such as atrial fibrillation can further exacerbate diastolic ventricular filling. Beta-blockers, amiodarone, and digitalis are also used.

4.3 Hypertrophic cardiomyopathy

Hypertrophic cardiomyopathy (HCM) is a congenital or acquired disorder of the heart muscle, which is clinically manifested by symmetrical or asymmetric hypertrophy of the left and/or right ventricle in the absence of any other cause of hypertrophy (Fig. 9). The symptoms are similar to those of most other types of heart disease: chest pain, shortness of breath, syncope, and sudden death. Some patients also develop arrhythmia, which is a risk factor for sudden cardiac death. Shortness of breath is a consequence of increased left ventricular filling pressure and increased pressure in the pulmonary veins. Several pathophysiological mechanisms may be involved in the development of syncope, including low diastolic filling, myocardial ischemia, cardiac arrhythmias, and ventricular outflow obstruction in obstructive disease. It is treated with β-blockers and sometimes the obstruction is treated with chemicals

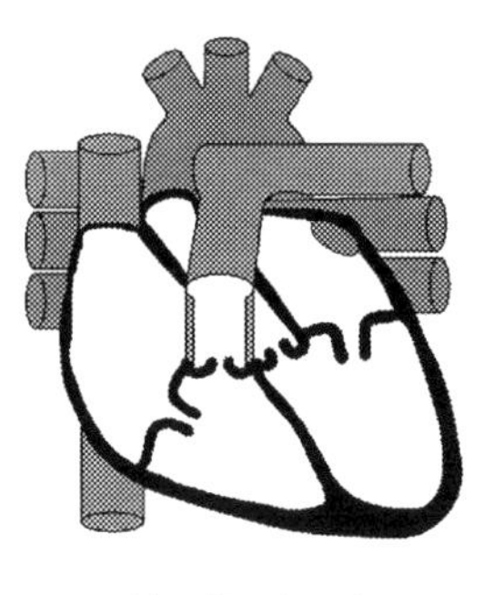

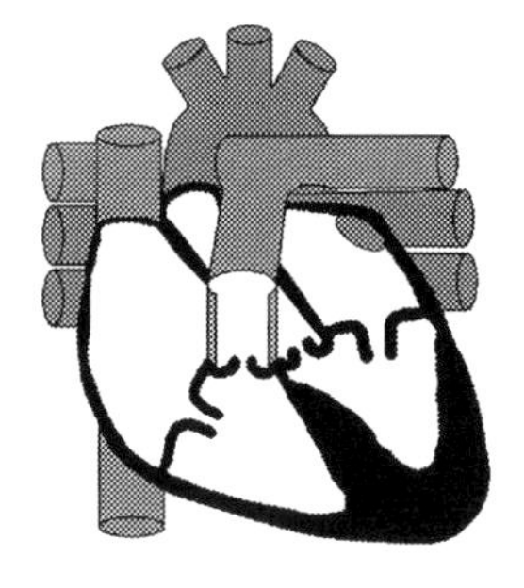

FIG. 9 Normal heart and heart with hypertrophic cardiomyopathy. *No permission required.*

or surgically to prevent sudden death. Patients usually develop hypertrophy during rapid physical growth in adolescence, but there are also known cases of HCM, which occur at the age of 60 (Maron, 2002).

Common to all cardiomyopathies is a disrupted organization of cells and muscle fibers. In the most common form of HCM, the upper part of the interventricular septum below the aortic valve is significantly hypertrophic and thickened, with simultaneous slight or no thickening of the posterior wall of the left ventricle. During systole, the septum thickens and sometimes the anterior leaflet of the mitral valve is attracted to the septum, which further narrows the outlet tract and reduces minute volume. Cardiac contractility usually does not change, but hypertrophy leads to stiffness and distension of the heart cavities. Resistance to diastolic filling is created, the pressure at the end of diastole increases and consequently the pressure in the pulmonary veins. Blood flow through the coronary arteries also changes, leading to angina pectoris, syncope, and arrhythmias without the presence of coronary disease itself. Due to insufficient blood flow and narrowed lumen of intramyocardial arteries due to hypertrophy and hyperplasia of the intima and media, part of the myocytes completely die (Maron et al., 2002). They are replaced by fibrous tissue and lead to the expansion of the hypertrophic myocardium and the development of systolic dysfunction. Blood pressure and heart rate are usually normal, and a fourth heart tone is often heard due to strong atrial contraction versus weak dilated left ventricle in diastole. The mortality rate is inversely proportional to the age at which symptoms occurred and is highest in patients with frequent, intermittent ventricular tachycardia or syncope (Maron et al., 2003). Treatment: Beta-blockers alleviate and reduce the incidence of retrosternal pain and syncope. Ca antagonists reduce the end-diastolic pressure of the LC, and may reduce the gradient in the outflow tract.

The use of verapamil has led to sudden death, so its use is debatable. Amiodarone protects against sudden death because it has a protective effect against ventricular arrhythmias. Intraventricular defibrillators are also used. Digitalis, diuretics, and nitrates of prolonged action should be avoided, especially in those who have dynamic obstruction of the left gastrointestinal tract.

4.4 Diagnostic techniques

The choice of imaging diagnostics depends on the stability of the patient, local expertise, and the availability of the technique itself. In approaching the diagnosis of HCM, it is essential to exclude hypertrophy arising from knowledge of causes such as hypertension, although HCM may coexist with hypertension or coronary heart disease (Gersh et al., 2011). It is necessary that patients with suspected HCM have an ECG, chest X-ray, and echocardiography examination. In the next part of the chapter, the basic HCM diagnostic techniques will be presented.

4.4.1 *Electrocardiography*

A 12-channel ECG is abnormal in 75%–95% of patients with HCM (Gersh et al., 2011). The ECM finding of HCM is dominated by the image of LV hypertrophy with deviation of the electrical axis to the left. The finding is difficult to distinguish from hypertrophy due to secondary causes, such as hypertension, but also ECG changes in other cardiomyopathies. A regular ECG finding is possible, but it is rarely found. Hypertrophy can sometimes be found in drains where we do not expect it, such as the middle precordial (Lj & Birtić, 1993) (Cerovec, 2014). The presence of septal hypertrophy can be seen in the ECG such as the presence of pathological Q peaks in the left precordial and inferior leads and high R peaks in the right precordial leads. Deep Q peaks may be indicative of HCM. Localized hypertrophies may give a different ECG finding than we consider indicative of HCM (Wigle et al., 1995). Arrhythmias are a common finding in patients with HCM. Their presence is best revealed by 24-h ECG recordings. When reviewing ECG findings, it is important to analyze the QT interval, especially before planning the introduction of antiarrhythmic therapy (Johnson et al., 2011).

Arrhythmias are a common finding in patients with HCM. Their presence is best revealed by 24-h ECG recordings (Holter). Around 88% of patients have premature ventricular contractions, 12% have over 500 premature ventricular contractions in a 24-h ECG, and 37% have supraventricular tachyarrhythmias. In 14% of Holter images analyzed, sinus bradycardia is found (Adabag et al., 2005). Paroxysmal atrial fibrillation is present in about 20%–25% of patients with HCM (Adabag et al., 2005). A 24-h ECG monitoring reveals intermittent ventricular tachycardia in 25%–30% of patients (Adabag et al., 2005; Gersh et al., 2011). Although intermittent ventricular tachycardia is common, most patients have rare (<5 in 24h) episodes of intermittent ventricular tachycardia. A 1994 study [30] found that the relative risk of heart death in patients with intermittent ventricular tachycardia was 1.4 versus those without it, at a rate of 1.4% per year versus 0.9% per year. Conduction system abnormalities were present in 23% of the patients. First-degree AV block was most common in 17% of the patients, second-degree AV block in 3% of the patients, and sinus pause was longer than 2s in 7% of the patients (Spirito et al., 1994). The presence of preexcitation in young patients with HCM casts suspicion that it is Danon's disease and not classical HCM (Arad et al., 2005).

A 12-channel ECG is recommended as part of screening relatives in the first knee of patients with HCM. In young relatives, with no signs of hypertrophy, a 12-channel ECG should be recorded every 12–18 months. A 24-h Holter ECG is recommended to be performed annually or biennially in stable patients who do not have manifest arrhythmias or an implanted cardioverter defibrillator (Gersh et al., 2011).

4.4.2 *Echocardiography*

Echocardiography, or heart ultrasound, is a noninvasive diagnostic method that uses ultrasound waves to show the heart and blood vessels. Echocardiography is crucial for the diagnosis and monitoring of HCM. In most patients, hypertrophy primarily involves the interventricular septum in the basal LV segments but often extends to the lateral wall, posterior septum, and LV apex. Since increased ventricular wall thickness can be found anywhere (including the right ventricle), the presence, distribution, and severity of hypertrophy should be documented using a standardized protocol for cross-sectional imaging from several projections. Proper orientation and slope of the beam along orthogonal planes are essential to avoid concealed parts and overestimation of wall thickness.

Echocardiography has been used in cardiology since 1953. There is one-dimensional (M-mode), two-dimensional, and three-dimensional echocardiography. In addition to the above, a Doppler echocardiography measures blood flow in the heart, while echocardiography shows blood flow in the heart and blood vessels in blue and red to determine the direction of flow. An indicative finding for HCM is LV hypertrophy with wall thickness greater than 15 mm. In 28% of patients, only one segment of the LV was affected as hypertrophy, in 38% two segments are affected, while in 34% of cases hypertrophy diffusely affects three or more segments. In only 2% of patients, the septum is not affected by hypertrophy. The number of affected segments correlates with the maximum LV wall thickness, which is thicker the more segments are affected (Cerovec, 2014).

Systolic anterior motion (SAM) is also more common in patients, who have multiple segments involved and according to a thicker LV wall (24 ± 5 mm) [32]. The more dystrophied the hypertrophy from the base of the heart, the less likely it is that systolic anterior motion and obstruction of the left ventricular outflow tract (LVOT) will occur (Maron, 2002). The cusps of the mitral valve are abnormally large and elongated, which is associated with impaired hemodynamics due to pressure gradients. At the point of contact of the cusp of the mitral valve with the septum, there is a thickening of the endocardium. SAM and pressure gradient lead to the appearance of mitral regurgitation, which in obstructive HCM has a jet directed posteriorly. Mitral regurgitation caused by mitral valve disease usually has an anterior-directed jet (Yu et al., 2000). In the patients with apical HCM, the LV has a markedly reduced apical portion of the cavity (spade shaped). In patients with "true apical" HCM, only the apical part is affected in isolation, while in the mixed form the hypertrophy from the apex extends to other segments of the left ventricle. A complication of apical HCM is the formation of an apical aneurysm. Echocardiographic presentation of apical hypertrophy and aneurysms is often difficult (Stainback, 2012).

Transthoracic echocardiography (TTE) is a method that determines the morphological appearance and function of the heart muscle, valve, and pericardium, as well as the ascending aorta. The transducer works with a probe in the 1–5 MHz band. It is necessary for the patient to take off his waistband and lie on his left side. Then, the probe, on which the gel was previously placed, searches for the image of the heart in the predicted positions/sections. Transthoracic echocardiography is performed in standard sections: parasternal, apical, and subcostal.

Transesophageal echocardiography (TEE) uses an ultrasound probe that, unlike transthoracic, is inserted into the esophagus through an endoscope, so as only the esophageal wall separates the probe from the heart and blood vessels; the heart and blood vessels show much better than transthoracic echocardiography. TEE works with a 3.5 and 7 MHz probe (Klues et al., 1995). Novel TEE probes allow recording in one, two, or more cross sections, which allows the heart to be viewed from any angle, so that images obtained at different angles can be used for 3D reconstruction.

4.4.3 *Magnetic resonance*

Magnetic resonance (MR) is a noninvasive and accurate diagnostic method that provides a picture of the health of individual organs. It is painless and harmless for the patient; so, the examination can be repeated several times. MR devices record signals originating from hydrogen nuclei found in molecules of the human body, which were previously placed in a strongly homogeneous magnetic field. This technique can show focal and massive hypertrophy, provide better estimation of the magnitude of hypertrophy and the display is not as susceptible to anatomical disorders as echocardiography. It also provides a better view of HCM RV, which may be difficult to visualize clearly with echocardiography, allows the view of narrowing of the RV outlet tract and its evaluation, as well as the assessment of the need for surgical correction of narrowing (Maron, 2002). It is particularly useful for the diagnosis of apical HCM and apical aneurysms, which are more difficult to visualize echocardiographically. The MR technique makes it possible to distinguish HCM from metabolic and infiltrative cardiomyopathies, which are phenocopies of HCM, as well as hypertrophy due to other causes (hypertension, sports heart, and LV noncompaction syndrome). In addition to the anatomical characterization of hypertrophy, MR can be used to assess hemodynamics, plan and evaluate cardiac surgery, and assess risk for sudden heart death (Maron, 2012).

4.4.4 *Computed tomography angiography*

Computed Tomography Angiography (CTA) is a radiological method of imaging which, in addition to X-rays, also uses tomography, a method based on the mathematical procedure of image processing using modern computers. First-generation devices (rough image and scan length of about 30min) provide data for each X-ray tube cycle used to reconstruct axial images, while newer versions (third-generation devices) provide accurate images and shorter imaging procedures, and also have the ability to simultaneously move the table on which the patient lies and exposes himself to X-rays. High-resolution CT with contrast enables clear separation of the myocardium and precise measurement of wall thickness, ventricular volumes, ejection fraction, and LV mass. Cardiovascular CT allows simultaneous imaging of coronary arteries and ventricles and can be used to guide catheters in supraventricular arrhythmia ablation. Data on myocardial tissue characterization in small cohorts suggest that contrast-enhanced CT may be useful in detecting myocardial fibrosis, but this requires further research. CT should be considered in patients with inadequate echocardiographic imaging and contraindications for MR (Schroeder et al., 2008; Shiozaki et al., 2013).

4.4.5 *Heart catheterization*

Catheterization is approached when there is conflicting data from physical findings and echocardiography, unclear diagnosis, or the need for more accurate hemodynamic assessment and as part of treatment in indicated surgery (Gersh et al., 2011). Cardiac catheterization allows us to determine the pressures within the heart and is most useful for measuring the pressure gradient in the LV. The pressure gradient can be highly variable and have a range of 0–175mmHg in the same patient under different conditions (Braunwald, 1997). Fake elevated pressures may occur if the catheter is trapped between hypertrophic LV trabeculae. Proper catheter selection and careful catheterization reduces the likelihood of falsely elevated pressures. Stroke volume is mostly normal, but may be reduced in patients with a long-term elevated pressure gradient. One-fourth of patients with HCM have pulmonary hypertension resulting from elevated LA pressure and decreased LV compliance. The pressure gradient in the RV appears in 15% of the patients with LVOT obstruction and is due to RV myocardial hypertrophy (Kroeker et al., 2006).

4.5 Differential diagnosis

Many conditions, such as hypertension, mitral regurgitation, ventricular septal defect, and discrete variants of aortic stenosis, can present a similar clinical picture of HCM. A problem in differential diagnosis is the distinction between HCM and physiological hypertrophy of the athlete's heart. Aerobic sports and static sports, with pronounced isometric muscle contraction, stimulate an increase in cardiac muscle mass and remodeling of the heart in the direction of increasing ventricular size and sometimes ventricular wall thickness. It is a physiological adaptation to sports activity without threatening cardiovascular consequences. The biggest changes are in sports such as swimming, cycling, rowing, and cross-country skiing. Athlete's hypertrophy can be accompanied by ECG changes that we associate with certain heart diseases. A special problem is the echocardiographic finding of LV wall thickness > 12 mm and LV size >60 mm in athletes, which resembles HCM with mild hypertrophy. Noninvasive procedures such as echocardiography, DNA analysis, and deconditioning can determine whether it is the so-called athletic heart or HCM (Maron et al., 2003).

In the elderly, secondary hypertrophy due to hypertension as well as changes in the septum and mitral valve due to aging may resemble HCM. Secondary hypertrophy due to hypertension is more often symmetrical, whereas in HCM it is asymmetric. The deep Q teeth found in the ECG in HCM can be easily mistaken for an ECG finding in ischemic heart disease. The presence of high R teeth and inverted giant T waves speak more in favor of HCM (Wigle & Baron, 1966). Some patients with HCM do not actually have "classic" HCM but phenocopies such as Nooan syndrome, mitochondrial myopathy, delay disease, amyloidosis, Friedreich's ataxia, Fabry's disease, or AMP kinase mutations (PRKAG2) (Maron et al., 2003). LV noncompaction may resemble apical HCM on echocardiography, but CMR can successfully differentiate these two conditions (Maron, 2012). Danon's disease (X-linked LAMP2 deficiency) may have the same phenotypic picture as HCM, and the finding of early onset of massive hypertrophy with ventricular preexcitation in male patients should raise suspicion of this disease. Gene analysis leads to a definitive diagnosis (Arad et al., 2005).

4.6 Treatment

4.6.1 Drug treatment

The approach to treatment depends on the intensity of the symptoms and the presence of obstruction as well as the complications of the disease. It is unclear whether asymptomatic patients should be treated and the decision to treat should be based on the risk factors present for sudden cardiac death (Wigle et al., 1995). Patients should avoid physical exertion, dehydration, factors that lead to vasodilation, and adopt a healthy lifestyle that will prevent the onset of coronary artery disease. Treatment goals are to reduce symptoms and prevent complications (Gersh et al., 2011).

β-Blockers are the first line of choice for patients, who have shortness of breath in exertion, presyncope, angina pectoris, or intolerance to exertion, regardless of obstruction. They reduce heart rate and contractility, prolong diastole and passive ventricular filling, prevent arrhythmias, can reduce myocardial oxygen demand, reduce the LVOT pressure gradient and increase it after physical exertion. They have little effect on obstruction during rest as well as on preventing the growth of the gradient provoked by the Valsalva maneuver, PVC, and amyl nitrite. Patients with latent obstructions in whom physical activity provokes obstruction are thought to benefit most from β-blockers. β-blockers also have a beneficial effect on reducing ventricular and supraventricular arrhythmias (Marian, 2009). Their effect on reducing the LVOT pressure gradient is based on slowing down the acceleration of the LV ejection jet and allowing the jet to reach its peak velocity in the second half of the systole rather than the first. This results in lower initial flow rates in systole that exert a proportionally smaller force on the mitral valve and thus delay the onset of systolic anterior motion (SAM). The SAM delay generates a much smaller pressure gradient. In addition, the later onset of SAM leaves more time for the papillary muscles to retract the cusp by their shortening and thus reduce the force acting on the cusp (Sherrid et al., 1998).

Calcium channel blockers or calcium antagonists are drugs that block the entry of calcium ions into a cell through calcium channels. Calcium antagonists include three chemically distinct classes: phenylalkylamines (verapamil), dihydropyridines (nifedipine, amlodipine), and benzothiazepines (diltiazem). Drugs from each of the above three chemical classes bind to the alpha1 subunit of the L-type cardiac calcium channel, but in different places, entering into allosteric interactions and interactions with the channel closure mechanism, and indirectly prevent the diffusion of potassium ions through the pores into the open channel. Most calcium antagonists cause use-dependent channel blockade (Bottinelli et al., 1998). The main effect of calcium antagonists, in therapeutic application, is focused on the heart and smooth muscle. Verapamil primarily affects the heart, while most dihydropyridines have a stronger effect on smooth muscle than on the heart. Diltiazem is, in its action, in the middle. If cardiac effects are observed, calcium antagonists can cause AV block and slow down the work of the heart by acting on the conducting tissue, but this is alleviated by a reflex increase in sympathetic activity that occurs due to their vasodilatory effect. For example, nifedipine typically causes reflex tachycardia; diltiazem causes only mild changes in beats per minute, or does not change heart rate, while verapamil slows heart rate. Calcium antagonists also have a negative inotropic effect, which is a consequence of the inhibition of the slow input current during the plateau of the action potential. Nevertheless, the minute volume generally remains unchanged or increases due to a decrease in peripheral resistance. Verapamil has the most pronounced negative inotropic effect and is, therefore, contraindicated in heart failure, as are most other calcium channel blockers. Calcium antagonists cause generalized dilation of arteries and arterioles and thus lower blood pressure, but do not have a great effect on veins. They act on all vascular beds, although the regional effects vary significantly with different drugs. They cause coronary vasodilation and are used in patients with coronary artery spasm (vasospastic angina). Calcium antagonists also relax other types of smooth muscle, but these effects are of less therapeutic importance than their effect on vascular smooth muscle, although they can cause side effects.

Calcium channel blockers are used in arrhythmias (to slow down the speed of ventricular work in rapid atrial fibrillation and to prevent the recurrence of supraventricular tachycardia), hypertension, and prophylaxis of angina pectoris. Calcium antagonists in clinical use are well absorbed from the gastrointestinal tract and are given orally, except for some specific indications, such as use after subarachnoid hemorrhage, where intravenous preparations are used. They are metabolized intensively. Most of the side effects of calcium antagonists are related to the length of their pharmacological action. They can cause redness of the face and headache, swelling of the joints, constipation, etc. (Rosing et al., 1985).

Antiarrhythmics—The continuous work of the heart is enabled by the constant creation of stimuli in the part of the right atrium of the heart called the sinus node. The stimulus spreads from that place through the atria, a special muscle bundle of fibers called the His bundle, all the way to the heart muscle cells of the left and right ventricles. These electrical impulses generated in the heart allow the heart to work in the form of convulsions (systole) or relaxation of the heart (diastole) 60–80 times per minute. Accelerated heart rate over 100 beats per minute is called sinus tachycardia and is most often the result of extracardiac factors such as: physical exertion, mental tension, fever, increased thyroid activity, and excessive intake of caffeine, alcohol, and nicotine. Slowed heart rate below 50 beats per minute is called sinus bradycardia and is most often the result of good physical training, reduced thyroid function, jaundice, aging, and the effects of some drugs (digitalis, beta-blockers, verapamil, diltiazem, etc.) (Connolly et al., 2000).

Cardiac arrhythmias occur due to disorders in the creation or conduction of stimuli in the heart or both disorders at the same time. The causes that lead to cardiac arrhythmias can be divided into three groups: heart disease, diseases of other organs, and general disorders. Heart diseases are: coronary heart disease (angina pectoris, myocardial infarction), inflammatory processes in the heart, heart muscle weakness, heart defects, etc. Diseases of other organs are: diseases of the lungs, diseases of the central nervous system, diseases of the kidneys, diseases of the endocrine glands (most often the thyroid gland), and diseases of the gastrointestinal tract. General disorders are various infections and toxic conditions, loss of minerals or fluids from the body, the effects of some drugs, especially digitalis, diuretics and some drugs for the treatment of cardiac arrhythmias (quinidine, flecainide, etc.).

The significance of cardiac arrhythmias is twofold. First, they can reduce heart rate by up to 30%. This refers to cardiac arrhythmias with accelerated heart rate. Second, they can cause sudden cardiac death in the event of ventricular fibrillation (ventricular fibrillation). According to the origin, cardiac arrhythmias can be from the atria (atrial) and ventricles (ventricular). From the point of view of treatment, this difference is significant for two reasons: (1) atrial arrhythmias are less dangerous for the patient, and (2) the approach to treatment is different. The most common heart rhythm disorders are premature heartbeats (extrasystoles). They can occur in the atria or ventricles. Atrial extrasystoles often occur in healthy people. They can be a precursor to flickering and fluttering of the atria (Marian, 2009).

Ventricular extrasystoles also occur not only in healthy people, but also in patients with coronary heart disease, heart muscle weakness, arterial hypertension, potassium loss, and because of side effects of some drugs (digitalis, antiarrhythmics). Atrial fibrillation (arrhythmia absoluta, fibrilatio atriorum) is one of the most common cardiac arrhythmias. It occurs in people with coronary heart disease, heart muscle weakness, mitral valve disease (especially mitral stenosis), increased thyroid function, and heart disease (pericardium). In a number of patients, the cause cannot be detected (Hamada et al., 2005). The most severe heart rhythm disorders are ventricular acceleration (ventricular tachycardia) and ventricular fibrillation (ventricular fibrillation). They most often occur in acute myocardial infarction and unstable angina pectoris.

4.6.2 Electrostimulation of the heart

Electrostimulation of the heart is a part of the electrostimulation system which, by generating impulses, repairs the weakened chronotropic function of the heart muscle. The electrostimulation system consists of a pulse generator and electrodes that are introduced into the heart muscle (myocardium). By generating impulses through the heart, the process of depolarization expands, which causes the heart muscle to contract and thus enables a normal heart rhythm. The implantation of a dual-chamber pacemaker may benefit some patients with LVOT obstruction and severe symptoms who are not candidates for surgery and do not respond to drug therapy (Marian, 2009), or in those who have bradycardia that requires electrostimulation (Nishimura & Holmes, 2004). Properly adjusted electrostimulation of RV leads to paradoxical movement of the septum leading to reduction of blood flow velocity, which reduces SAM and MR. Most patients reported improvement in symptoms, and a reduction in the LVOT pressure gradient was achieved in the long term in 57% of the patients. Some of the advantages of cardiac pacemaker treatment, especially in elderly patients, are lower risk than surgical procedures, easier availability (Maron, 2002).

4.6.3 Surgical treatment

The most commonly performed surgical procedure to treat LVOT is an anterior ventricular septum myectomy (Morrow procedure), in which a rectangular canal extends from the basal septum below the aorta distally to above the point of septal contact with the mitral leaflet. This reverses or significantly reduces the exit tract gradient LV in over 90% of the cases, reduces mitral regurgitation caused by SAM, and improves functional capacity and symptoms. Long-term benefit in terms of symptomatology has been achieved in 70%–80% of patients with long-term survival comparable to that in the general population. Preoperative determinants of a good long-term outcome are age < 50 years, left atrium size < 46 mm, absence of atrial fibrillation, and male gender. The main surgical complications are AV block, ventricular septal defect and, aortic regurgitation (AR), but they are rare in experimental centers when the operation is guided by intraoperative transesophageal echocardiography. When left ventricular body obstruction is present at the same time, standard myectomy may be extended distally to the middle ventricle around the base of the papillary muscles, but data on the efficacy and long-term outcomes of this approach are limited (Desai et al., 2013).

5 Anatomical model

An individual 3D anatomical model of a patient's torso is developed from CT scans. The data are obtained on a CT scanner of the Siemens SOMATOM Dual Source 256 brand, with a pixel dimension of 0.33 mm and a layer thickness of 0.3 mm. Several stages make up this whole process:

- acquisition of CT images,
- segmentation of CT images and contour detection,

- generating 3D models,
- processing the 3D models,
- creating a mesh of finite elements, and
- setting boundary conditions.

The first phase involves the application of a semiautomatic algorithm of the segmentation CT images to get an axial array of contours, and then a creation of a 3D view using the Materialize Mimics 10.01 program. Adjusting the contrast of the loaded images further simplifies the reconstruction process because the application of the mentioned option highlights the bright areas of the scanned structures.

The next stage is the application of the Geomagic Studio 10 program in order to obtain the volume model presented as a mesh of triangular surfaces. The creation of a finite element mesh is done using the TetGen program, which represents the final phase in the preparation of the model for simulation. The last phase involves adding boundary conditions and use of PAKSF solver. These programs are widely accepted and for this reason, they are used in this study.

3D reconstruction starts with the software Materialize Mimics 10.01, developed at the KU Leuven in Belgium. This software is specially developed for processing medical images, such as CT, microCT, MR, ultrasound, and others. With this software, it is possible to convert anatomical images into 3D models and this process is known as segmentation (Fig. 10).

The anatomical model consists of human torso, spine, ribs, costal cartilage, sternum, lungs, and heart (Fig. 11 left).

The finite element mesh was created using TetGen software. A tetrahedral three-dimensional finite element with four nodes was used. The PAKSF solver used in the next chapter gives excellent results with octahedral elements, so it is necessary to convert the obtained tetrahedral elements into octahedral ones. Each edge of the tetrahedral elements

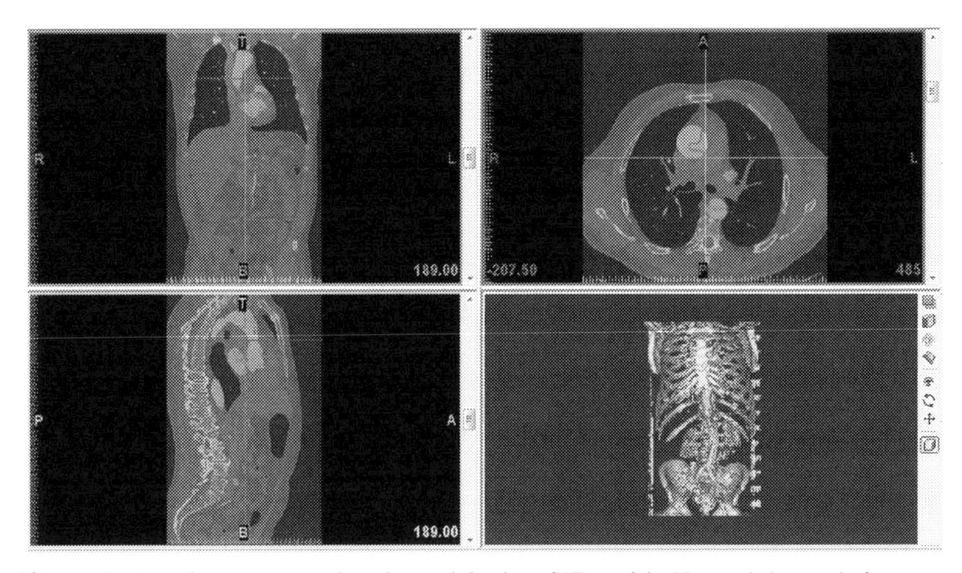

FIG. 10 Application of segmentation algorithm and display of 3D models. *No permission required.*

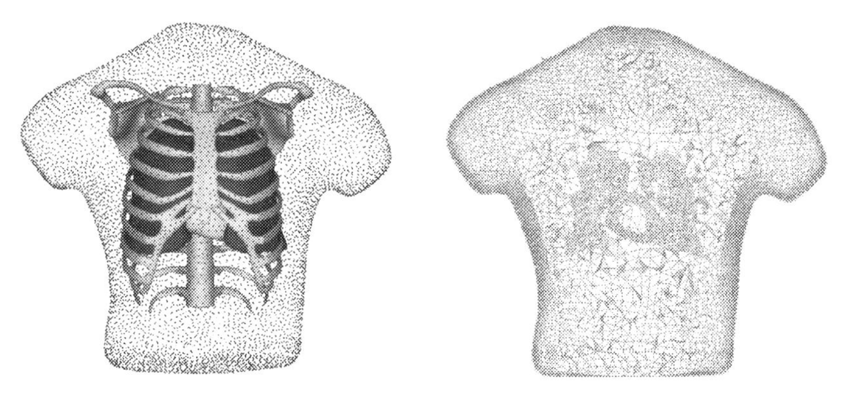

FIG. 11 Anatomical model (left) and cut view of the heart-torso computational mesh (right). Anatomical model—human torso, spine, ribs, costal cartilage, sternum, lungs, and heart. *No permission required.*

is assigned another node in the middle of the edge, while in the center of gravity of each page, one node is assigned. By connecting these nodes, new eight-node elements are obtained. Four new eight-node elements are obtained from each tetrahedron. Finally, the following values of nodes and elements were obtained: the torso has 27,315 nodes and 139,186 elements, lungs 12,184 nodes and 60,955 elements, heart 8445 nodes and 43,200 elements, and bones 92,374 nodes and 402,244 elements (Fig. 11 right).

6 Numerical model

The goal of this research was to develop a finite element (FE) model of the electrical activity of the heart embedded in the torso and use it to simulate healthy and ischemic event. It is very important to use detailed and complex, high-resolution anatomically accurate model of whole-heart electrical activity, which requires extensive computation times, dedicated software, and even the use of supercomputers (Trudel et al., 2004). In order to simulate the electrical model of the heart, it requires an adequate model of the transfer between cardiac electrical activity and the ECG signals measured on the torso surface. Six electrodes (V1, V2, V3, V4, V5, and V6) were positioned at the chest to model the precordial leads. It also requires solving a mathematical inverse problem for which no unique solution exists. Clinical validation in humans is very limited since simultaneous whole heart electrical distribution recordings are inaccessible for both practical and ethical reasons (Kojic et al., 2019). In this chapter, we describe the methodology and concepts used to generate an electro-mechanical model of the heart and also present results of the numerical simulation.

Cardiac cells are filled with and surrounded by an ionic solution, mostly sodium $Na+$, potassium $K+$, and calcium Ca^{2+}. These charged atoms move between the inside and the outside of the cell through proteins called ion channels. Cells are connected through gap junctions that form channels that allow ions to flow from one cell to another. An accurate numerical model is needed for better understanding of heart behavior in cardiomyopathy, heart failure, cardiac arrhythmia, and other heart diseases. These numerical models usually include drug transport, electrophysiology, and muscle mechanics (Wilhelms et al., 2011).

Heart geometry has seven different regions: (1) sinoatrial node; (2) atria; (3) atrioventricular node; (4) His bundle; (5) bundle fibers; (6) Purkinje fibers; (7) ventricular myocardium (Figs. 12 and 13).

The monodomain model of the modified FitzHugh-Nagumo model of the cardiac cell was used (FitzHugh, 1961; Nagumo et al., 1962; Sovilj et al., 2013).

$$\frac{dV}{dt} = -kc_1(V_m - B)\left(-\frac{(V_m - B)}{A} + a\right)\left(-\frac{(V_m - B)}{A} + 1\right) - kc_2R(V_m - B) \quad (1)$$

$$\frac{dR}{dt} = -ke\left(\frac{(V_m - B)}{A} - R\right) \quad (2)$$

where V_m is the membrane potential, R is the recovery variable a, which is relating to the excitation threshold e is relating to the excitability A, which is the action potential amplitude, B is the resting membrane potential and c_1, c_2 and k are membrane-specific parameters.

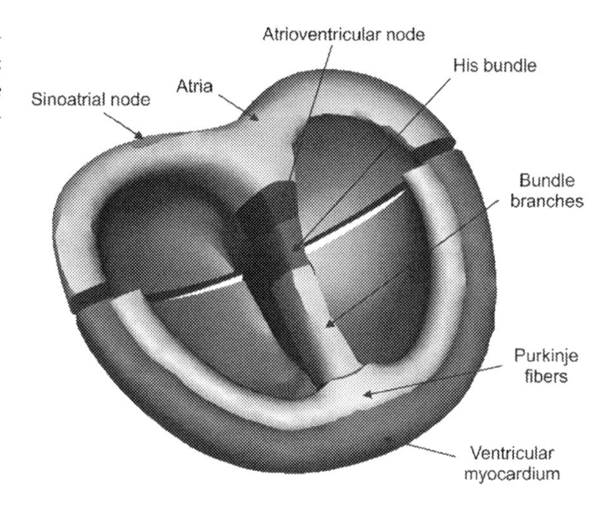

FIG. 12 Heart geometry and seven different regions of the model. (1) Sinoatrial node; (2) atria; (3) atrioventricular node; (4) His bundle; (5) bundle fibers; (6) Purkinje fibers; and (7) ventricular myocardium. *No permission required.*

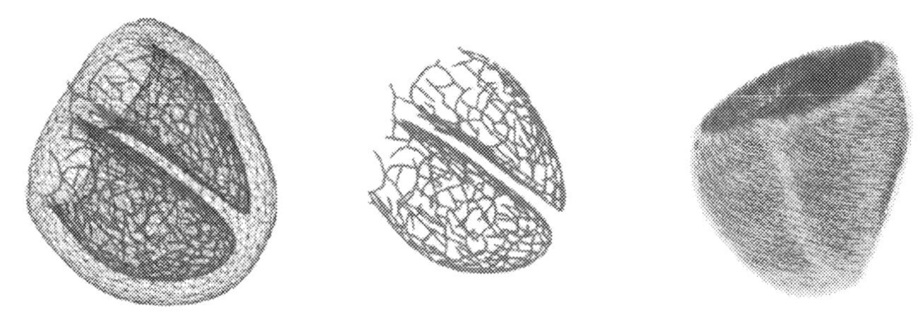

FIG. 13 Model of the left and right ventricles (left) with Purkinje fibers (middle); Directions of muscle fibers (right). *No permission required.*

The monodomain model (Nagumo et al., 1962; Sovilj et al., 2013) with the incorporated modified FitzHugh-Nagumo equations are:

$$\frac{\partial V_m}{\partial t} = \frac{1}{\beta C_m}(\nabla \bullet (\sigma \nabla V_m) - \beta(I_{ion} - I_s)), D = \frac{\sigma}{\beta C_m} \tag{3}$$

where β is the membrane surface-to-volume ratio, C_m is the membrane capacitance per unit area, σ is the tissue conductivity, I_{ion} is the ionic transmembrane current density per unit area, and I_s is the stimulation current density per unit area.

Parameters for the monodomain model with the modified FitzHugh-Nagumo equations are presented in Table 1 (Sovilj et al., 2013).

Computer simulations were conducted using the fully coupled heart-torso monodomain equations including a detailed description of human ventricular cellular electrophysiology. Myocardial and torso conductivities were based on the literature (Sovilj et al., 2013), as presented in Table 2.

Boundary conditions on all interior boundaries in contact with the torso, lungs, and cardiac cavities are zero flux for m; therefore, $-\mathbf{n} \cdot \mathbf{\Gamma} = 0$ where $\mathbf{n}$ is the unit outward normal vector on the boundary and $\mathbf{\Gamma}$ is the flux vector through that boundary for the intracellular voltage, equal to $\mathbf{\Gamma} = - \cdot \mathrm{m}/\mathbf{n}$. For the variable m, the inward flux on these boundaries is equal to

TABLE 1 Parameters for the monodomain model with the modified FitzHugh-Nahumo equations.

Parameter	SAN	Atria	AVN	His	BNL	Purkinje	Ventricles
a	−0.60	0.13	0.13	0.13	0.13	0.13	0.13
b	−0.30	0	0	0	0	0	0
c_1 ($\mathrm{AsV^{-1}\,m^{-3}}$)	1000	2.6	2.6	2.6	2.6	2.6	2.6
c_2 ($\mathrm{AsV^{-1}\,m^{-3}}$)	1.0	1.0	1.0	1.0	1.0	1.0	1.0
D	0	1	1	1	1	1	1
e	0.066	0.0132	0.0132	0.005	0.0022	0.0047	0.006
A (mV)	33	140	140	140	140	140	140
B (mV)	−22	−85	−85	−85	−85	−85	−85
k	1000	1000	1000	1000	1000	1000	1000
σ ($\mathrm{mSm^{-1}}$)	0.5	8	0.5	10	15	35	8

From https://doi.org/10.1155/2013/134208.

TABLE 2 Conductivity parameters for lungs, bone, and remaining regions.

Parameter	Lungs	Bone	Remaining regions
σ ($\mathrm{Scm^{-1}}$)	2.4×10^{-4}	4×10^{-5}	6×10^{-4}

From https://doi.org/10.1155/2013/134208.

the outward current density J from the torso/chamber volume conductor; therefore, −·m / n = n · J (Nagumo et al., 1962).

In the second part, the performance of classical approaches for solving the ECG inverse problem using the epicardial potential formulation was implemented. The studied methods are the family of Tikhonov methods and L regularization-based methods (Wang & Rudy, 2006; Van Oosterom, 1999; Oosterom, 2001, 2003, 2010). ECG measurement was performed at the Clinical Center Kragujevac, University of Kragujevac, on a healthy volunteer.

7 Results

In this section, the results of simulations for two cases are presented. The first is a simulation of the transmembrane potential on the epicardial surface of a normal heart, while the second simulation shows the ECG results of the heart with apical myocardial ischemia. Myocardial ischemia or infarction is an imbalance between the supply of oxygenated blood and the oxygen requirements of the myocardium. The ischemia or infarction can alter the myocardial action potentials. The principal changes are decreases in the magnitude of the resting potential and in the action potential duration.

Whole healthy heart activation simulation from leads' ECG signal at various time points on the ECG signal is shown in Fig. 14. There are 1–9 activation sequences corresponding to the ECG signal. The color bar denotes mV of the transmembrane potential.

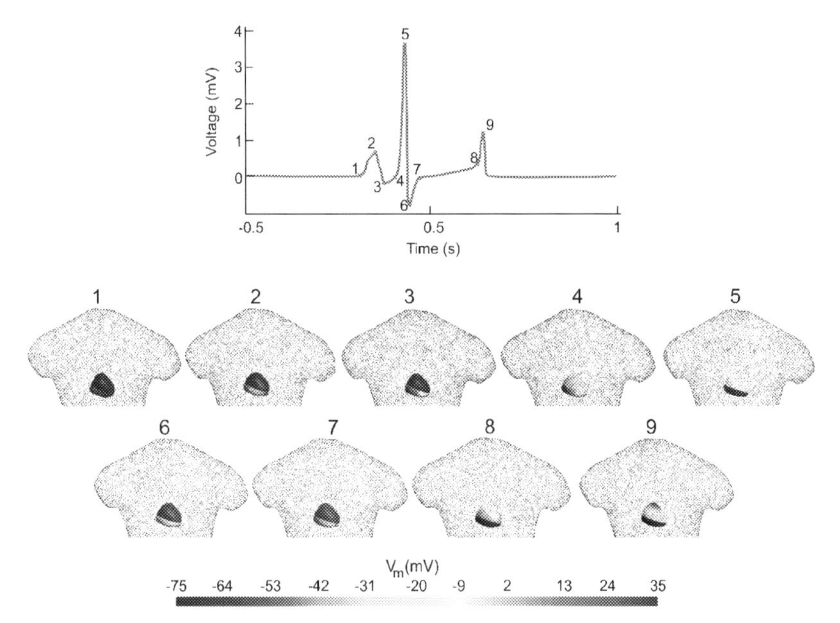

FIG. 14 Simulated lead II ECG signal and corresponding whole healthy heart activation sequences. *No permission required.*

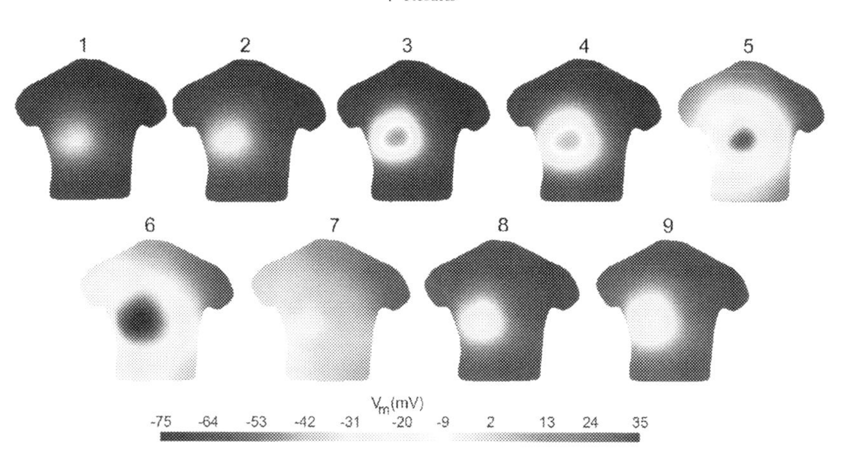

FIG. 15 Body surface potential maps in a healthy subject during progression of ventricular activation in nine sequences corresponding to an ECG signal. *No permission required.*

Body surface potential maps in a healthy subject during the progression of ventricular activation in nine sequences, which are corresponding to a measured ECG signal, are presented in Fig. 15.

Cardiac ischemia introduces several electrophysiological changes in the myocardium. First, the membrane potential, regulated by parameters A, B, and e, is raised together with the extracellular potassium concentration. To change the conduction velocity, the conductivity was reduced to 0.003 in the ischemic region. Figs. 16–18 show six standard precordial leads for a healthy subject.

The ECG in the presence of apical ischemia is shown in Figs. 19–21.

It can be noticed that the R wave in the presence of apical ischemia has a slightly lower value than in a healthy heart. The QRS complex has a similar response. The T wave in the presence of apical ischemia has a significantly different response. The maximum values of T waves in this case at leads V4, V5, and V6 are 0.2, 0.15, and 0.09 mV, respectively. In a healthy heart, these values are higher and amount to 0.97, 0.55, and 0.32 mV, respectively.

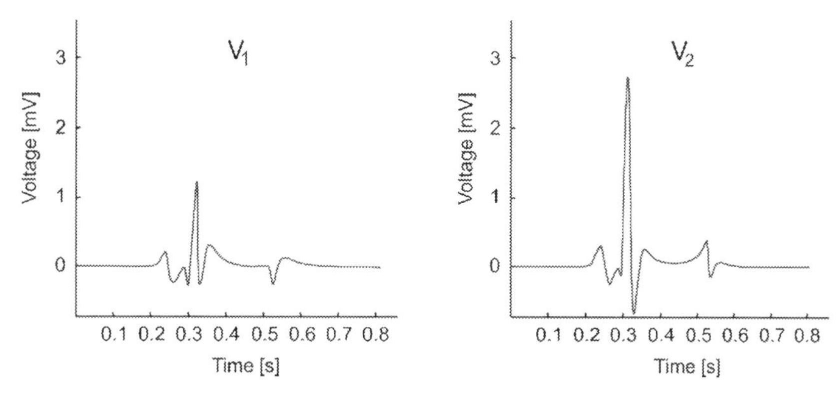

FIG. 16 ECG leads for a normal heart (V1, V2). *No permission required.*

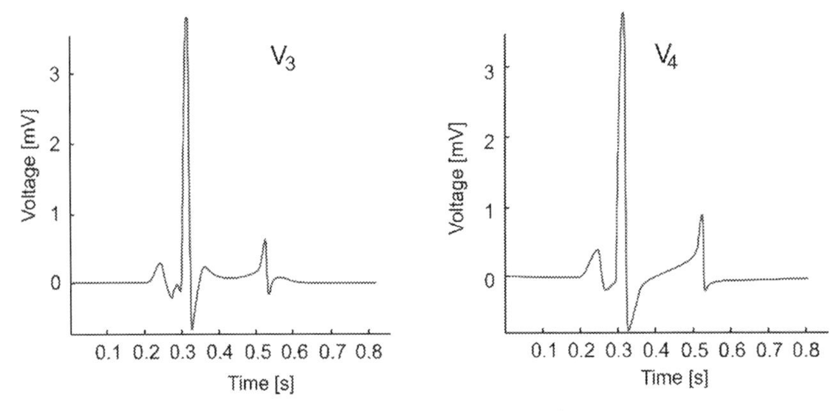

FIG. 17 ECG leads for a normal heart (V3, V4). *No permission required.*

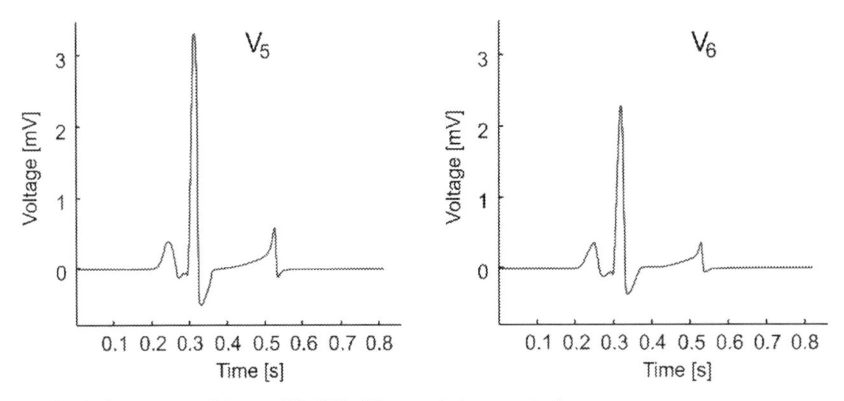

FIG. 18 ECG leads for a normal heart (V5, V6). *No permission required.*

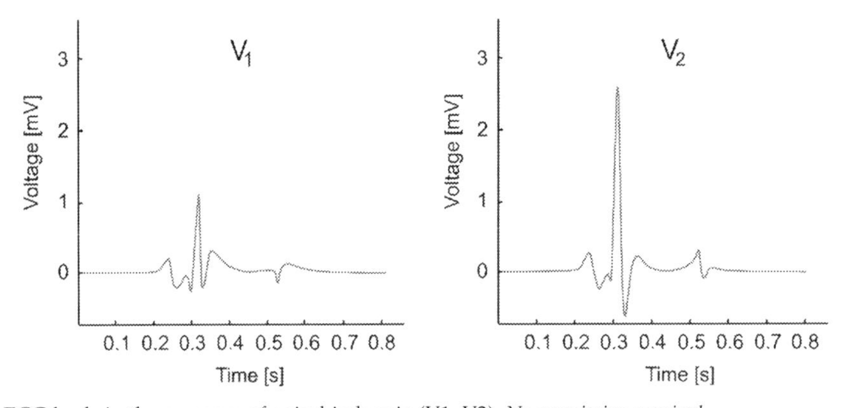

FIG. 19 ECG leads in the presence of apical ischemia (V1, V2). *No permission required.*

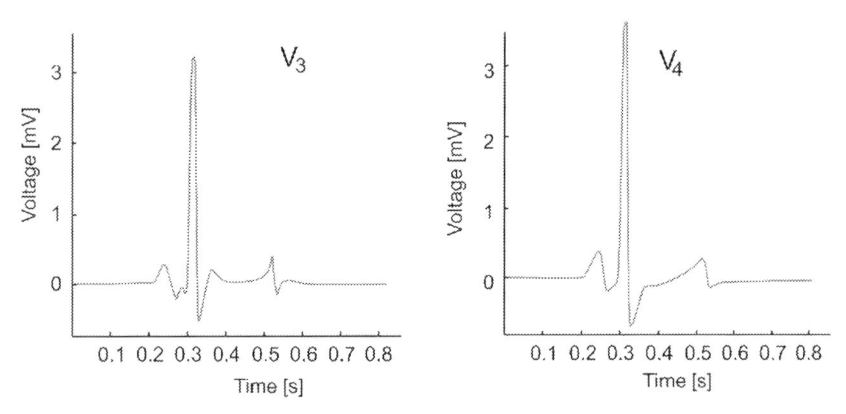

FIG. 20 ECG leads in the presence of apical ischemia (V3, V4). *No permission required.*

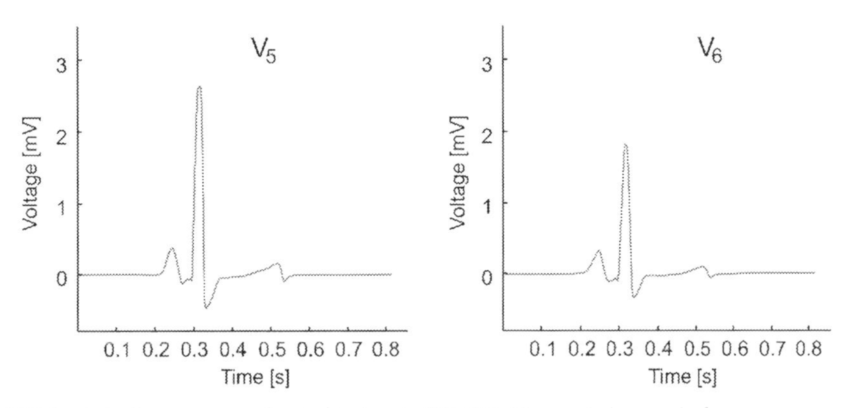

FIG. 21 ECG leads in the presence of apical ischemia (V5, V6). *No permission required.*

In this way, by applying finite elements and setting boundary conditions that can simulate heart disease, we can fully monitor heart behavior.

8 Conclusions

In this research, we developed a finite element model of the electrical activity of the heart embedded in the torso in order to simulate healthy and ischemic events. An individual 3D anatomical model of a patient's torso was developed based on CT scans. The anatomical model consists of human torso, spine, ribs, costal cartilage, sternum, lungs, and heart. Heart geometry consists of seven different regions: sinoatrial node, atria, atrioventricular node, His bundle, bundle fibers, Purkinje fibers, and ventricular myocardium. The monodomain model based on the modified FitzHugh-Nagumo model of the cardiac cell was used. Computer simulations were conducted using the fully coupled heart-torso monodomain equations including the detailed description of human ventricular cellular electrophysiology. Parameters for the monodomain model with the modified FitzHugh-Nagumo equations as well as myocardial and torso conductivities were based on data from the literature.

In the second part, the performance of classical approaches for solving the ECG inverse problem using the epicardial potential formulation was implemented. The studied methods are the family of Tikhonov methods and L regularization-based methods.

The results of the simulations for the two cases are presented. The first was the simulation of the transmembrane potential on the epicardial surface of the normal heart, while the second simulation showed ECG results of the heart with apical myocardial ischemia. Cardiac ischemia introduces several electrophysiological changes in the myocardium. First, the membrane potential, regulated by parameters A, B, and e, was raised together with the extracellular potassium concentration. To change the conduction velocity, the conductivity was reduced to 0.003 in the ischemic region. The results of the simulations showed that the R wave in the presence of apical ischemia had a slightly lower value than in the healthy heart. The QRS complex had a similar response. The T wave in the presence of apical ischemia had a significantly different response. The maximum values of T waves in this case at leads V4, V5, and V6 were 0.2, 0.15, and 0.09 mV, respectively. Observing the healthy heart, these values were higher and amounted to 0.97, 0.55, and 0.32 mV, respectively, for these ECG leads.

Echocardiography is the main method for the initial evaluation of patients with suspected hypertrophic cardiomyopathy. Echocardiography allows the clinician to evaluate the presence and severity of LV wall thickness, the presence of LVOT obstruction, mitral regurgitation, diastolic dysfunction, and assist with therapeutic and surgical interventions. The presented 3D model offers a wide range of possibilities for modeling various cardiomyopathies and electrical abnormalities in order to explore and develop new ECG-based diagnostic methods.

References

Adabag, A. S., Casey, S. A., Kuskowski, M. A., Zenovich, A. G., & Maron, B. J. (2005). Spectrum and prognostic significance of arrhythmias on ambulatory Holter electrocardiogram in hypertrophic cardiomyopathy. *Journal of the American College of Cardiology, 45*(5), 697–704. https://doi.org/10.1016/j.jacc.2004.11.043.

AlGhatrif, M., & Lindsay, J. (2012). A brief review: history to understand fundamentals of electrocardiography. *Journal of Community Hospital Internal Medicine Perspectives, 2*. https://doi.org/10.3402/jchimp.v2i1.14383.

Arad, M., Maron, B. J., Gorham, J. M., Johnson, W. H., Saul, J. P., Perez-Atayde, A. R., Spirito, P., Wright, G. B., Kanter, R. J., Seidman, C. E., & Seidman, J. G. (2005). Glycogen storage diseases presenting as hypertrophic cardiomyopathy. *New England Journal of Medicine, 352*(4), 362–372. https://doi.org/10.1056/NEJMoa033349.

Bottinelli, R., Coviello, D. A., Redwood, C. S., Pellegrino, M. A., Maron, B. J., Spirito, P., Watkins, H., & Reggiani, C. (1998). A mutant tropomyosin that causes hypertrophic cardiomyopathy is expressed in vivo and associated with an increased calcium sensitivity. *Circulation Research, 82*(1), 106–115. https://doi.org/10.1161/01.RES.82.1.106.

Brandenburg, I., Robert, O., Valentini, F., & Giuliani, E. R. (1989). *Cardiology fundamentals and practice* (pp. 45–164). Year Book Medical Pub.

Braunwald, E. (1997). *Heart disease: A textbook of cardiovascular medicine* (5th ed., pp. 1414–1426). W.B. Saunders Company, str.

Carley, S. D. (2003). Beyond the 12 lead: Review of the use of additional leads for the early electrocardiographic diagnosis of acute myocardial infarction. *Emergency Medicine, 15*(2), 143–154. https://doi.org/10.1046/j.1442-2026.2003.00431.x.

Cerovec, D. (2014). *Hipertrofijska kardiomiopatija*. Diplomski rad Sveučilište u Zagrebu, Medicinski fakultet.

Connolly, S. J., Hallstrom, A. P., Cappato, R., Schron, E. B., Kuck, K. H., Zipes, D. P., Greene, H. L., Boczor, S., Domanski, M., Follmann, D., Gent, M., & Roberts, R. S. (2000). Metal-analysis of the implantable cardioverter defibrillator secondary prevention trials. *European Heart Journal, 21*(24), 2071–2078. https://doi.org/10.1053/euhj.2000.2476.

Desai, M. Y., Bhonsale, A., Smedira, N. G., Naji, P., Thamilarasan, M., Lytle, B. W., & Lever, H. M. (2013). Predictors of long-term outcomes in symptomatic hypertrophic obstructive cardiomyopathy patients undergoing surgical relief of left ventricular outflow tract obstruction. *Circulation, 128*(3), 209–216. https://doi.org/10.1161/CIRCULATIONAHA.112.000849.

Ebrard, G., Fernández, M. A., Gerbeau, J. F., Rossi, F., & Zemzemi, N. (2009). From intracardiac electrograms to electrocardiograms: Models and metamodels. In *Vol. 5528. Lecture Notes in Computer Science (Including Subseries Lecture Notes in Artificial Intelligence and Lecture Notes in Bioinformatics)* (pp. 524–533). https://doi.org/10.1007/978-3-642-01932-6_56.

FitzHugh, R. (1961). Impulses and physiological states in theoretical models of nerve membrane. *Biophysical Journal*, 445–466. https://doi.org/10.1016/s0006-3495(61)86902-6.

Gersh, B. J., Maron, B. J., Bonow, R. O., Dearani, J. A., Fifer, M. A., Link, M. S., … Yancy, C. W. (2011). ACCF/AHA Guideline for the Diagnosis and Treatment of Hypertrophic Cardiomyopathy: A report of the American College of Cardiology Foundation/American Heart Association Task Force on Practice Guidelines. Developed in collaboration with the American Association for Thoracic Surgery. *Heart Failure Society of America, 58*(25), 212–260.

Gregory, A. (2001). *Harvey's heart, the discovery of blood circulation*. Icon Books.

Guyton. (1991). *Textbook of medical physiology*. Harcourt College Pub.

Hamada, M., Shigematsu, Y., Inaba, S., Aono, J., Ikeda, S., Watanabe, K., Ogimoto, A., Ohtsuka, T., Hara, Y., & Higaki, J. (2005). Antiarrhythmic drug cibenzoline attenuates left ventricular pressure gradient and improves transmitral doppler flow pattern in patients with hypertrophic obstructive cardiomyopathy caused by midventricular obstruction. *Circulation Journal, 69*(8), 940–945. https://doi.org/10.1253/circj.69.940.

Harvey, W. (1889). *On the motion of the heart and blood in animals*. Prometheus Books.

Jean, R.-A. (2013). La médecine égyptienne, Médecine cardiaque: le coeur, l'infectiologie, Pharaon Magazine. *Juin, 13*, 42–46.

Johnson, J. N., Grifoni, C., Bos, J. M., Saber-Ayad, M., Ommen, S. R., Nistri, S., Cecchi, F., Olivotto, I., & Ackerman, M. J. (2011). Prevalence and clinical correlates of QT prolongation in patients with hypertrophic cardiomyopathy. *European Heart Journal, 32*(9), 1114–1120. https://doi.org/10.1093/eurheartj/ehr021.

Klues, H. G., Schiffers, A., & Maron, B. J. (1995). Phenotypic spectrum and patterns of left ventricular hypertrophy in hypertrophic cardiomyopathy: Morphologic observations and significance as assessed by two-dimensional echocardiography in 600 patients. *Journal of the American College of Cardiology, 26*(7), 1699–1708. https://doi.org/10.1016/0735-1097(95)00390-8.

Kojic, M., Milosevic, M., Simic, V., Geroski, V., Ziemys, A., Filipovic, N., & Ferrari, M. (2019). Smeared multiscale finite element model for electrophysiology and ionic transport in biological tissue. *Computers in Biology and Medicine*, 288–304. https://doi.org/10.1016/j.compbiomed.2019.03.023.

Kroeker, C. A. G., Adeeb, S., Tyberg, J. V., & Shrive, N. G. (2006). A 2D FE model of the heart demonstrates the role of the pericardium in ventricular deformation. *American Journal of Physiology, 291*(5), H2229–H2236.

Lj, B., & Birtić, K. (1993). *Elektrokardiografija u Praksi*. University of Zagreb, School of Medicine.

Marian, A. J. (2009). Contemporary treatment of hypertrophic cardiomyopathy. *Texas Heart Institute Journal, 36*(3), 194–204. http://www.pubmedcentral.nih.gov/picrender.fcgi?artid=2696493&blobtype=pdf.

Maron, B. J. (2002). Hypertrophic cardiomyopathy: A systematic review. *Journal of the American Medical Association, 287*(10), 1308–1320.

Maron, M. S. (2012). Clinical utility of cardiovascular magnetic resonance in hypertrophic cardiomyopathy. *Journal of Cardiovascular Magnetic Resonance, 14*(1). https://doi.org/10.1186/1532-429X-14-13.

Maron, B. J., McKenna, W. J., Danielson, G. K., Kappenberger, L. J., Kuhn, H. J., Seidman, C. E., Shah, P. M., Spencer, W. H., Spirito, P., Ten Cate, F. J., Wigle, E. D., Vogel, R. A., Abrams, J., Bates, E. R., Brodie, B. R., Danias, P. G., Gregoratos, G., Hlatky, M. A., Hochman, J. S., … Torbicki, A. (2003). American College of Cardiology/European Society of Cardiology clinical expert consensus document on hypertrophic cardiomyopathy: A report of the American College of Cardiology Foundation Task Force on clinical expert consensus documents and the European Society of Cardiology Committee for practice guidelines. *Journal of the American College of Cardiology, 42*(9), 1687–1713. Elsevier Inc https://doi.org/10.1016/S0735-1097(03)00941-0.

Maron, B. J., Olivotto, I., Bellone, P., Conte, M. R., Cecchi, F., Flygenring, B. P., Casey, S. A., Gohman, T. E., Bongioanni, S., & Spirito, P. (2002). Clinical profile of stroke in 900 patients with hypertrophic cardiomyopathy. *Journal of the American College of Cardiology, 39*(2), 301–307. https://doi.org/10.1016/S0735-1097(01)01727-2.

Nagumo, J., Arimoto, S., & Yoshizawa, S. (1962). An active pulse transmission line simulating nerve axon. *Proceedings of the IRE*, 2061–2070. https://doi.org/10.1109/JRPROC.1962.288235.

Nishimura, R. A., & Holmes, D. R. (2004). Hypertrophic obstructive cardiomyopathy. *New England Journal of Medicine*, *350*(13), 1320–1327. https://doi.org/10.1056/NEJMcp030779.

Noble, R., Hillis, J., & Rothbaum, D. (1990). Electrocardiography. In *Clinical methods: The history, physical, and laboratory examinations*. Boston: Butterworths.

Nutton, V. (1973). The chronology of Galen's early career. *The Classical Quarterly*, *23*(1), 158–171. https://doi.org/10.1017/s0009838800036600.

Oosterom, A. V. (1999). The spatial covariance used in computing the pericardial potential. *IEEE Transactions on Biomedical Engineering*, *46*(7), 778–787. https://doi.org/10.1109/10.771187.

Oosterom, A. V. (2003). Source models in inverse electrocardiography. *International Journal of Bioelectromagnetism*, *5*, 211–214.

Oosterom, A. V. (2010). The equivalent double layer: source models for repolarization. In *Comprehensive electrocardiology* (pp. 227–246).

Podrid, P. J., & Kowey, P. R. (2010). *Cardiac arrhythmia, mechanism, diagnosis and management*. Lippincott Williams and Wilkins Publication.

Pullan, A. J., Buist, M. L., & Cheng, L. K. (2005). *Mathematically modelling the electrical activity of the heart—From cell to body surface and back again*. World Scientific Publishing Company.

Rosing, D., Idanpaan-Heikkila, U., Maron, B., Bonow, R., & Epstein, S. E. (1985). Use of calciumchannel blocking drugs in hypertrophic cardiomyopathy. *The American Journal of Cardiology*, *55*.

Schalick, W. O. (2006). Jerome J. Bylebyl: An appreciation. *Bulletin of the History of Medicine*, vii–x. https://doi.org/10.1353/bhm.2006.0116.

Schroeder, S., Achenbach, S., Bengel, F., Burgstahler, C., Cademartiri, F., De Feyter, P., George, R., Kaufmann, P., Kopp, A. F., Knuuti, J., Ropers, D., Schuijf, J., Tops, L. F., & Bax, J. J. (2008). Cardiac computed tomography: Indications, applications, limitations, and training requirements—Report of a Writing Group deployed by the Working Group Nuclear Cardiology and Cardiac CT of the European Society of Cardiology and the European Council of Nuclear Cardiology. *European Heart Journal*, *29*(4), 531–556. https://doi.org/10.1093/eurheartj/ehm544.

Sherrid, M. V., Pearle, G., & Gunsburg, D. Z. (1998). Mechanism of benefit of negative inotropes in obstructive hypertrophic cardiomyopathy. *Circulation*, *97*(1), 41–47. https://doi.org/10.1161/01.CIR.97.1.41.

Shiozaki, A. A., Senra, T., Arteaga, E., Martinelli Filho, M., Pita, C. G., Ávila, L. F. R., Parga Filho, J. R., Mady, C., Kalil-Filho, R., Bluemke, D. A., & Rochitte, C. E. (2013). Myocardial fibrosis detected by cardiac CT predicts ventricular fibrillation/ventricular tachycardia events in patients with hypertrophic cardiomyopathy. *Journal of Cardiovascular Computed Tomography*, *7*(3), 173–181. https://doi.org/10.1016/j.jcct.2013.04.002.

Sovilj, S., Magjarević, R., Lovell, N. H., & Dokos, S. (2013). A simplified 3D model of whole heart electrical activity and 12-lead ECG generation. *Computational and Mathematical Methods in Medicine*, *2013*. https://doi.org/10.1155/2013/134208.

Spirito, P., Rapezzi, C., Autore, C., Bruzzi, P., Bellone, P., Ortolani, P., Fragola, P. V., Chiarella, F., Zoni-Berisso, M., Branzi, A., Cannata, D., Magnani, B., & Vecchio, C. (1994). Prognosis of asymptomatic patients with hypertrophic cardiomyopathy and nonsustained ventricular tachycardia. *Circulation*, *90*(6), 2743–2747. https://doi.org/10.1161/01.cir.90.6.2743.

Stainback, R. F. (2012). Apical hypertrophic cardiomyopathy. *Texas Heart Institute Journal*, *39*(5), 747–749.

Standring, S., Borley, N. R., & Gray, H. (2008). *Gray's anatomy: The anatomical basis of clinical practice* (40th ed., anniversary ed.). Edinburgh: Churchill/Elsevier Livingstone/Elsevier.

Trudel, M.-C., Dube, B., Potse, M., Gulrajani, R. M., & Leon, L. J. (2004). Simulation of QRST integral maps with a membrane-based computer heart model employing parallel processing. *IEEE Transactions on Biomedical Engineering*, *51*(8), 1319–1329. https://doi.org/10.1109/TBME.2004.827934.

Van Oosterom, A. (1999). The use of the spatial covariance in computing pericardial potentials. *IEEE Transactions on Biomedical Engineering*, *46*(7), 778–787. https://doi.org/10.1109/10.771187.

Wang, Y., & Rudy, Y. (2006). Application of the method of fundamental solutions to potential-based inverse electrocardiography. *Annals of Biomedical Engineering*, *34*(8), 1272–1288. https://doi.org/10.1007/s10439-006-9131-7.

Wigle, E. D., & Baron, R. H. (1966). The electrocardiogram in muscular subaortic stenosis. Effect of a left septal incision and right bundle-branch block. *Circulation*, *34*(4), 585–594. https://doi.org/10.1161/01.CIR.34.4.585.

Wigle, E. D., Rakowski, H., Kimball, B. P., & Williams, W. G. (1995). Hypertrophic cardiomyopathy: Clinical spectrum and treatment. *Circulation*, *92*(7), 1680–1692. https://doi.org/10.1161/01.CIR.92.7.1680.

Wilhelms, M., Dossel, O., & Seemann, G. (2011). In silico investigation of electrically silent acute cardiac ischemia in the human ventricles. *IEEE Transactions on Biomedical Engineering*, *58*(10), 2961–2964. https://doi.org/10.1109/tbme.2011.2159381.

Yu, E. H. C., Omran, A. S., Wigle, E. D., Williams, W. G., Siu, S. C., & Rakowski, H. (2000). Mitral regurgitation in hypertrophic obstructive cardiomyopathy: Relationship to obstruction and relief with myectomy. *Journal of the American College of Cardiology*, *36*(7), 2219–2225. https://doi.org/10.1016/S0735-1097(00)01019-6.

CHAPTER

5

Simulation of carotid artery plaque development and treatment

Tijana Djukic and Nenad Filipovic

Bioengineering Research and Development Center (BioIRC), Kragujevac, Serbia

1 Introduction

One of the diseases of the human cardiovascular system is carotid artery stenosis (CAS). This disease is characterized by the constriction of the arterial wall, caused by the atherosclerotic plaque. This constriction causes anomalies in the blood flow to the brain and this can further cause severe cerebrovascular events, such as transient ischemic attack (TIA) or stroke. There are several noninvasive imaging techniques that are applied in clinical examination of the potential CAS patients. These include magnetic resonance imaging (MRI), computed tomography (CT), and high-resolution ultrasound (US). US, as a noninvasive and inexpensive technique, is the one that is most commonly applied first when a symptomatic or an asymptomatic patient comes to the clinic for diagnostics. In clinical diagnosis of CAS there are three parameters that are usually considered: the degree of stenosis, the presence of symptoms, and the recency of the symptoms (Grotta, 2013). A detected stenosis that is larger than 70% is considered as high risk. These patients (either symptomatic or asymptomatic) are considered to be at high risk of cerebrovascular events and are therefore directed to surgical intervention (carotid endarterectomy or stenting). In contrast, patients with <70% stenosis are considered at low-intermediate risk when asymptomatic, and unless other confounding factors exist, they are subjected to medical treatment alone (Abbott et al., 2015). The stenosis cutoff for high-risk patients is lowered to 50% when a symptomatic patient with recent events is treated. However, this classification is too general and often causes either unnecessary surgical treatment or undertreatment in patients with lower levels of stenosis who can also be at high risk if all other patient-specific factors are considered. Early and accurate prediction of individuals at high risk of myocardial infarction and stroke would allow for preventive, therapeutic, or

https://doi.org/10.1016/B978-0-12-823956-8.00004-3

surgical (e.g., stenting) measures to be applied to the patient before any of these life-threatening events take place (Lekadir et al., 2017; Paraskevas et al., 2018).

Stenosis in the carotid arteries is generated through the buildup of atherosclerotic carotid plaque formation (accumulation of lipid, protein, and cholesterol esters in the blood vessel wall) and it reduces blood flow significantly. That is why plaque image analysis (Vancraeynest et al., 2011) has the potential to extract valuable information about plaques, identify atherosclerotic plaque components, and ultimately to classify patients as high or low risk (Boi et al., 2018). It is well established that tissue composition plays a central role in the stability and vulnerability of atherosclerotic plaques. Therefore, it is important to develop computational techniques that can automatically and objectively extract the shape of the carotid artery, determine the atherosclerotic plaque constituents from imaging data, and afterward use the obtained data to simulate plaque progression. Also, computer simulations could give valuable insight and quantitative information about blood flow through the arteries. Another application of numerical simulations is the prediction of the outcome of the stenting intervention, providing medical staff a useful tool to analyze possible positions of the stent and the state of the arterial wall after the intervention.

US examination has the advantage of being simple, safe, and inexpensive. There are two possibilities when examining blood vessels using US: traditional B-mode (gray-scale) and color Doppler. Using the traditional B-mode, images of the considered vessels are created at rest. Using the color Doppler, the blood flow is also analyzed and there is a possibility to measure the blood velocity. In both cases, the two-dimensional (2D) cross-sectional images are obtained and these images are analyzed by the clinicians during clinical examination. Computer tools could be useful in this area too, to help extract the vessel geometry using segmentation techniques and then to perform three-dimensional (3D) visualization. This would enable better diagnostics and more detailed analysis.

The attempts to segment medical US images have had limited success in comparison with the success in segmenting images from other medical imaging modalities. Evaluation of a carotid US requires segmentation of the vessel wall, lumen, and plaque of the carotid artery. Convolutional neural networks (CNN) are state of the art in image segmentation and there are some papers in the literature dealing with segmentation of carotid US images. In Xie et al. (2019), the U-Net convolutional neural network for lumen segmentation from US images of the entire carotid system was used. There are many other U-Net modifications used to segment medical US images. The segmentation results using a U-Net-like CNN with noisy rectified linear unit (NReLU) functions, noisy hard sigmoid (NHSigmoid) functions, and noisy hard tanh (NHTanh) function on a small dataset were presented in Li (2016). A similar method for detecting blood vessels in B-mode ultrasound images was presented in a study by Smistad and Løvstakken (2016). In another study, the authors proposed a semiautomatic segmentation method based on deep learning (U-Net) to segment the MAB and LIB from carotid three-dimensional ultrasound (3DUS) images (Zhou, Fenster, Xia, Spence, & Ding, 2019). The 3D reconstruction was also discussed in the literature, where 2D transversal US images were used for the 3D reconstruction of the carotid bifurcation (Rosenfield et al., 1991; Yeom et al., 2014). In this chapter, a similar computer-based automated 3D reconstruction was performed, but due to the limited number of transversal US images obtained during clinical examination, instead of performing the reconstruction only using US images, a generalized model of the CA was used and updated with extracted data from US images, to create a patient-specific reconstructed CA geometry. The extraction of data from US images is performed using a complete and automatic computer-aided diagnosis tool that segments images using a deep learning technique (U-net convolutional neural network). Within this tool, the identification of atherosclerotic plaque components is also performed. This reconstructed geometrical data is then used to perform blood flow simulations and determine regions of possible further plaque growth. Also, this same data is used to perform stent implantation simulation and subsequently blood flow simulation, to determine the restoration of blood flow after this intervention.

The chapter is organized as follows. Section 2 presents the computational approaches and numerical models used to obtain results within this chapter. The results of the simulations for a particular patient are presented and discussed in Section 3. A discussion of other relevant findings from the literature and a short conclusion are given in Section 4.

2 Materials and methods

In this section, the numerical methods and approaches used within this chapter will be described. First, the used clinical dataset in presented in Section 2.1. Then the approach applied for the segmentation of US images, including the plaque classification, is described in Section 2.2. Section 2.3 discusses the 3D reconstruction procedure. The numerical model for blood flow and plaque progression is presented in Section 2.4 and the numerical model used for stent implantation is presented in Section 2.5.

2.1 Clinical dataset

Clinical data from US examination from the Serbian clinical center was used in this chapter. Examination of the carotid arteries was performed in B-mode and color-Doppler mode, in longitudinal and transversal projection. The common carotid artery (CCA), internal (ICA), and external (ECA) carotid arteries are monitored. The dataset that is further used for development and validation of image segmentation and atherosclerotic plaque characterization contained original and annotated US images. The annotated images have been used for development, while the original images (ground truth) have been used for the validation. The dataset for segmentation of the lumen and wall consisted of 108 patients and the dataset for plaque characterization consisted of 44 patients who underwent the US examination. All images were anonymized in order to respect the data protection policies and safety. US images were annotated by two observers (clinical experts). Image annotation included labeling the lumen regions, arterial wall boundaries, and atherosclerotic plaque components such as the lipid core, fibrous and calcified tissue. All patients included in the dataset had the fibrous plaque component as the dominant one among other components.

2.2 Segmentation and plaque classification using deep learning

The deep learning approach was used in this chapter to perform three tasks:

1. Segmentation of images to obtain the shape of the lumen of the patient's carotid artery.

2. Segmentation of images to obtain the shape of the wall of the patient's carotid artery.
3. Classification of the plaque components inside the carotid artery.

The segmentation of the lumen area from US images was presented (Djukic, Arsic, Koncar, & Filipovic, 2020) and validated in the literature (Djukic, Arsic, Djorovic, Filipovic, & Koncar, 2020). In this chapter, this same approach is used for the segmentation of the wall and further enhanced to enable the classification of plaque types.

The US images from the clinical dataset described in Section 2.1 were first preprocessed by performing automated isolation of the region of interest, i.e., by extracting a static window with dimensions 512 × 512 pixels. This step was necessary to exclude segments of the US image that are irrelevant for the segmentation (the view of the plane, quantitative information that the medical device includes for the medical staff, etc.). Afterward, all images were labeled twice—once to denote the lumen region and the second time to denote the wall region. For the classification, a special variant of the U-net CNN (Zhou & Yang, 2019) was used. This model was presented in detail in Ronneberger et al. (2015). The whole dataset was divided into three parts such that 80% of the images were used for the training phase, while 10% were used for the validation phase and 10% were used for the testing phase.

The loss function that was used during the training of the model is given by

$$Loss = binary\_crossentropy\left(y_{true}, y_{pred}\right) + 1 - dice\_coeff\left(y_{true}, y_{pred}\right) \quad (1)$$

where y_{pred} and y_{true} represent the flattened predicted probabilities and the flattened ground truths of the image, respectively.

For the plaque classification task, the preprocessing of US images has included the following steps:

1. Overlapping the previously detected and segmented carotid lumen and wall areas in transversal cross sections
2. Extraction of plaque area
3. Creation of four different masks corresponding to plaque components (plus background color)

The CNN outputs for the lumen and wall contours are used to extract the "ring" consisting of the tissue of the arterial wall that contains plaque inside it. Those "ring" images were then used to automatically detect and classify plaque. In order to accurately perform the classification task, a team of clinical experts manually annotated three different types of plaque (fibrous, lipid, and calcified plaque). The dataset created in this way was used for the training of the neural network. For this multiclass image segmentation, two architectures were tested: U-net and SegNET. In the case of the plaque classification considered in this chapter, the problem can be defined as the multiclass segmentation model, where four classes should be detected: background (area outside the ring), fibrous, calcified, and lipid atherosclerotic plaque components. Background (denoted by 0) is annotated with black color, fibrous plaque (denoted by 1) with yellow, calcified (denoted by 2) with blue, and lipid (denoted by 3) with green color.

For the training phase, 100 epochs were performed with batch size 2. In order to handle highly imbalanced classes, different loss functions were tested, including categorical crossentropy loss, focal loss, categorical focal dice loss, and generalized dice loss. The best performances were achieved by using weighted categorical cross-entropy loss function with manually selected weights for each class [0.2, 2, 3, 3]. The reasoning behind this is that the goal was to increase the importance of nonbackground classes.

As a result of the plaque classification task, a map of pixels is obtained, where pixels representing each individual plaque component are denoted with numbers 0–3, as previously mentioned.

2.3 Three-dimensional reconstruction

The data obtained using the deep learning approach is used to perform 3D reconstruction of the patient-specific carotid bifurcation. This data consists of contours of the lumen and wall area and the map denoting plaque types for a specific transversal cut of the ICA. One of the main drawbacks of the dataset used in this chapter is that the number of transversal cuts obtained during US examination is limited. This problem was overcome by combining the available data with the generalized model of the CA proposed in the literature (Perktold, Peter, et al., 1991; Perktold, Resch, & Peter, 1991). The reconstruction process was presented in the literature (Djukic, Arsic, Koncar, & Filipovic, 2020). The characteristics of the generalized model are shown in Fig. 1 on the left. This model assumes that the cross-sectional shapes of the CA are circular. Another assumption that was combined with the generalized model is to consider that the outer diameter of the wall is 25% greater than the diameter of the lumen. These assumptions are not true in a real patient-specific case. Therefore, the contours obtained using the deep learning approach for the lumen and wall are used to define the cross sections of the segments of CA—CCA, ICA, and ECA. The most sclerotic segment of the CA for the patients considered within this study was the ICA. For that reason, the main focus of 3D reconstruction was on the ICA. Consequently, the centerlines of the CCA and ECA are considered to be straight lines and their lengths are equal to the ones used in the generalized model, but the longitudinal cut of the ICA is used to extract the centerline of the ICA and the change of lumen and wall diameters in this segment. Fig. 1 on the right shows the CA model used for reconstruction, with a combination of generalized and extracted parameters. The rectangles show segments that are extracted from US images. A similar process of 3D

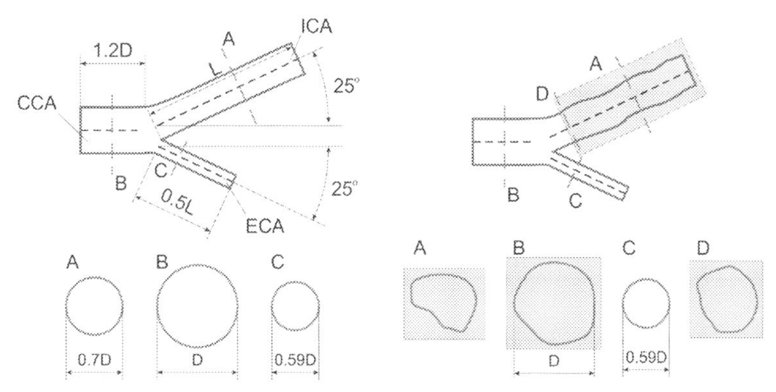

FIG. 1 The adaptation of the generalized model of the carotid artery using US images obtained for a specific patient.

reconstruction was applied in the literature (Vukicevic et al., 2018) to reconstruct the geometry of the coronary arteries from angiography images. It was adapted to the case considered in this chapter. This approach to 3D reconstruction of the lumen of the CA was successfully validated in the literature (Djukic, Arsic, Djorovic, et al., 2020).

The contours obtained from the deep learning approach are smoothed and converted to nonuniform B-splines. In order to perform this conversion, it was first necessary to define the basis functions. These functions are calculated using the Cox-de Boor recursive algorithm (Vukicevic et al., 2014) and are described using the following equations:

$$N_{i,1}(t) = \begin{cases} 1, & \text{if } s_i \le t < s_{i+1} \\ 0, & \text{otherwise} \end{cases} \tag{2}$$

$$N_{i,k}(t) = \frac{(t-s_i)N_{i,k-1}(t)}{s_{i+k-1}-s_i} + \frac{(s_{i+k}-t)N_{i+1,k-1}(t)}{s_{i+k}-s_{i+1}} \tag{3}$$

In Eqs. (2) and (3), s represents the knot vector calculated using the Chord length algorithm (Bastl et al., 2012).

The nonuniform B-spline curve is defined using the following equation:

$$\vec{P}(l) = \sum_{i=1}^{q} \mathbf{B}_i N_{i,k}(t) \tag{4}$$

where q denotes the number of control points and $\mathbf{B}_i$ denotes the control points that are actually extracted within the deep learning approach. In Eq. (4), subscript i denotes the coordinate index ($i \in [0,1]$) and the subscript k has values in the interval $[2, i+1]$.

Using a similar approach, the centerlines for all the three branches are parameterized by using the same equation already used for the nonuniform B-spline curve [Eq. (4)]. These centerlines can be denoted by $\vec{c}^{\,j}(t)$, where the subscript j takes values 0–2, indicating the index of each branch (CCA—0, ICA—1, ECA—2).

The branches of the CA are considered to be tube-like and in order to connect the cross sections to the centerlines, it is necessary to define the orientation of the centerlines. This is done by using the Frenet–Serret formulas (Vukicevic et al., 2014). For this purpose, the tangent $\vec{T}^{\,i}(t)$, normal $\vec{N}^{\,i}(t)$, and binormal $\vec{B}^{\,i}(t)$ of the parameterized centerlines are defined by performing the interpolation and calculation of the first and second derivatives for the B-spline curves. The cross sections are then projected onto the trihedron normal-binormal plane, in the appropriate positions. In this way, the patches of the final surface for all branches are obtained. In the process of forming patches, if the points are not properly defined, the twisting of some patches may occur. In order to prevent this and to ensure the regularity and continuity of the lines, all control points of the contours obtained using deep learning were converted to polar coordinates in the normal-binormal plane and sorted in a circular direction. The obtained patches in 3D space can be represented by using NUBRS surfaces. These surfaces can be defined using the following equation:

$$\vec{S}^{\,i}(u,v) = \sum_{i=1}^{q} \sum_{j=1}^{w} \mathbf{B}_{i,j} N_{i,k}(u) M_{j,l}(v) \tag{5}$$

In Eq. (5), the basis functions defined in Eqs. (2) and (3) are again used and denoted by $N_{i,k}$ and $M_{j,l}$ (and calculated for the kth and lth order). The control points are denoted by $\mathbf{B}_{i,j}$, and indexes u and v can have values in the interval [0,1].

In this way, the parameterized branches are fully defined with quantities $\vec{c}^{\,i}(t)$ and $\vec{S}^{\,i}(u,v)$. Now it is possible to perform the discretization and create the mesh of finite elements (FE). Two quantities are used to define the density of the FE mesh—the number of nodes in the longitudinal and circular directions. The procedure used for the creation of the FE mesh is already described in Vukicevic et al. (2018). When the hexahedral elements are created, the three branches of the CA are connected at the bifurcation (with additional smoothing of the centerlines and the obtained elements).

The 3D reconstruction software is additionally improved to include the separation of plaque from the arterial wall and to reconstruct three plaque types, according to the data obtained from the plaque classification segmentation. In consultation with the clinicians who performed the US examination, it was concluded that the plaque separation from the arterial wall should be performed such that a layer of the plaque should be present radially within the entire wall. This was ensured in the reconstruction software in two steps:

1. The plaque was computationally located along the longitudinal direction of the ICA.
2. A layer of elements representing the plaque was extracted from the previously created finite element (FE) mesh of the ICA wall in the radial direction.

In order to perform the automatic calculation of the plaque in the longitudinal direction, it was necessary to calculate the ratio of the wall and lumen diameters along the longitudinal direction. The cross sections where this ratio is greater than 15% are considered stenotic and in these cross sections the plaque layer was separated, as illustrated in Fig. 2.

After separating the plaque from the arterial wall, an overlap of the FE mesh with the map of plaque components is performed to determine the elements of the FE mesh belonging to the fibrous, lipid, and calcified plaque.

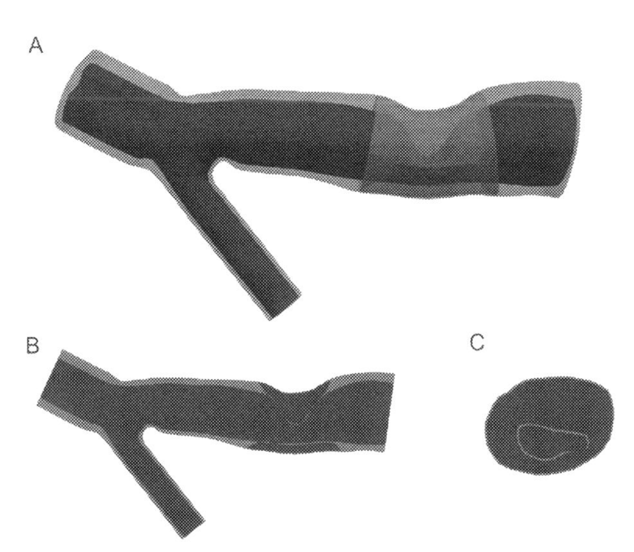

FIG. 2 The reconstructed model with separated plaque: (A) whole 3D geometry, (B) cross section in longitudinal direction, and (C) cross section in radial direction.

2.4 Blood flow and plaque progression simulation

The numerical model for the simulation of blood flow and plaque progression was developed within the international FP7 project ARTreat and presented in the literature (Filipovic et al., 2013; Parodi et al., 2012; Rakocevic et al., 2013). The developed software uses the finite element method to perform the simulations using this numerical model. In this section, a detailed explanation of this model is given.

If the blood velocity is denoted by u_l, pressure is denoted by p_l, and the dynamic viscosity of the blood and density of the blood are denoted by μ and ρ, respectively, then the basic equations used for the simulation of blood flow (Navier-Stokes equation and the continuity equation) can be written as

$$-\mu \nabla^2 u_l + \rho (u_l \cdot \nabla) u_l + \nabla p_l = 0 \tag{6}$$

$$\nabla u_l = 0 \tag{7}$$

Within the blood, there are also other elements that flow along. These other substances can be considered as a solute. If the concentration of this solute is denoted by c_l and the diffusivity of the solute in the lumen is denoted by D_l, then the mass transfer of the solute within the blood lumen can be described using the convection-diffusion equation that is given by

$$\nabla \cdot (-D_l \nabla c_l + c_l u_l) = 0 \tag{8}$$

The solute can also pass through the arterial wall. If the concentration of the solute in the arterial wall is denoted by c_w and the diffusivity of the solute in the arterial wall is denoted by D_w, then the mass transfer of the solute in the arterial wall can be described using the convection-diffusion-reaction equation that is given by

$$\nabla \cdot (-D_w \nabla c_w + k c_w u_w) = r_w c_w \tag{9}$$

where k represents the solute lag coefficient. In this case, the mass transfer is coupled with the transmural flow, and hence in Eq. (9) r_w represents the consumption rate constant.

The Kedem-Katchalsky equations are used to model the transport of low-density lipoprotein (LDL) in the lumen. If the hydraulic conductivity of the endothelium is denoted by L_p, the solute concentration difference across the endothelium is denoted by Δc, the mean endothelial concentration is denoted by c, the pressure drop across the endothelium is denoted by Δp, and the oncotic pressure difference across the endothelium is denoted by $\Delta \pi$, then the Kedem-Katchalsky equations can be written as

$$J_v = L_p (\Delta p - \sigma_d \Delta \pi) \tag{10}$$

$$J_s = P \Delta c + (1 - \sigma_f) J_v c \tag{11}$$

where σ_d represents the osmotic reflection coefficient, σ_f represents the solvent reflection coefficient, and P represents the solute endothelial permeability.

If the solute is considered to be actually oxidized LDL, then it is possible to model the inflammatory process within the arterial wall. If the concentration of macrophages is denoted by M and the concentration of cytokines is denoted by S, then the additional reaction-diffusion partial differential equations can be written as

$$\partial_t c_w = d_2 \Delta c_w - k_1 c_w \cdot M \tag{12}$$

$$\partial_t M + \text{div}\,(v_w M) = d_1 \Delta M - k_1 c_w \cdot M + \frac{S}{1-S} \tag{13}$$

$$\partial_t S = d_3 \Delta S - \lambda S + k_1 c_w \cdot M + \gamma \left(c_w - c_w^{thr}\right) \tag{14}$$

where the corresponding diffusion coefficients are denoted by d_1, d_2, and d_3, l, and g represent the coefficients of degradation and LDL oxidized detection. In Eq. (13), v_w represents the inflammatory velocity of plaque growth. If the pressure in the arterial wall is denoted by p_w, then it is possible to write the equations that represent Darcy's law and the continuity equation for the wall domain (Filipovic et al., 2010):

$$v_w - \nabla \cdot (p_w) = 0 \tag{15}$$

$$\nabla v_w = 0 \tag{16}$$

In this chapter, the blood flow is not modeled as motion of simple fluid, where the boundaries (arterial walls) are treated as impermeable. Instead, it is considered that there is a transport of gases (e.g., oxygen) and macromolecules (e.g., LDL) both through the blood and also through the permeable walls. In the numerical model used in this chapter, it is considered that this mixture of gases and macromolecules within the blood can be modeled as a diluted mixture, i.e., that the transported gases and/or macromolecules do not affect the blood flow. The mass transport through the walls is actually a diffusion process and can be modeled by using convection-diffusion equations. If a concentration of a particular gas or macromolecule is denoted by c, this equation can be written as

$$\frac{\partial c}{\partial t} + v_x \frac{\partial c}{\partial x} + v_y \frac{\partial c}{\partial y} + v_z \frac{\partial c}{\partial z} = D \left(\frac{\partial^2 c}{\partial x^2} + \frac{\partial^2 c}{\partial y^2} + \frac{\partial^2 c}{\partial z^2} \right) \tag{17}$$

where v_x, v_y, and v_z represent the components of blood velocity and D represents the diffusion coefficient of the transported material. The diffusion coefficients are considered constant.

A macromolecule that is directly responsible for the process of atherosclerosis and formation of plaque is LDL. The motion of LDL particles through the bloodstream was analyzed experimentally and numerically in Filipovic et al. (2014). In this chapter, the transfer of LDL within the arterial wall is analyzed. In the literature, the permeability of the arterial wall to LDL has been studied and it was concluded that the coefficient of permeability is of the order of 10^{-8} cm/s (Bratzler et al., 1977).

As already defined, c_w denotes the surface concentration of LDL. The quantity from Eq. (13) can also be described as the filtration velocity of LDL transport through the wall. In that case, the following equation is valid at the arterial wall:

$$c_w v_w - D \frac{\partial c}{\partial n} = K c_w \tag{18}$$

where n is the normal vector in that particular segment of the wall, D represents the diffusivity of LDL, and K represents the overall mass transfer coefficient of LDL at the arterial wall. This equation practically defines the conversion of the mass of LDL, during the passage through a semipermeable wall, the movement toward the arterial wall by a filtration flow, and the diffusion back to the mainstream at the wall.

The presented numerical model is valid for a continuum and it is necessary to apply numerical procedures to adapt them and ensure easier numerical solving. The finite element method

(FEM) was used for this purpose. But in the presented equations, the convection terms dominate due to the low values of diffusion coefficients (Kojic et al., 2008), and due to this fact it was necessary to perform a specific technique during the transformation of equations that ensures a stable numerical solution. In this case, the streamline upwind/Petrov-Galerkin stabilizing technique (SUPG) (Brooks & Hughes, 1982) was applied in combination with the standard numerical integration scheme applied regularly within FEM. The diffusion equations are transformed into an incremental form and included in the incremental-iterative form of finite element equations of balance. This procedure provides the following system of equations:

$$\begin{bmatrix} \frac{1}{\Delta t}\mathbf{M}_v + {}^{n+1}\mathbf{K}_{vv}^{(i-1)} + {}^{n+1}\mathbf{K}_{iv}^{(i-1)} + {}^{n+1}\mathbf{J}_{vv}^{(i-1)} & {}^{n+1}\mathbf{K}_{vp}^{(i-1)} & 0 \\ \mathbf{K}_{vp}^{T} & 0 & 0 \\ {}^{n+1}\mathbf{K}_{cv}^{(i-1)} & 0 & \frac{1}{\Delta t}\mathbf{M}_c + {}^{n+1}\mathbf{K}_{cc}^{(i-1)} + {}^{n+1}\mathbf{J}_{cc}^{(i-1)} \end{bmatrix}$$

$$\times \begin{Bmatrix} \Delta\mathbf{V}^{(i)} \\ \Delta\mathbf{P}^{(i)} \\ \Delta\mathbf{C}^{(i)} \end{Bmatrix} = \begin{Bmatrix} {}^{n+1}\mathbf{F}_v^{(i-1)} \\ {}^{n+1}\mathbf{F}_p^{(i-1)} \\ {}^{n+1}\mathbf{F}_c^{(i-1)} \end{Bmatrix} \quad (19)$$

where the matrices in the system are defined as

$$\begin{aligned} (\mathbf{M}_v)_{jjKJ} &= \int_V \rho N_K N_J \, dV; (\mathbf{M}_c)_{jjKJ} = \int_V N_K N_J \, dV \left({}^{n+1}\mathbf{K}_{cc}^{(i-1)}\right)_{jjKJ} \\ &= \int_V D N_{K,j} N_{J,j} \, dV; \left({}^{n+1}\mathbf{K}_{\mu v}^{(i-1)}\right)_{jjKJ} = \int_V \mu N_{K,j} N_{J,j} \, dV \left({}^{n+1}\mathbf{K}_{cv}^{(i-1)}\right)_{jjKJ} \\ &= \int_V N_K \, {}^{n+1}c_j^{(i-1)} N_J \, dV; \left({}^{n+1}\mathbf{K}_{vv}^{(i-1)}\right)_{jjKJ} = \int_V \rho N_K \, {}^{n+1}v_j^{(i-1)} N_{J,j} \, dV \left({}^{n+1}\mathbf{J}_{cc}^{(i-1)}\right)_{jjKJ} \\ &= \int_V \rho N_K \, {}^{n+1}v_j^{(i-1)} N_{J,j} \, dV; \left({}^{n+1}\mathbf{K}_{vp}^{(i-1)}\right)_{jjKJ} = \int_V \rho N_{K,j} \widehat{N}_J \, dV \left({}^{n+1}\mathbf{J}_{vv}^{(i-1)}\right)_{jkKJ} \\ &= \int_V \rho N_K \, {}^{n+1}v_{j,k}^{(i-1)} N_J \, dV \end{aligned} \quad (20)$$

and the vectors are given by

$$\begin{aligned} {}^{n+1}\mathbf{F}_c^{(i-1)} &= {}^{n+1}\mathbf{F}_{\mathrm{q}} + {}^{n+1}\mathbf{F}_{\mathrm{sc}}^{(\mathrm{i}-1)} - \frac{1}{\Delta t}\mathbf{M}_c \left\{ {}^{n+1}\mathbf{C}^{(\mathbf{i-1})} - {}^{n}\mathbf{C} \right\} - {}^{n+1}\mathbf{K}_{cv}^{(i-1)} \left\{ {}^{n+1}\mathbf{V}^{(i-1)} \right\} \\ &\quad - {}^{n+1}\mathbf{K}_{cc}^{(i-1)} \left\{ {}^{n+1}\mathbf{C}^{(i-1)} \right\} \left({}^{n+1}\mathbf{F}_q\right)_{\mathbf{K}} \\ &= \int_V N_K q^B \, dV \, {}^{n+1}\mathbf{F}_{sc}^{(i-1)} = \int_S D N \nabla \, {}^{n+1}\mathbf{c}^{(i-1)} \cdot \mathbf{n} dS \end{aligned} \quad (21)$$

In Eqs. (20) and (21), the interpolation functions for velocities are denoted by N_I, while the interpolation functions for pressure are denoted by $\widehat{N}_J$. It should be noted that the interpolation functions for pressure are one order of magnitude lower than the interpolation functions for velocities, due to the applied integration scheme.

2.5 Stent deployment simulation

The stent implantation procedure involves an expansion of a stent (a metal cylindrical structure) within the artery. During this process, the stent pushes the arterial wall outward, and accordingly blood flow through the stenosed segment is restored after the implantation. The numerical model used in this chapter for the simulation of stent implantation within CA consists of three segments that are simulated at the same time: modeling the expansion of the stent, modeling the interaction of the stent with the arterial wall, and modeling the deformation of the arterial wall. This model was presented and used for the stent implantation in coronary arteries in the literature (Djukic et al., 2019), where it was also validated against clinical data.

The stent is modeled by using a mesh of interconnected nodes. These nodes are grouped in rings, such that nodes corresponding to each ring have the same distance from the beginning of the stent, measured along the axis of the stent. The stent expansion is modeled using equations of motion of the nodes and the internal expansion force that are defined in the simplex deformable model, as proposed in the literature (Larrabide et al., 2012). The equation of motion of a particular stent node with coordinates $\mathbf{p}_i$ is given by

$$\mathbf{p}_i^{t+1} = \mathbf{p}_i^t + (1-\gamma)\cdot\left(\mathbf{p}_i^t - \mathbf{p}_i^{t-1}\right) + \alpha\cdot\mathbf{f}_{int}\left(\mathbf{p}_i^t\right) + \beta\cdot\mathbf{f}_{ext}\left(\mathbf{p}_i^t\right) \quad (22)$$

where the upper index denotes the time point, and the coefficients α, β, and γ are equal to 0.018, 0.018, and 0.5, respectively, and are defined according to the values provided in the literature (Larrabide et al., 2012; Paliwal et al., 2016).

The internal force $\mathbf{f}_{int}$ in Eq. (22) can be defined as an expanding force and can be written as

$$\mathbf{f}_{int}\left(\mathbf{p}_i\right) = \mathbf{n}_i k\left(r_0 - \left|\mathbf{p}_i - \mathbf{c}_s\right|\right) \quad (23)$$

where $\mathbf{c}_s$ is the center of mass of the stent ring to which the particular stent node belongs, $\mathbf{n}_i$ is the normal vector at the considered stent node, k is the stiffness coefficient, and r_0 is the predefined radius.

The external force $\mathbf{f}_{ext}$ in Eq. (22) can be defined as the force with which the arterial wall opposes the expansion of the stent. In the model proposed in Larrabide et al. (2012), this force was defined for the rigid arterial wall. For each stent node, it was checked whether the distance between that node and the closest segment of the arterial wall is smaller than half of the stent radial thickness. If this is the case, the external force was included in the system of equations and is taken to have the intensity of the expansion force $\mathbf{f}_{int}$, defined in Eq. (23), but acting in the opposite direction. This model of external force from Larrabide et al. (2012) was expanded in Djukic et al. (2019) to include the interaction of the stent nodes with the deformable arterial wall, so it can be represented as a sum of two components:

$$\mathbf{f}_{ext} = \mathbf{f}_{ext}^{(1)} + \mathbf{f}_{ext}^{(2)} \quad (24)$$

This additional component to the external force $\mathbf{f}_{ext}^{(2)}$ is defined as the internal force that appears within the arterial wall as a consequence of the deformation caused by the stent in the previous time step.

In order to model the interaction of the stent and arterial wall, it would be necessary to perform a pairing of nodes from the mesh representing the stent and the mesh representing the arterial wall. Since the discretization of the simulated entities is not the same (the stent is modeled as a shell, with interconnected 1D elements, and the arterial wall is modeled with a mesh of tetrahedral elements), it is not possible to exactly pair the nodes. Instead, this problem is solved by interpolating the forces. This interpolation over different discretization meshes is resolved such that for each stent node the influence of several nodes from the arterial wall is considered and vice versa. The interpolation is performed using the Dirac delta function proposed in the literature (Peskin, 1977). This function is given by

$$\delta\left(\mathbf{p}-\mathbf{x}^{A}(t)\right)=\delta\left(p_{x}-x^{A}\right)\delta\left(p_{y}-y^{A}\right)\delta\left(p_{z}-z^{A}\right) \tag{25}$$

where the $\mathbf{x}_i^A$ vector denotes the coordinates of a particular node of the arterial wall.

The value of function $\delta(r)$ is given by

$$\delta(r)=\begin{cases}\frac{1}{4h}\left(1+\cos\left(\frac{\pi r}{2}\right)\right), & |r|\leq 2\\ 0, & |r|>2\end{cases} \tag{26}$$

The stent acts on the arterial wall with a force that is equal to the force with which the artery opposes further stent expansion, as defined in Newton's third law. This external force $\mathbf{f}_{ext}$ from Eq. (22) was already discussed in previous paragraphs. This force from a particular stent node is interpolated to a particular node of the arterial wall, by applying the following equation:

$$\mathbf{f}_{ext}^{A}(t)=\sum_{l=1}^{m}\mathbf{f}_{ext}\left(\mathbf{p}_{l},t\right)\delta\left(\mathbf{p}_{l}-\mathbf{x}_{i}^{A}(t)\right) \tag{27}$$

The equation of motion of a particular node of the arterial wall is given by

$$\mathbf{x}_{i}^{A}\mathbf{t+1}=\mathbf{x}_{i}^{A}t+(1-\gamma)\cdot\left(\mathbf{x}_{i}^{A}t-\mathbf{x}_{i}^{A}t-1\right)+\alpha\cdot\mathbf{f}_{int}^{A}\left(\mathbf{x}_{i}^{A}t\right)+\beta\cdot\mathbf{f}_{ext}^{A}\left(\mathbf{x}_{i}^{A}t\right) \tag{28}$$

This equation is practically identical to the equation used for stent nodes [Eq. (22)], but the forces are calculated differently. The term for the calculation (interpolation) of external force is already given in Eq. (27). The internal force is calculated using FEM (Kojic et al., 2008). Specifically, during the simulation, the stent causes a deformation of the arterial wall. This deformation further causes an internal force that is the reaction of the wall that is opposing further deformation. Since the deformations and displacements of nodes in these simulations are large, it means that geometric nonlinearity has to be considered and this is done by applying the so-called updated Lagrange method (Kojic et al., 2008).

If the internal stress is denoted by σ, the expression for the calculation of this force can be written as

$$\mathbf{f}_{int}^{A}=\int_{V}\mathbf{B}_{h}^{T}\boldsymbol{\sigma}dV \tag{29}$$

where $\mathbf{B}_h$ represents the matrix of derivatives of interpolation functions for the finite element considered in simulation. The exact expression for this matrix will be given in the sequel.

The internal force from Eq. (29) is then interpolated to stent nodes, using an equation similar to Eq. (27):

$$\mathbf{f}_{ext}^{(2)}(t)=\sum_{l=1}^{m}\mathbf{f}_{int}^{A}\left(\mathbf{x}_{i}^{A},t\right)\delta\left(\mathbf{p}_{l}-\mathbf{x}_{i}^{A}(t)\right) \tag{30}$$

In the literature (Djukic et al., 2019), the material of the arterial wall is considered to be linearly elastic and the material properties are assumed to be isotropic. In Djukic et al. (2021), a different mathematical model was employed to simulate the behavior of the arterial wall and this approach will be used in this Chapter as well. Similar approaches to the modeling of deformable bodies were already successfully applied to model the motion of highly deformable red blood cells (RBCs) (Djukic, 2015; Djukic et al., 2016) and circulating tumor cells (CTCs) (Djukic et al., 2015; Djukic & Filipovic, 2019).

The arterial wall is modeled as an incompressible and isotropic material. In the literature, there are several material models used to model the behavior of human blood vessels, as discussed in Noble et al. (2020). In this chapter, a material model based on the Mooney-Rivlin model is used. This model considers the material as hyperelastic and defines a hyperelastic strain energy density function in terms of the invariants of the Cauchy-Green deformation tensor I_1 and I_2.

The strain energy density function is given by

$$W = C_{10}(I_1 - 3) + C_{01}(I_2 - 3) + C_{20}(I_1 - 3)^2 + C_{11}(I_1 - 3)(I_2 - 3) + C_{30}(I_1 - 3)^3 \tag{31}$$

where C_{ij} are the material coefficients. These coefficients were measured in the literature (Bennetts et al., 2013) and these values are used in this chapter as well and are listed in Table 1.

The material function from Eq. (31) is used to define the relation between stress and deformation and this is afterwards used to calculate the internal stress within each element of the arterial wall that is then inserted in Eq. (29).

The internal stress from Eq. (29) is actually the Cauchy stress tensor and it can be expressed as (Holzapfel, 2000; Živkovic, 2011)

$$\sigma_{ij} = \frac{1}{J} T_{im} F_{jm} \tag{32}$$

where T_{im} represents the components of the first Piola-Kirchhoff stress tensor, and J is the determinant of the deformation gradient.

The first Piola-Kirchhoff stress tensor can be expressed in terms of the strain energy density and deformation gradient as follows:

$$T_{im} = \frac{\partial W}{\partial F_{im}} \tag{33}$$

If Eq. (33) is substituted into Eq. (32), the following expression is obtained:

TABLE 1 Material coefficients for the considered material model.

	C_{10} (kPa)	C_{01} (kPa)	C_{20} (kPa)	C_{11} (kPa)	C_{30} (kPa)
Arterial wall	18.90	18.90	590.4	857.0	0
Fibrous plaque	1.4	0	7.3	0	5.8
Calcified plaque	2.5	0	16.4	0	7
Lipid plaque	1.6	0	9.3	0	11

$$\sigma_{ij} = \frac{1}{J}\frac{\partial W}{\partial F_{im}} F_{jm} \tag{34}$$

The partial derivative of the strain energy density with respect to the deformation gradient from Eq. (34) can be transformed using the chain rule:

$$\frac{\partial W}{\partial F_{im}} = \frac{\partial W}{\partial I_1}\frac{\partial I_1}{\partial F_{im}} + \frac{\partial W}{\partial I_2}\frac{\partial I_2}{\partial F_{im}} + \frac{\partial W}{\partial J}\frac{\partial J}{\partial F_{im}} \tag{35}$$

The expressions for the calculation of partial derivatives of the determinant and invariants with respect to the deformation gradients are defined in the literature (Holzapfel, 2000; Živkovic, 2011) and when they are substituted back into Eq. (34), the following equation for the Cauchy stress tensor can be obtained:

$$\sigma_{ij} = \frac{1}{J}\left(\frac{\partial W}{\partial I_1}2B_{ij} + \frac{\partial W}{\partial I_2}(2I_1 B_{ij} - B_{ik}B_{kj}) + \frac{\partial W}{\partial J}J\delta_{ij}\right) == \frac{2}{J}\left(\frac{\partial W}{\partial I_1} + I_1\frac{\partial W}{\partial I_2}\right)B_{ij} + \frac{\partial W}{\partial J}\delta_{ij} - \frac{2}{J}\frac{\partial W}{\partial I_2}B_{ik}B_{kj} \tag{36}$$

where δ_{ij} is the Kronecker delta symbol and **B** represents the left Cauchy-Green deformation tensor.

The left Cauchy-Green deformation tensor is defined (in index notation) as

$$B_{ij} = F_{im}F_{jm} \tag{37}$$

where **F** represents the deformation gradient tensor that is calculated as

$$F_{kj} = \delta_{kj} + \frac{\partial u_i}{\partial^t x_j} \tag{38}$$

In Eq. (38), **u** represents the displacement of a material point that is given by

$$\mathbf{u} = {}^t\mathbf{x} - {}^0\mathbf{x} \tag{39}$$

The tensor invariants can be expressed in terms of the Cauchy-Green deformation tensor **B** (Živković, 2006):

$$I_1 = tr\mathbf{B} = B_{11} + B_{22} + B_{33} = B_{nn} \tag{40}$$

$$I_2 = \frac{1}{2}\left((tr\mathbf{B})^2 - tr\mathbf{B}^2\right) \tag{41}$$

It is now possible to define the stress-strain relationship for the material model considered in this chapter [in Eq. (31)].

The partial derivatives of strain energy density with respect to the determinant and invariants for the first considered material are given by

$$\frac{\partial W}{\partial I_1} = C_{10} + 2C_{20}(I_1 - 3) + C_{11}(I_2 - 3) + 3C_{30}(I_1 - 3)^2; \frac{\partial W}{\partial I_2} = C_{01} + C_{11}(I_1 - 3); \frac{\partial W}{\partial J} = 0 \tag{42}$$

By substituting the partial derivatives from Eq. (42) into Eq. (36), the final expression that defines the stress-strain relationship for the first considered material is obtained:

$$\sigma_{ij} = \frac{2}{J}\left(C_{10} + 2C_{20}(I_1 - 3) + C_{11}(I_2 - 3) + 3C_{30}(I_1 - 3)^2 + I_1(C_{01} + C_{11}(I_1 - 3))\right) B_{ij}$$
$$-\frac{2}{J}(C_{01} + C_{11}(I_1 - 3)) B_{ik} B_{kj} \quad (43)$$

Eq. (43) is valid for the continuum. In order to use these relations in the simulations of stent deployment [in order to include them in Eq. (29) for the calculation of internal force of the arterial wall], it is necessary to transform these relations to a more appropriate form using FEM. The whole discussion about the transformation of relations and the calculations of quantities in FEM can be found in the literature (Bathe, 1996; Kojić et al., 1998), and in the sequel only a part of this process will be discussed, in order to list all relevant equations needed for the calculations of quantities in the numerical model for stent deployment.

All quantities in FEM simulations are calculated only in nodal points. In all other points of the discretized domain, quantities are calculated by interpolating the quantities in nodal points. If coordinates of the kth node are denoted by X_i^k, where i denotes the index of the coordinate axis ($i=1$ corresponds to the x axis, $i=2$ corresponds to the y axis, and $i=3$ corresponds to the z axis), then the coordinate of an arbitrary point inside the considered finite element can be calculated using the following interpolation:

$$x_i = \sum_{k=1}^{N} h_k X_i^k \quad (44)$$

where N is the overall number of nodes of the finite element (e.g., $N=4$ for the tetrahedral finite element, $N=8$ for the isoparametric three-dimensional (hexagonal) finite element, etc.), and h_k are the interpolation functions that are defined for the specific type of finite element.

Similar to Eq. (44), using the values of the displacement of nodal points, it is possible to calculate the displacement of an arbitrary point of the considered finite element:

$$u_i = \sum_{k=1}^{N} h_k U_i^k \quad (45)$$

In FEM simulations, all required quantities are first determined in the so-called natural (local) coordinate system, which is related to each element separately, and then equations for each element are grouped together in a system of equations of the entire domain. This system is then solved and eventually the calculated quantities are transformed back to the global coordinate system. The natural (local) coordinate system is defined using the coordinate axes r, s, and t that are mutually perpendicular, just like the axes of the global coordinate system and the center of the local coordinate system are connected to the center of gravity of the finite element. Actually, the finite element is mapped from a space defined by the x, y, and z coordinates to a space where the coordinates of nodes have to satisfy the following conditions: $-1 \leq r \leq 1, -1 \leq s \leq 1, -1 \leq t \leq 1$. The interpolation functions are defined for each particular type of finite element.

The strain vector **e** (which has six components: e_{11}, e_{22}, e_{33}, e_{23}, e_{31}, e_{12}) in a particular material point within a finite element can be defined as

$$\mathbf{e} = \mathbf{B}_h \mathbf{U} \tag{46}$$

where $\mathbf{B}_\mathrm{h}$ is the matrix that contains the derivatives of the interpolation function. This matrix, just as the interpolation functions, has a different form depending on the type of finite element. For a triangular finite element, for the kth node, the matrix of derivatives of interpolation functions is given by

$$B_h^k = \begin{bmatrix} h_{k,1} & 0 & 0 \\ 0 & h_{k,2} & 0 \\ 0 & 0 & h_{k,3} \\ h_{k,2} & h_{k,1} & 0 \\ 0 & h_{k,3} & h_{k,2} \\ h_{k,3} & 0 & h_{k,1} \end{bmatrix} \tag{47}$$

where the derivative of the interpolation function with respect to the coordinate is denoted by $h_{k,j} = \frac{\partial h_k}{\partial x_j}$.

Of course, it should be described how to calculate the derivatives of interpolation functions with respect to global coordinates. Since the interpolation functions are defined in terms of natural coordinates, the chain rule can be applied:

$$\frac{\partial h_k}{\partial x_j} = \frac{\partial h_k}{\partial r}\frac{\partial r}{\partial x_j} + \frac{\partial h_k}{\partial s}\frac{\partial s}{\partial x_j} + \frac{\partial h_k}{\partial t}\frac{\partial t}{\partial x_j} \tag{48}$$

The Jacobian of the transformation between the global and local coordinate system is introduced:

$$\mathbf{J} = \left[\frac{\partial \mathbf{x}}{\partial \mathbf{r}}\right] = \begin{bmatrix} \frac{\partial x}{\partial r} & \frac{\partial y}{\partial r} & \frac{\partial z}{\partial r} \\ \frac{\partial x}{\partial s} & \frac{\partial y}{\partial s} & \frac{\partial z}{\partial s} \\ \frac{\partial x}{\partial t} & \frac{\partial y}{\partial t} & \frac{\partial z}{\partial t} \end{bmatrix} \tag{49}$$

and it is also possible to define the inverse Jacobian:

$$\mathbf{J}^{-1} = \left[\frac{\partial \mathbf{r}}{\partial \mathbf{x}}\right] = \begin{bmatrix} \frac{\partial r}{\partial x} & \frac{\partial s}{\partial x} & \frac{\partial t}{\partial x} \\ \frac{\partial r}{\partial y} & \frac{\partial s}{\partial y} & \frac{\partial t}{\partial y} \\ \frac{\partial r}{\partial z} & \frac{\partial s}{\partial z} & \frac{\partial t}{\partial z} \end{bmatrix} \tag{50}$$

Now the derivatives in Eq. (48) can be rewritten as

$$\frac{\partial h_k}{\partial x_j} = J_{j1}^{-1}\frac{\partial h_k}{\partial r} + J_{j2}^{-1}\frac{\partial h_k}{\partial s} + J_{j3}^{-1}\frac{\partial t}{\partial x_j} \tag{51}$$

Elements of the Jacobian matrix can be calculated by substituting Eq. (44) into Eq. (49):

$$J_{ij} = \sum_{k=1}^{N}\frac{\partial h_k}{\partial r_i}X_j^k \tag{52}$$

The deformation gradient for a specific element can be calculated as follows:

$$F_{ij} = \delta_{ij} + \sum_{k=1}^{N} U_i^k \frac{\partial h_k}{\partial x_j} \tag{53}$$

The integrals defined for each finite element are numerically calculated using the Gaussian quadrature. In this dissertation, the problem of strain in the membrane plane is considered, and hence the integrals should be calculated over a two-dimensional domain. In this case, applying the Gaussian quadrature, the following expression is obtained:

$$\int_{-1}^{1}\int_{-1}^{1}\int_{-1}^{1} f(r,s,t)\,dr\,ds\,dt = \sum_{i=1}^{n_i}\sum_{j=1}^{n_j}\sum_{k=1}^{n_k} f\left(r_i, s_j, t_k\right) w_i w_j w_k \tag{54}$$

where n_i, n_j, and n_k represent the numbers of integration points along the axes of the natural coordinate system, w_i are the weight coefficients, and (r_i, s_j, t_k) defines the position of a Gaussian point.

Coordinates of Gaussian points and the corresponding weight coefficients are defined for each particular type of finite element.

All equations of the numerical model for stent deployment are solved repeatedly in iterations. The simulation is stopped when the convergence criteria are satisfied. The convergence criteria are defined such that the maximum absolute change in the nondimensional displacement in a time step must be less than 10^{-3}.

3 Results

In this section, the results of the simulations for a particular patient are going to be presented. After performing the segmentation procedure and extracting the relevant information about plaque types, the 3D reconstruction for the specific patient is performed. The obtained FE mesh is shown in Fig. 3. Fig. 3A shows the complete FE mesh, comprising the lumen, wall, and plaque components. In Fig. 3B, the wall segment is shown to be transparent, in order to better illustrate the position of the plaque within the wall, and in Fig. 3C the plaque is augmented and different plaque components are visible. It should be noted that yellow color represents a fibrous plaque, whereas blue color represents a calcified plaque and green color represents a lipid plaque.

The reconstructed geometry shown in Fig. 3 is used to perform simulations of stent deployment. Two simulations were performed in order to analyze the effect of different plaque components on the outcome. The first one (denoted as Simulation #1) was performed without considering the plaque components, i.e., the whole arterial wall was considered as a single material, with material properties defined in the first row of Table 1. The second simulation (denoted as Simulation #2) considered all plaque components separately, i.e., the components were considered as different materials and the material properties of plaque components are set to be equal to the ones listed in the entire Table 1. The chosen patient had a diagnosed stenosis of 65% in the ICA branch, so the stent is positioned within this branch. The initial position of the stent is shown in Fig. 4A and it was the same in both simulations. Fig. 4B

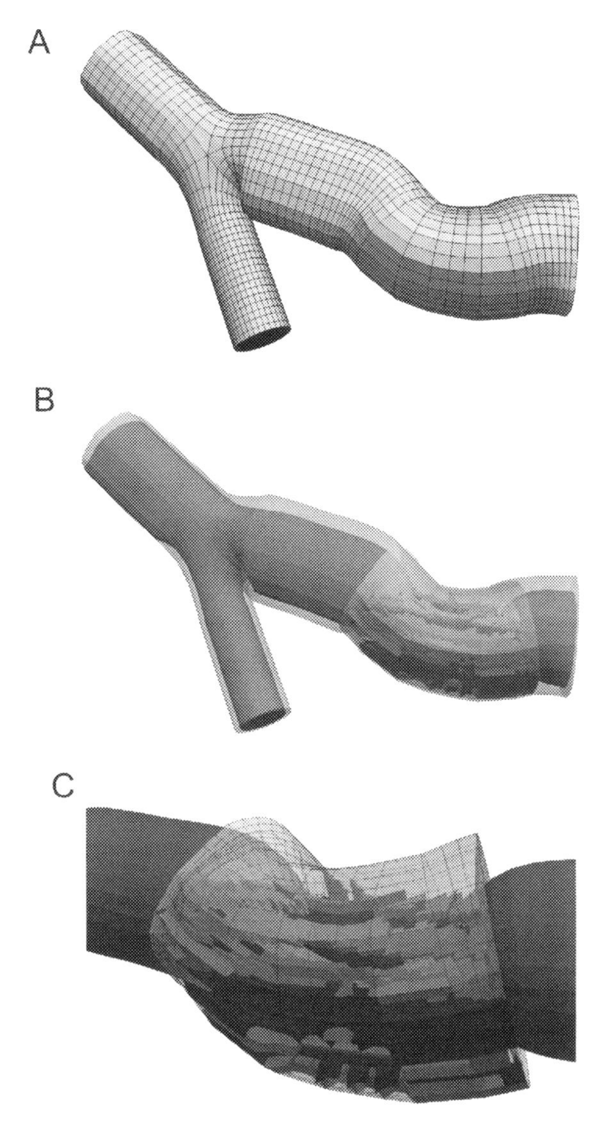

FIG. 3 The reconstructed 3D FE mesh for a particular patient: (A) full model; (B) model with a transparent wall and colored plaque components; and (C) augmented plaque segment.

and C show the final shape of the carotid bifurcation for both simulations. The change of diameter of the ICA branch at baseline and after both simulations is shown in Fig. 5. As can be observed from Figs. 4 and 5, the increase of diameter is smaller in Simulation #1, which is logical since the arterial wall is much stiffer in this simulation setup. The distribution of force with which the stent is acting on the arterial wall is shown in Fig. 6A and B for Simulation #1 and Simulation #2, respectively. Fig. 7A and B show the distribution of internal stress within the arterial wall obtained at the end of both simulations. The overall distributions within the whole ICA, as well as distributions in several chosen cross sections, are shown in Figs. 6 and 7, in order to provide a better overview of the results within the vessel wall. The distributions in

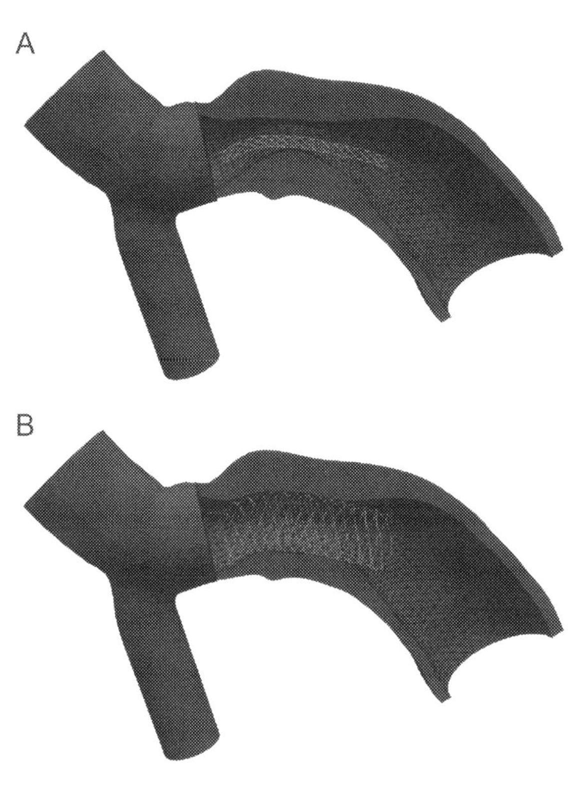

FIG. 4 Results of the stent deployment simulation for two considered cases: (A) initial stent position; (B) final shape of the carotid bifurcation after Simulation #1 (arterial wall modeled as a single material); and (C) final shape of the carotid bifurcation after Simulation #2 (arterial wall modeled with plaque components).

cross sections are shown separately for three cases—at the end of Simulation #1, at the end of Simulation #2, and for Simulation #2 at the iteration when Simulation #1 ended (iteration #3000). These moments in time are chosen to enable the comparison of obtained results. The overall internal stress and stent forces are higher at the end of Simulation #2 because the expansion of the vessel is greater in Simulation #2, but when values for Simulation #2 at iteration 3000 are compared with values at the end of Simulation #1, the values are greater for the latter.

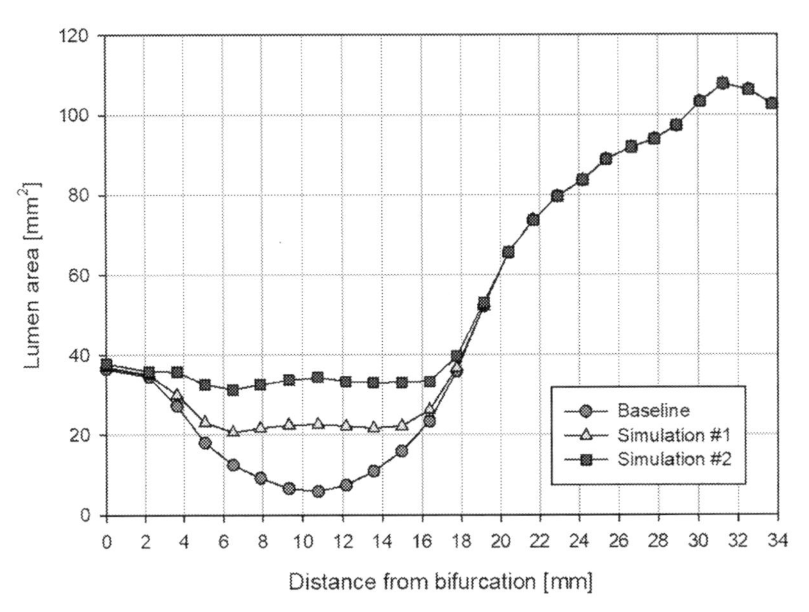

FIG. 5 Change of lumen area along the length of the ICA at baseline and after both stent deployment simulations: Simulation #1—arterial wall modeled as a single material; Simulation #2—arterial wall modeled with plaque components.

The simulations of blood flow were performed together with the simulations of plaque progression within the arterial wall. In these simulations, it is considered that the blood can be represented as an incompressible Newtonian fluid. The density of the blood is set to $1.05\,g/cm^3$ and blood viscosity is set to $0.035\,cm^2/s$. There are several other techniques proposed in the literature that are used to model the viscosity of the blood (Johnston et al., 2004; Shibeshi & Collins, 2005), including the Power-law, Carreau-Yasuda, and Casson models. These models use nonlinear equations to describe the dependency of viscosity of other parameters, among others on the shear rate, hematocrit level, etc. On the other hand, there are authors in the literature who claim that in large arteries the effect of the mentioned factors on viscosity can be neglected (Evju et al., 2013; Steinman, 2012). Since in this chapter the simulations are performed within the CA, the assumption of constant viscosity was considered acceptable. An additional assumption that is introduced within the numerical model used in this chapter is to consider the arterial wall as rigid and nondeformable. Again, since CA is a large artery, the assumption of the rigid and nondeformable arterial wall is considered valid. The velocity of all nodes on the arterial wall is set to zero as part of the boundary condition in the blood flow simulation. The initial condition is the prescribed velocity profile at the inlet of the carotid bifurcation. The value of the velocity is defined according to the values measured for the specific patient. In the simulations of plaque progression, the additional boundary condition is included for the concentration of LDL—the uniform constant concentration is defined at the inlet.

The simulations are performed for two states of the carotid bifurcation—before stent implantation (the state reconstructed from US examination) and after the stent implantation (the

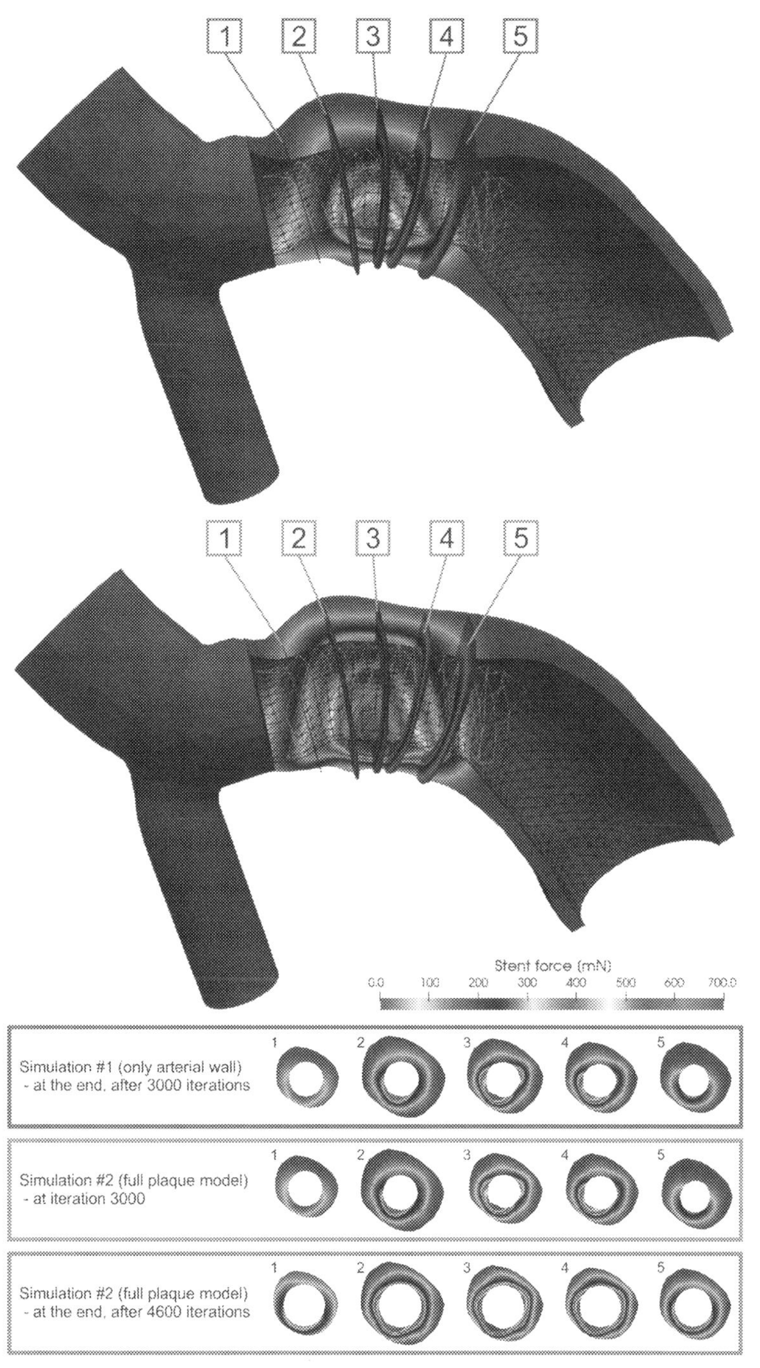

FIG. 6 Results of the stent implantation simulations—Distribution of overall force with which the stent is acting on the arterial wall: Simulation #1 (top)—arterial wall modeled as a single material; Simulation #2 (bottom)—arterial wall modeled with plaque components.

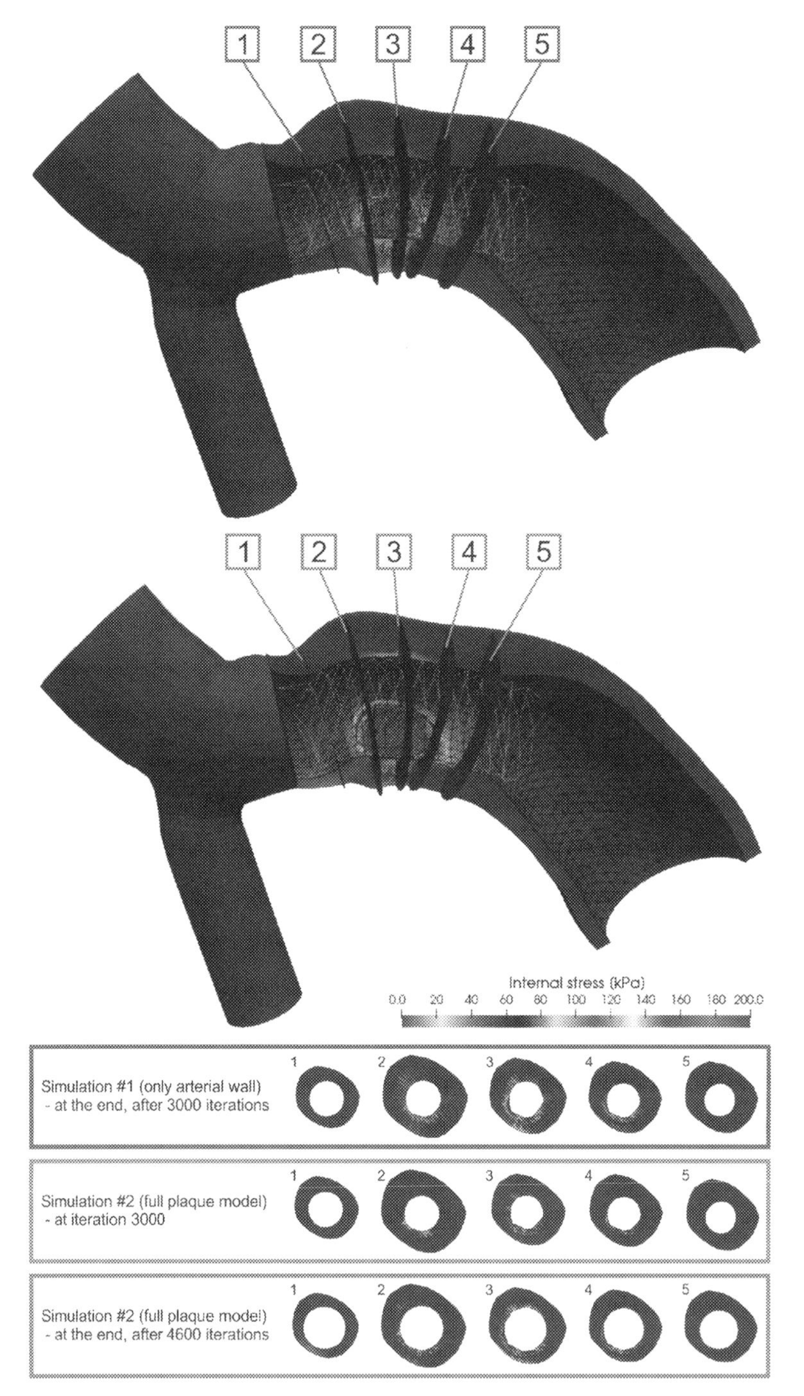

FIG. 7 Results of the stent implantation simulations—Distribution of the von-Mises stress within the arterial wall: Simulation #1 (top)—arterial wall modeled as a single material; Simulation #2 (bottom)—arterial wall modeled with plaque components.

state obtained using the previously discussed Simulation #2). The results for blood flow for both states are shown comparatively in Fig. 8. It can be observed that the stent implantation procedure caused a significant decrease of all relevant quantities—pressure was reduced from 910 Pa to approximately 370 Pa (as shown in Fig. 8A and B), velocity was reduced from 1.9 m/s to 1.1 m/s (as shown in Fig. 8C and D), and WSS was reduced from 200 to 50 dyne/cm^2 (as shown in Fig. 8E and F).

The biological mechanism behind the plaque progression that is considered within this chapter consists of the analysis of the behavior of several macromolecules, as discussed in Section 2.4. The LDL molecules enter the arterial wall and it is assumed within the numerical model that they are immediately oxidized. The number of these molecules then increases until it reaches a certain threshold. Afterward, the recruitment of monocytes occurs and these

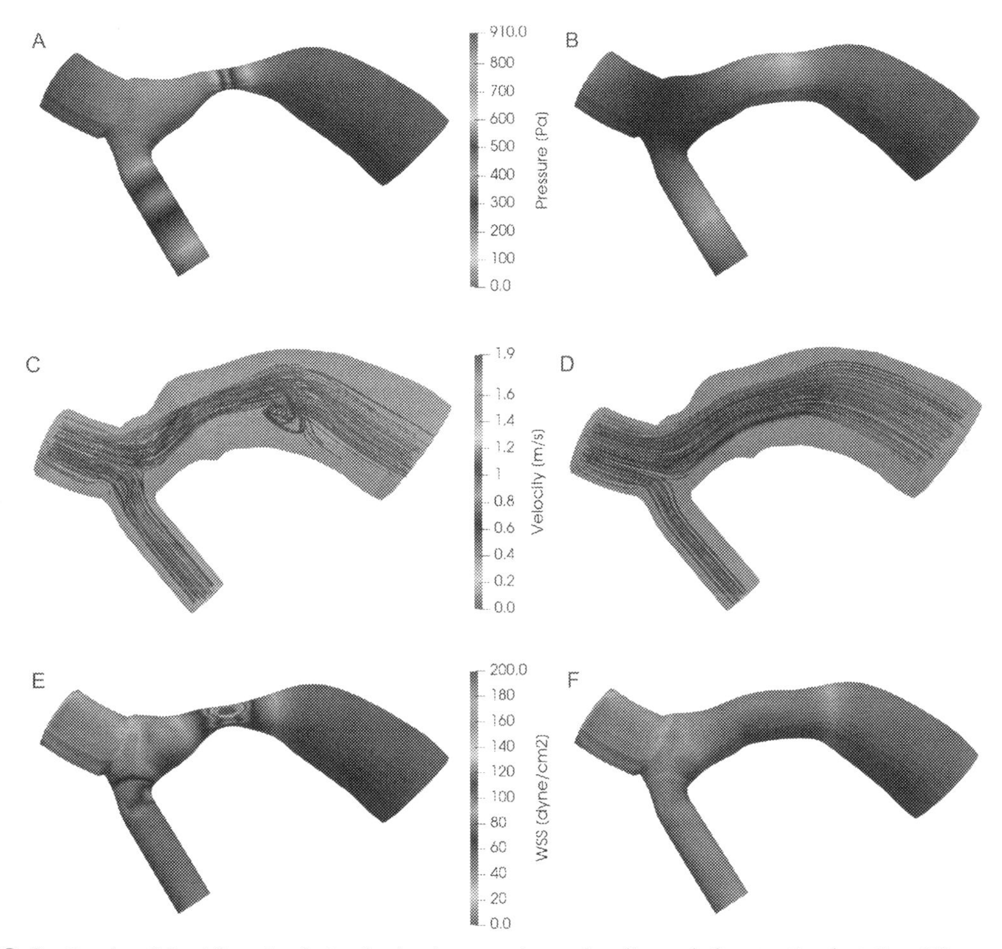

FIG. 8 Results of blood flow simulation for the chosen patient, at baseline and after stent implantation; (A) pressure distribution at baseline; (B) pressure distribution after stent implantation; scale bar for pressure is between A and B; (C) velocity streamlines at baseline; (D) velocity streamlines after stent implantation; scale bar for velocity is between C and D; (E) WSS distribution at baseline; and (F) WSS distribution after stent implantation; scale bar for WSS is between E and F.

monocytes differentiate into macrophages. This further causes the transformation of macrophages into foam cells. At the same time, new monocytes are recruited and secretion of a proinflammatory signal occurs, causing the appearance of cytokines, as part of the inflammatory reaction. The foam cells that are created from macrophages are actually responsible for the increase of the plaque volume. This approach was presented in the literature and validated in both experimental data (in vitro studies and experiments with pigs) and human clinical data (Rakocevic et al., 2013).

The results for the transfer of macromolecules through the arterial wall (within the simulations of plaque progression) are shown in Figs. 9–11, with distributions of macrophages, cytokines, and oxidized LDL, respectively. Again, the results are comparatively shown for both states—before and after stent implantation. Also, several cross sections of the carotid bifurcation are extracted and shown, in these figures, to demonstrate the alterations that stent implantation caused on the distribution of the considered macromolecules. As expected, the concentrations remained of the same magnitude in the CCA and ECA and the greatest change is visible in the treated segment of the ICA. It can be observed that the concentration of macrophages close to the stenosis is reduced after stent implantation and that the LDL concentration is significantly smaller close to the stenosis. The distribution of cytokines is also different; it is more regularly distributed along the considered cross sections after stent implantation, in comparison with a higher concentration in one part of the wall at baseline.

4 Discussion and conclusions

The capabilities of CFD simulations to accurately predict the blood flow parameters have been studied in the literature (Cibis et al., 2016; Long et al., 2002), where the results of simulations were compared with MRI measurements and the good agreement of results was obtained. In clinical practice, the hemodynamic assessment of significant coronary stenosis is commonly performed using a well-established technique called FFRct. In analogy with this technique, the authors of a study introduced a pressure-based carotid arterial functional assessment (CAFA) index (Zhang et al., 2018). This index is calculated in CFD simulations that are performed using geometry extracted from digital subtracted angiography (DSA). The pressure gradient ratio obtained in CFD simulation and using invasive measurement for a chosen patient was compared to demonstrate the feasibility of the proposed technique.

The authors of another study performed simulations of blood flow through an anatomically realistic CA and analyzed the distribution of relevant quantities with only CFD simulations and the simulations considering the deformable walls and the interaction of the blood with walls (Lopes et al., 2019). They concluded that the carotid sinus (in the ICA branch) has the lowest values of WSS and that it is the region with the most probable plaque accumulation. It was also concluded in other previous studies (Samady et al., 2011; Stone et al., 2012) that there is a correlation between low WSS regions and plaque progression. There is also a correlation between high concentration of LDL and plaque progression, as discussed in the literature (A. Sakellarios et al., 2017; A. I. Sakellarios et al., 2017). Time resolved phase-contrast MRI with three-directional velocity encoding (so-called flow-sensitive 4D MRI) was used in (Markl et al., 2010) to study the distribution of WSS in vivo in patient-specific CA. The authors

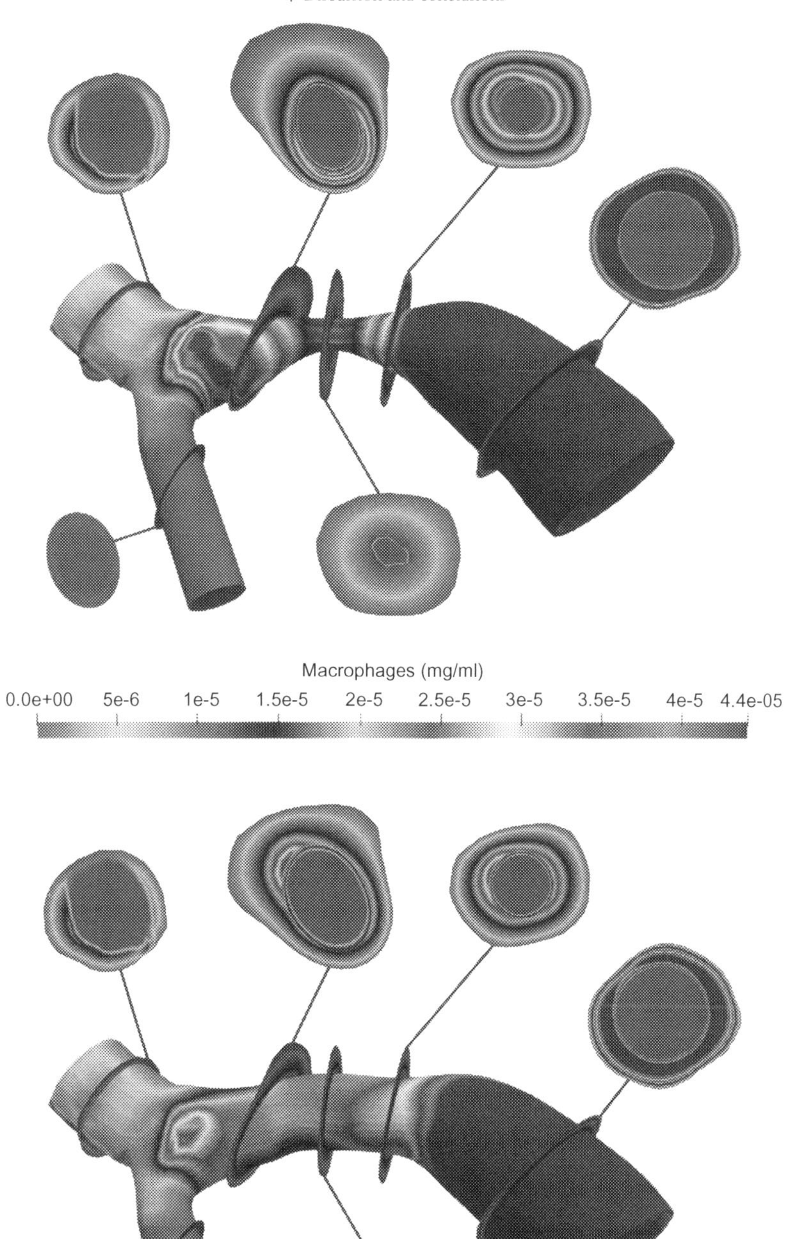

FIG. 9 Results of blood flow simulation for the chosen patient, at baseline (top) and after stent implantation (bottom); macrophages distribution within the arterial wall.

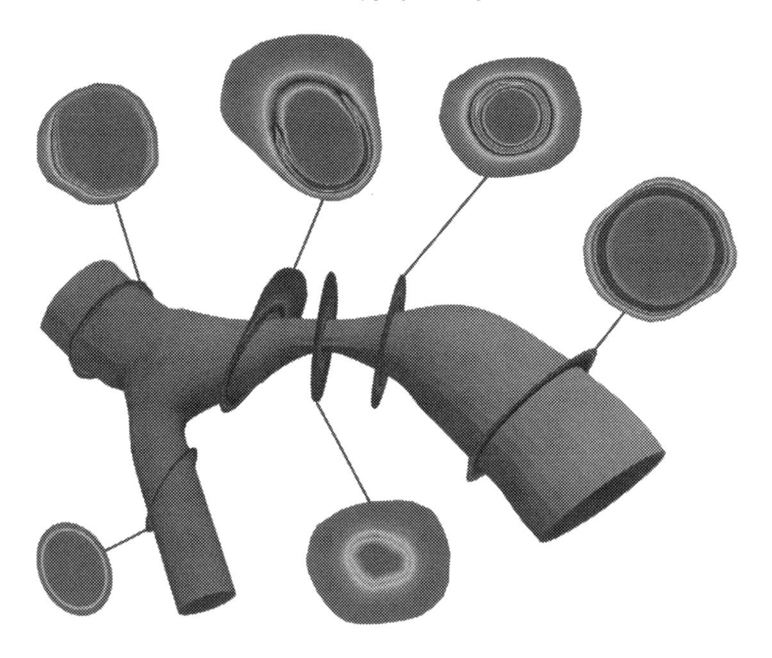

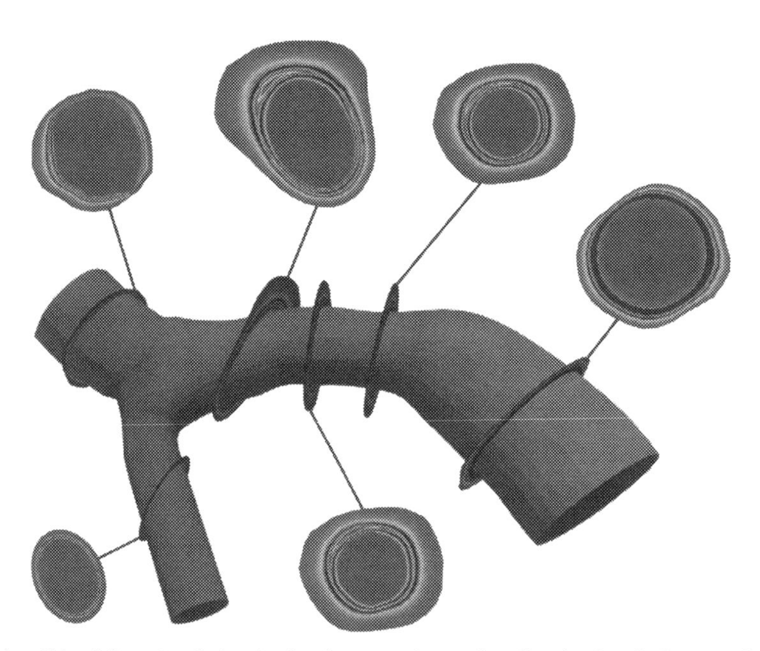

FIG. 10 Results of blood flow simulation for the chosen patient, at baseline (top) and after stent implantation (bottom); cytokines distribution within the arterial wall.

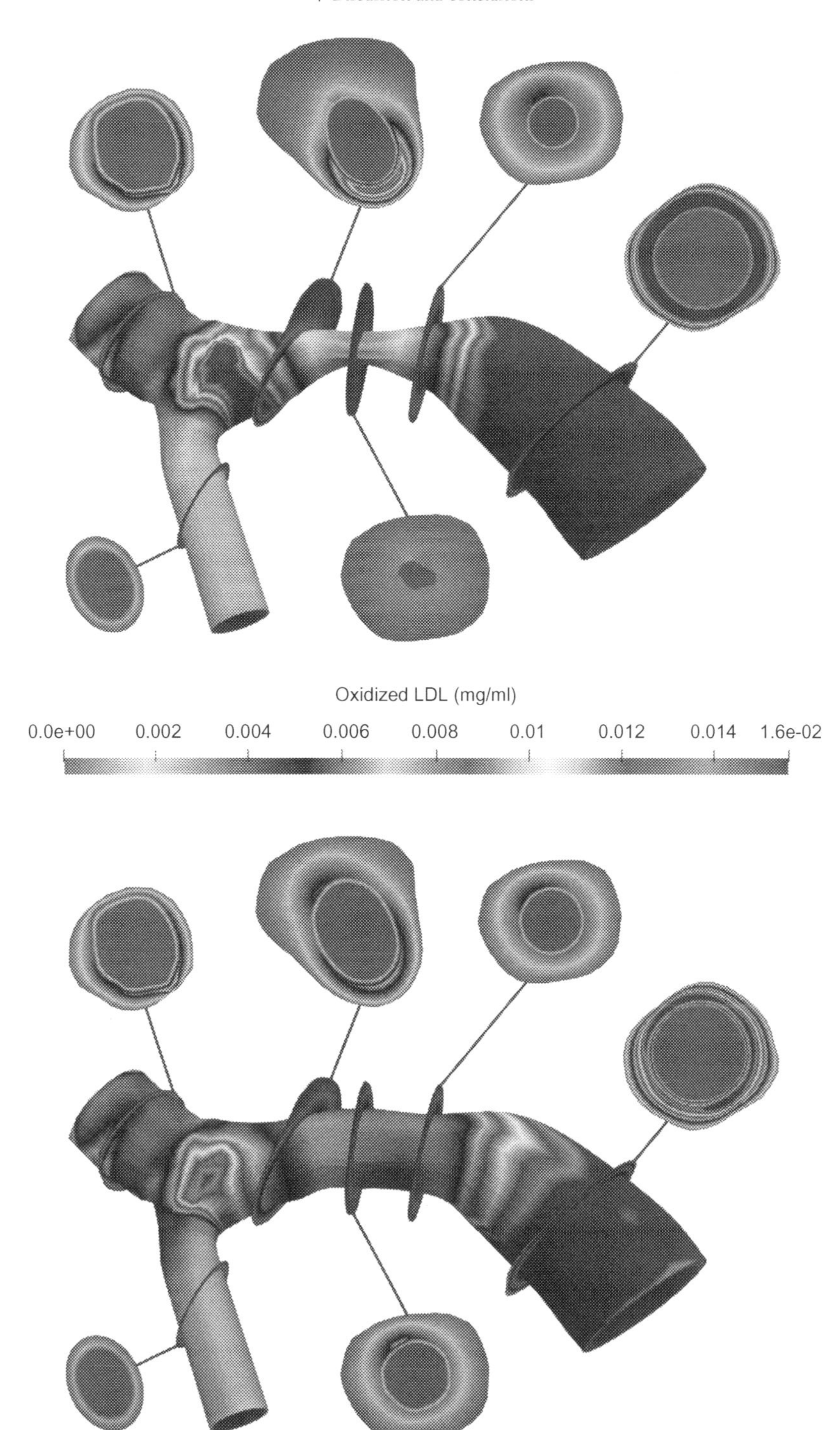

FIG. 11 Results of blood flow simulation for the chosen patient, at baseline (top) and after stent implantation (bottom); oxidized LDL distribution within the arterial wall.

demonstrated that it is very important to analyze the velocity and WSS distributions in 3D, because the estimation of the state of blood flow within the patient's vessel, which is based only on unidirectional velocity measurement or axial WSS, may severely underestimate the real situation. All the aforementioned demonstrate the importance of CFD simulations. In this chapter, a complex numerical model is used for the simulation of blood flow and plaque progression within the reconstructed patient-specific geometry. This model provided results that can be used to analyze the distribution of relevant blood flow parameters and macromolecules important for the process of inflammation and plaque within the arterial wall.

The change of morphological and hemodynamic parameters was analyzed in the literature (Morales et al., 2010) to observe the effects of two clinical treatments of carotid stenoses—endarterectomy with patch closure and stenting. The US examination was performed after stenting and patch closure and the relevant parameters were analyzed 24 h and 12 months after the treatment. The authors concluded that the mean velocities normalized after stenting and also remained fairly stable after 12 months, and there was a low percentage of restenosis occurrences. The effects of three different implantation methods within bifurcations of coronary arteries on the blood flow were also studied in the literature (Brindise et al., 2017). Particle image velocimetry was used to perform the measurements in vitro. Several blood flow parameters were analyzed, including WSS, oscillatory shear index (OSI), and relative residence time (RRT). The findings suggest that different stenting treatments can have very different effects on the blood flow. In this Chapter, the numerical model for blood flow and plaque progression was also used to analyze the state of the patient-specific CA after stent implantation treatment. It was concluded that the applied stenting treatment improved the blood flow parameters.

The capabilities of the computer methods to accurately predict the results of stent implantation treatment have also been studied in the literature. The results of simulations using serially linked beam elements were compared with experimental micro-CT data in Krewcun et al. (2019). Stent implantation in peripheral arteries was considered in another study (Yang et al., 2017). Linear theoretical and computational approaches were used to determine the stent's mechanical properties. Auricchio et al. (2011) investigated the influence of stent design on the stress induced in the arterial wall, the lumen gain, and the vessel straightening of the patient-specific carotid artery. Perrin et al. (2015) developed a numerical method to predict the deployed shapes of the stent-graft for treatment of abdominal aortic aneurysms. The simulations were performed on three clinical cases, using preoperative scans to extract patient-specific vessel geometries. The numerical results were validated against the geometries of the stent-grafts that were extracted from postoperative scans.

Stent implantation was also performed within arteries whose arterial wall had separated plaque components. Wei et al. (2019)considered a blood vessel as a straight single-layered cylindrical tube and included four different plaque components—general plaque, fibrous cap, lipid pool, and calcified zone. The plaque components were placed at five different positions. They used the second-order Mooney-Rivlin constitutive equation to simulate the behavior of plaque components. The analysis of effects of these different configurations showed that there is a great difference in the obtained values of maximum plastic strain of stent, stent foreshortening, maximum von Mises stress, and stress distributions of plaque. Stent implantation in idealized and symmetrical geometries with separate plaque components was also the topic of other studies in the literature (Migliavacca et al., 2004; Wu et al., 2007). In this chapter, the plaque components that were extracted from US images are incorporated in the arterial wall and stent implantation was simulated taking into consideration different material characteristics of these components that were experimentally measured in the literature (Bennetts et al., 2013). The comparison of results with and without considering separate plaque components showed that they play an important role in the deformation of the arterial wall and that more realistic results are obtained when they are considered.

The results obtained using a complex numerical model presented in this chapter demonstrated that this approach can be very useful in providing more information to the clinicians during US examination. With these simulations, the clinicians can analyze the state of the patient's CA in more detail, perceive regions of low WSS and high LDL accumulation and possible plaque growth, and plan possible stent implantation treatment. Also, they can analyze the outcome of the stenting intervention and again analyze the relevant blood flow and plaque progression parameters prior to the intervention. This whole approach can help to achieve more detailed and patient-specific diagnostics and treatment planning.

Acknowledgments

The research presented in this study was part of the project that has received funding from the European Union's Horizon 2020 research and innovation programme under Grant agreement No. 755320-2—TAXINOMISIS. This article reflects only the author's view. The Commission is not responsible for any use that may be made of the information it contains. This research is also funded by the Serbian Ministry of Education, Science, and Technological Development [451-03-68/2020-14/200378 (Institute for Information Technologies, University of Kragujevac)]. We acknowledge that the results of this research have been achieved using the DECI resource KAY based in Ireland at ICHEC (Irish Center for High-End Computing) with support from the PRACE aisbl.

References

Abbott, A. L., Paraskevas, K. I., Kakkos, S. K., Golledge, J., Eckstein, H. H., Diaz-Sandoval, L. J., Cao, L., Fu, Q., Wijeratne, T., Leung, T. W., Montero-Baker, M., Lee, B. C., Pircher, S., Bosch, M., Dennekamp, M., & Ringleb, P. (2015). Systematic review of guidelines for the management of asymptomatic and symptomatic carotid stenosis. *Stroke*, *46*(11), 3288–3301. https://doi.org/10.1161/STROKEAHA.115.003390.

Auricchio, F., Conti, M., De Beule, M., De Santis, G., & Verhegghe, B. (2011). Carotid artery stenting simulation: From patient-specific images to finite element analysis. *Medical Engineering and Physics*, *33*(3), 281–289. https://doi.org/10.1016/j.medengphy.2010.10.011.

Bastl, B., Jüttler, B., Lávička, M., & Šír, Z. (2012). Curves and surfaces with rational chord length parameterization. *Computer Aided Geometric Design*, *29*(5), 231–241. https://doi.org/10.1016/j.cagd.2011.04.003.

Bathe, K. J. (1996). *Finite element procedures*. Prentice-Hall, Inc.

Bennetts, C. J., Erdemir, A., & Young, M. (2013). Surface stiffness of patient-specific arterial segments with varying plaque compositions. In *ASME 2013 conference on frontiers in medical devices: Applications of computer modeling and simulation, FMD 2013* American Society of Mechanical Engineers. https://doi.org/10.1115/FMD2013-16132.

Boi, A., Jamthikar, A. D., Saba, L., Gupta, D., Sharma, A., Loi, B., Laird, J. R., Khanna, N. N., & Suri, J. S. (2018). A survey on coronary atherosclerotic plaque tissue characterization in intravascular optical coherence tomography. *Current Atherosclerosis Reports*, *20*(7). https://doi.org/10.1007/s11883-018-0736-8.

Bratzler, R. L., Chisolm, G. M., Colton, C. K., Smith, K. A., & Lees, R. S. (1977). The distribution of labeled low-density lipoproteins across the rabbit thoracic aorta in vivo. *Atherosclerosis*, *28*(3), 289–307. https://doi.org/10.1016/0021-9150(77)90177-0.

Brindise, M. C., Chiastra, C., Burzotta, F., Migliavacca, F., & Vlachos, P. P. (2017). Hemodynamics of stent implantation procedures in coronary bifurcations: An in vitro study. *Annals of Biomedical Engineering*, *45*(3), 542–553. https://doi.org/10.1007/s10439-016-1699-y.

Brooks, A. N., & Hughes, T. J. R. (1982). Streamline upwind/Petrov-Galerkin formulations for convection dominated flows with particular emphasis on the incompressible Navier-stokes equations. *Computer Methods in Applied Mechanics and Engineering, 32*(1–3), 199–259. https://doi.org/10.1016/0045-7825(82)90071-8.

Cibis, M., Potters, W. V., Selwaness, M., Gijsen, F. J., Franco, O. H., Arias Lorza, A. M., de Bruijne, M., Hofman, A., van der Lugt, A., Nederveen, A. J., & Wentzel, J. J. (2016). Relation between wall shear stress and carotid artery wall thickening MRI versus CFD. *Journal of Biomechanics, 49*(5), 735–741. https://doi.org/10.1016/j.jbiomech.2016.02.004.

Djukic, T. (2015). *Modeling motion of deformable body inside fluid and its application in biomedical engineering*. PhD dissertation, University of Kragujevac, Serbia.

Djukic, T., Arsic, B., Djorovic, S., Filipovic, N., & Koncar, I. (2020). Validation of the machine learning approach for 3D reconstruction of carotid artery from ultrasound imaging. In *Proceedings—IEEE 20th international conference on bioinformatics and bioengineering, BIBE 2020* (pp. 789–794). Institute of Electrical and Electronics Engineers Inc. https://doi.org/10.1109/BIBE50027.2020.00134.

Djukic, T., Arsic, B., Koncar, I., & Filipovic, N. (2020). 3D reconstruction of patient-specific carotid artery geometry using clinical ultrasound imaging. In *Workshop computational biomechanics for medicine XV, 23rd international conference on medical image computing & computer assisted intervention (MICCAI)*.

Djukic, T., & Filipovic, N. (2019). Numerical modeling of cell separation in microfluidic chips. In *Computational modeling in bioengineering and bioinformatics* (pp. 321–352). Elsevier.

Djukic, T., Karthik, S., Saveljic, I., Djonov, V., & Filipovic, N. (2016). Modeling the behavior of red blood cells within the caudal vein plexus of zebrafish. *Frontiers in Physiology, 7*. https://doi.org/10.3389/fphys.2016.00455.

Djukic, T., Saveljic, I., Pelosi, G., Parodi, O., & Filipovic, N. (2019). Numerical simulation of stent deployment within patient-specific artery and its validation against clinical data. *Computer Methods and Programs in Biomedicine, 175*, 121–127. https://doi.org/10.1016/j.cmpb.2019.04.005.

Djukic, T., Saveljic, I., Pelosi, G., Parodi, O., & Filipovic, N. (2021). Improved numerical model of the arterial wall applied for simulations of stent deployment within patient-specific coronary arteries. *Journal of Applied Engineering Science, 19*(1), 109–113. https://doi.org/10.5937/jaes0-27805.

Djukic, T., Topalovic, M., & Filipovic, N. (2015). Numerical simulation of isolation of cancer cells in a microfluidic chip. *Journal of Micromechanics and Microengineering, 25*(8). https://doi.org/10.1088/0960-1317/25/8/084012.

Evju, Ø., Valen-Sendstad, K., & Mardal, K. A. (2013). A study of wall shear stress in 12 aneurysms with respect to different viscosity models and flow conditions. *Journal of Biomechanics, 46*(16), 2802–2808. https://doi.org/10.1016/j.jbiomech.2013.09.004.

Filipovic, N., Meunier, N., & Kojic, M. (2010). *Technical solution: PAK-Athero, Specialized three-dimensional PDE software for simulation of plaque formation and development inside the arteries*. Kragujevac, Serbia: University of Kragujevac.

Filipovic, N., Teng, Z., Radovic, M., Saveljic, I., Fotiadis, D., & Parodi, O. (2013). Computer simulation of three-dimensional plaque formation and progression in the carotid artery. *Medical and Biological Engineering and Computing, 51*(6), 607–616. https://doi.org/10.1007/s11517-012-1031-4.

Filipovic, N., Zivic, M., Obradovic, M., Djukic, T., Markovic, Z., & Rosic, M. (2014). Numerical and experimental LDL transport through arterial wall. *Microfluidics and Nanofluidics, 16*(3), 455–464. https://doi.org/10.1007/s10404-013-1238-1.

Grotta, J. C. (2013). Carotid stenosis. *New England Journal of Medicine, 369*(12), 1143–1150. https://doi.org/10.1056/NEJMcp1214999.

Holzapfel, G. (2000). *Nonlinear solid mechanics a continuun approach for engineering*. Wiley.

Johnston, B. M., Johnston, P. R., Corney, S., & Kilpatrick, D. (2004). Non-Newtonian blood flow in human right coronary arteries: Steady state simulations. *Journal of Biomechanics, 37*(5), 709–720. https://doi.org/10.1016/j.jbiomech.2003.09.016.

Kojic, M., Filipovic, N., Stojanovic, B., & Kojic, N. (2008). *Computer modeling in bioengineering: Theoretical background, examples and software*. Chichester, England: John Wiley and Sons.

Kojić, M., Slavković, R., Živković, M., & Grujović, N. (1998). *Metod konačnih elemenata 1*. Kragujevac: Mašinski fakultet Kragujevac.

Krewcun, C., Sarry, L., Combaret, N., & Pery, E. (2019). Fast simulation of stent deployment with plastic beam elements. In *Proceedings of the annual international conference of the IEEE engineering in medicine and biology society, EMBS* (pp. 6968–6974). Institute of Electrical and Electronics Engineers Inc. https://doi.org/10.1109/EMBC.2019.8857179.

Larrabide, I., Kim, M., Augsburger, L., Villa-Uriol, M. C., Rüfenacht, D., & Frangi, A. F. (2012). Fast virtual deployment of self-expandable stents: Method and in vitro evaluation for intracranial aneurysmal stenting. *Medical Image Analysis, 16*(3), 721–730. https://doi.org/10.1016/j.media.2010.04.009.

Lekadir, K., Galimzianova, A., Betriu, A., Del Mar Vila, M., Igual, L., Rubin, D. L., Fernandez, E., Radeva, P., & Napel, S. (2017). A convolutional neural network for automatic characterization of plaque composition in carotid ultrasound. *IEEE Journal of Biomedical and Health Informatics, 21*(1), 48–55. https://doi.org/10.1109/JBHI.2016.2631401.

Li, Y. (2016). *Segmentation of medical ultrasound images using convolutional neural networks with Noisy activating functions.* Project final report, Stanford Edu.

Long, Q., Xu, X. Y., Köhler, U., Robertson, M. B., Marshall, I., & Hoskins, P. (2002). Quantitative comparison of CFD predicted and MRI measured velocity fields in a carotid bifurcation phantom. *Biorheology, 39*(3–4), 467–474.

Lopes, D., Puga, H., Teixeira, J. C., & Teixeira, S. F. (2019). Influence of arterial mechanical properties on carotid blood flow: Comparison of CFD and FSI studies. *International Journal of Mechanical Sciences, 160*, 209–218. https://doi.org/10.1016/j.ijmecsci.2019.06.029.

Markl, M., Wegent, F., Zech, T., Bauer, S., Strecker, C., Schumacher, M., Weiller, C., Hennig, J., & Harloff, A. (2010). In vivo wall shear stress distribution in the carotid artery effect of bifurcation eometry, internal carotid artery stenosis, and recanalization therapy. *Circulation. Cardiovascular Imaging, 3*(6), 647–655. https://doi.org/10.1161/CIRCIMAGING.110.958504.

Migliavacca, F., Petrini, L., Massarotti, P., Schievano, S., Auricchio, F., & Dubini, G. (2004). Stainless and shape memory alloy coronary stents: A computational study on the interaction with the vascular wall. *Biomechanics and Modeling in Mechanobiology, 2*(4), 205–217. https://doi.org/10.1007/S10237-004-0039-6.

Morales, M. M., Anacleto, A., Buchdid, M. A., Simeoni, P. R. B., Ledesma, S., Cêntola, C., Anacleto, J. C., Aldrovani, M., & Piccinato, C. E. (2010). Morphological and hemodynamic patterns of carotid stenosis treated by endarterectomy with patch closure versus stenting: A duplex ultrasound study. *Clinics, 65*(12), 1315–1323. https://doi.org/10.1590/S1807-59322010001200015.

Noble, C., Carlson, K. D., Neumann, E., Dragomir-Daescu, D., Erdemir, A., Lerman, A., & Young, M. (2020). Patient specific characterization of artery and plaque material properties in peripheral artery disease. *Journal of the Mechanical Behavior of Biomedical Materials, 101*. https://doi.org/10.1016/j.jmbbm.2019.103453.

Paliwal, N., Yu, H., Xu, J., Xiang, J., Siddiqui, A. H., Yang, X., Li, H., & Meng, H. (2016). Virtual stenting workflow with vessel-specific initialization and adaptive expansion for neurovascular stents and flow diverters. *Computer Methods in Biomechanics and Biomedical Engineering, 19*(13), 1423–1431. https://doi.org/10.1080/10255842.2016.1149573.

Paraskevas, K. I., Veith, F. J., & Spence, J. D. (2018). How to identify which patients with asymptomatic carotid stenosis could benefit from endarterectomy or stenting. *Stroke and Vascular Neurology, 3*(2), 92–100. https://doi.org/10.1136/svn-2017-000129.

Parodi, O., Exarchos, T. P., Marraccini, P., Vozzi, F., Milosevic, Z., Nikolic, D., Sakellarios, A., Siogkas, P. K., Fotiadis, D. I., & Filipovic, N. (2012). Patient-specific prediction of coronary plaque growth from CTA angiography: A multiscale model for plaque formation and progression. *IEEE Transactions on Information Technology in Biomedicine, 16*(5), 952–965. https://doi.org/10.1109/TITB.2012.2201732.

Perktold, K., Peter, R. O., Resch, M., & Langs, G. (1991). Pulsatile non-newtonian blood flow in three-dimensional carotid bifurcation models: A numerical study of flow phenomena under different bifurcation angles. *Journal of Biomedical Engineering, 13*(6), 507–515. https://doi.org/10.1016/0141-5425(91)90100-L.

Perktold, K., Resch, M., & Peter, R. O. (1991). Three-dimensional numerical analysis of pulsatile flow and wall shear stress in the carotid artery bifurcation. *Journal of Biomechanics, 24*(6), 409–420. https://doi.org/10.1016/0021-9290(91)90029-M.

Perrin, D., Badel, P., Orgéas, L., Geindreau, C., Dumenil, A., Albertini, J. N., & Avril, S. (2015). Patient-specific numerical simulation of stent-graft deployment: Validation on three clinical cases. *Journal of Biomechanics, 48*(10), 1868–1875. https://doi.org/10.1016/j.jbiomech.2015.04.031.

Peskin, C. S. (1977). Numerical analysis of blood flow in the heart. *Journal of Computational Physics, 25*(3), 220–252. https://doi.org/10.1016/0021-9991(77)90100-0.

Rakocevic, G., Djukic, T., Filipovic, N., & Milutinović, V. (2013). Computational medicine in data mining and modeling. In *Vol. 9781461487852. Computational Medicine in Data Mining and Modeling* (pp. 1–376). New York: Springer. https://doi.org/10.1007/978-1-4614-8785-2.

Ronneberger, O., Fischer, P., & Brox, T. (2015). U-net: Convolutional networks for biomedical image segmentation. In *Vol. 9351. Lecture Notes in Computer Science (including subseries Lecture Notes in Artificial Intelligence and Lecture Notes in Bioinformatics)* (pp. 234–241). Springer Verlag. https://doi.org/10.1007/978-3-319-24574-4_28.

Rosenfield, K., Losordo, D. W., Ramaswamy, K., Pastore, J. O., Langevin, R. E., Razvi, S., Kosowsky, B. D., & Isner, J. M. (1991). Three-dimensional reconstruction of human coronary and peripheral arteries from images recorded during two-dimensional intravascular ultrasound examination. *Circulation, 84*, 1938–1956. https://doi.org/10.1161/01.cir.84.5.1938.

Sakellarios, A., Bourantas, C. V., Papadopoulou, S. L., Tsirka, Z., De Vries, T., Kitslaar, P. H., Girasis, C., Naka, K. K., Fotiadis, D. I., Veldhof, S., Stone, G. W., Reiber, J. H. C., Michalis, L. K., Serruys, P. W., De Feyter, P. J., & Garcia-Garcia, H. M. (2017). Prediction of atherosclerotic disease progression using LDL transportmodelling: A serial computed tomographic coronary angiographic study. *European Heart Journal Cardiovascular Imaging, 18*(1), 11–18. https://doi.org/10.1093/ehjci/jew035.

Sakellarios, A. I., Bizopoulos, P., Papafaklis, M. I., Athanasiou, L., Exarchos, T., Bourantas, C. V., Naka, K. K., Patterson, A. J., Young, V. E. L., Gillard, J. H., Parodi, O., Michalis, L. K., & Fotiadis, D. I. (2017). Natural history of carotid atherosclerosis in relation to the hemodynamic environment: A low-density lipoprotein transport modeling study with serial magnetic resonance imaging in humans. *Angiology, 68*(2), 109–118. https://doi.org/10.1177/0003319716644138.

Samady, H., Eshtehardi, P., McDaniel, M. C., Suo, J., Dhawan, S. S., Maynard, C., Timmins, L. H., Quyyumi, A. A., & Giddens, D. P. (2011). Coronary artery wall shear stress is associated with progression and transformation of atherosclerotic plaque and arterial remodeling in patients with coronary artery disease. *Circulation, 124*(7), 779–788. https://doi.org/10.1161/CIRCULATIONAHA.111.021824.

Shibeshi, S. S., & Collins, W. E. (2005). The rheology of blood flow in a branced arterial system. *Applied Rheology, 15*(6), 398–405.

Smistad, E., & Løvstakken, L. (2016). Vessel detection in ultrasound images using deep convolutional neural networks. In *Vol. 10008. Lecture notes in computer science (including subseries lecture notes in artificial intelligence and lecture notes in bioinformatics)* (pp. 30–38). Springer Verlag. https://doi.org/10.1007/978-3-319-46976-8_4.

Steinman, D. A. (2012). In D. Ambrosi, A. Quarteroni, & G. Rozza (Eds.), *Assumptions in modelling of large artery hemodynamics* (pp. 1–18). Springer. https://doi.org/10.1007/978-88-470-1935-5_1.

Stone, P. H., Saito, S., Takahashi, S., Makita, Y., Nakamura, S., Kawasaki, T., Takahashi, A., Katsuki, T., Nakamura, S., Namiki, A., Hirohata, A., Matsumura, T., Yamazaki, S., Yokoi, H., Tanaka, S., Otsuji, S., Yoshimachi, F., Honye, J., Harwood, D., … Feldman, C. L. (2012). Prediction of progression of coronary artery disease and clinical outcomes using vascular profiling of endothelial shear stress and arterial plaque characteristics: The PREDICTION study. *Circulation, 126*(2), 172–181. https://doi.org/10.1161/CIRCULATIONAHA.112.096438.

Vancraeynest, D., Pasquet, A., Roelants, V., Gerber, B. L., & Vanoverschelde, J. L. J. (2011). Imaging the vulnerable plaque. *Journal of the American College of Cardiology, 57*(20), 1961–1979. https://doi.org/10.1016/j.jacc.2011.02.018.

Vukicevic, A. M., Çimen, S., Jagic, N., Jovicic, G., Frangi, A. F., & Filipovic, N. (2018). Three-dimensional reconstruction and NURBS-based structured meshing of coronary arteries from the conventional X-ray angiography projection images. *Scientific Reports, 8*(1). https://doi.org/10.1038/s41598-018-19440-9.

Vukicevic, A. M., Stepanovic, N. M., Jovicic, G. R., Apostolovic, S. R., & Filipovic, N. D. (2014). Computer methods for follow-up study of hemodynamic and disease progression in the stented coronary artery by fusing IVUS and X-ray angiography. *Medical and Biological Engineering and Computing, 52*(6), 539–556. https://doi.org/10.1007/s11517-014-1155-9.

Wei, L., Chen, Q., & Li, Z. (2019). Influences of plaque eccentricity and composition on the stent–plaque–artery interaction during stent implantation. *Biomechanics and Modeling in Mechanobiology, 18*(1), 45–56. https://doi.org/10.1007/s10237-018-1066-z.

Wu, W., Wang, W. Q., Yang, D. Z., & Qi, M. (2007). Stent expansion in curved vessel and their interactions: A finite element analysis. *Journal of Biomechanics, 40*(11), 2580–2585. https://doi.org/10.1016/j.jbiomech.2006.11.009.

Xie, M., Li, Y., Xue, Y., Shafritz, R., Rahimi, S. A., Ady, J. W., & Roshan, U. W. (2019). Vessel lumen segmentation in internal carotid artery ultrasounds with deep convolutional neural networks. In *Proceedings—2019 IEEE International Conference on Bioinformatics and Biomedicine, BIBM 2019* (pp. 2393–2398). Institute of Electrical and Electronics Engineers Inc. https://doi.org/10.1109/BIBM47256.2019.8982980.

Yang, H., Fortier, A., Horne, K., Mohammad, A., Banerjee, S., & Han, H. C. (2017). Investigation of stent implant mechanics using linear analytical and computational approach. *Cardiovascular Engineering and Technology, 8*(1), 81–90. https://doi.org/10.1007/s13239-017-0295-0.

Yeom, E., Nam, K. H., Jin, C., Paeng, D. G., & Lee, S. J. (2014). 3D reconstruction of a carotid bifurcation from 2D transversal ultrasound images. *Ultrasonics*, *54*(8), 2184–2192. https://doi.org/10.1016/j.ultras.2014.06.002.

Zhang, D., Xu, P., Qiao, H., Liu, X., Luo, L., Huang, W., Zhang, H., & Shi, C. (2018). Carotid DSA based CFD simulation in assessing the patient with asymptomatic carotid stenosis: A preliminary study. *Biomedical Engineering Online*, *17*(1). https://doi.org/10.1186/s12938-018-0465-9.

Zhou, R., Fenster, A., Xia, Y., Spence, J. D., & Ding, M. (2019). Deep learning-based carotid media-adventitia and lumen-intima boundary segmentation from three-dimensional ultrasound images. *Medical Physics*, *46*(7), 3180–3193. https://doi.org/10.1002/mp.13581.

Zhou, X. Y., & Yang, G. Z. (2019). Normalization in training U-net for 2-D biomedical semantic segmentation. *IEEE Robotics and Automation Letters*, *4*(2), 1792–1799. https://doi.org/10.1109/LRA.2019.2896518.

Živković, M. (2006). *Nelinearna analiza konstrukcija*. Kragujevac, Srbija: Mašinski fakultet.

Živkovic, M. (2011). *Mehanika kontinuuma za analizu metodom konačnih elemenata*. Kragujevac, Srbija: WUS Austria.

CHAPTER

6

Myocardial work and aorta stenosis simulation

Smiljana Djorovic[a,b]

[a]Faculty of Engineering, University of Kragujevac, Kragujevac, Serbia [b]Bioengineering Research and Development Center, Kragujevac, Serbia

1 Introduction

The first objective of this chapter is to describe the background of myocardial work and aortic stenosis, and then present the model-based simulation of myocardial work for parametric left ventricle in the case of aortic stenosis. Previous works (Garcia et al., 2005; Korurek et al., 2010) have already shown that lumped-parameter models of ventricular-vascular coupling are able to provide a good agreement between the estimated and the measured left ventricular and aortic pressure waveforms. The second objective is to computationally simulate patient-specific aortic root with mild aortic stenosis. The used methodology relies on the implementation of the finite element (FE) procedure and the employment of an in-house FE PAK solver (Kojic et al., 2010). PAK is a high-performance FE software for solving complex coupled multiphysics/multiscale problems, with main application in the cardiovascular domain (Djorovic et al., 2019; Kojic et al., 2019) and with possibilities to interact with different computational solutions and solvers (Robnik-Šikonja et al., 2018).

In the case of heart malformations, complex changes in left ventricular geometry are in most cases caused by continuous exposure to cardiovascular risk factors and/or hemodynamic conditions, which usually start as a physiological response (Lavie et al., 2014; Marwick et al., 2015). Myocardial dysfunctions and heart failure are usually caused by alterations in left ventricular geometry, such as left ventricular hypertrophy and dilatation (Bluemke et al., 2008; de Simone et al., 2008; Inoko et al., 1994; Velagaleti et al., 2014). Despite intensive research, the mechanisms of myocardial dysfunctions are still not well understood. Invasive recording of pressure-volume (PV) loops as the reference standard provides real-time assessment of left ventricle loading conditions, contractility, and myocardial oxygen consumption (Bastos et al., 2020). However, its invasive nature restricts its use in daily clinical practice.

https://doi.org/10.1016/B978-0-12-823956-8.00010-9

A novel echocardiographic method has been introduced and validated against invasive measurements that noninvasively quantify active myocardial function, i.e., systolic and early diastolic active myocardial work (Russell et al., 2012). This approach allows differentiating constructive from wasted myocardial work, with the latter not contributing to left ventricle output. The concept of myocardial work measurement is based on speckle-tracking-derived longitudinal strain and systolic blood pressure and is widely applicable. On the other hand, echocardiography-derived myocardial work has to be differentiated from the definition of cardiac work derived from invasive PV loops (Schramm, 2010), as myocardial work approximates the work contributing to left ventricle output, i.e., constructive work, and quantifies energy loss due to uncoordinated left ventricular contractions. It results in stretching of individual left ventricle segments by the contraction of other left ventricle segments, i.e., wasted work (Russell et al., 2013). Further, myocardial work might allow profound insights into left ventricle performance and might also serve as a surrogate of regional and global myocardial metabolism (Russell et al., 2013, 2012). Left ventricle geometry patterns have been shown to be of prognostic relevance in community studies (Lieb et al., 2014; Lind & Sundstrom, 2019) and depend on the exposure to modifiable cardiovascular risk factors, such as hypertension and obesity (Bluemke et al., 2008; Cuspidi et al., 2016; Milani et al., 2006). Thus, the detailed evaluation of myocardial work in relation to left ventricle geometry might further advance the pathophysiological understanding of functional changes associated with abnormal left ventricle geometry. Therefore, we aimed to assess the association of left ventricle geometry with myocardial work in a parametric left ventricle model using FE analysis and simulation of heartbeat prescribing boundary conditions for aortic stenosis. Also, we performed the structural analysis of patient-specific aortic root with aortic stenosis in order to analyze the orifice area and leaflet stress distribution (Zakikhani et al., 2019).

The chapter is organized as follows: the basis of aortic stenosis and myocardial work is presented in Sections 2 and 3, respectively. Section 4 describes the computational simulation of myocardial work in aortic stenosis in a parametric model of the left ventricle. Section 5 is related to the computational simulation of the patient-specific aortic root with aortic stenosis. Finally, the main conclusions and future work are summarized in Section 6.

2 Stenosis of aorta

Aortic stenosis is a narrowing of the orifice of the aortic valve that causes increased resistance to blood flow from the ventricle into the systemic circulation (Dweck et al., 2012). The heart maintains its flow at the cost of increased pressure initiating pathophysiological processes that lead to unfavorable clinical outcomes. The usual focus of aortic stenosis assessments has been on the valve, but the disease process also reduces arterial compliance and alters the geometry of the left ventricle. For this reason, aortic stenosis is viewed as a systemic disease (Pibarot & Dumesnil, 2012). In addition, aortic stenosis is associated with different malformations. For example, aortic stenosis affecting a bicuspid valve is the most common indication for surgical aortic valve replacement in patients < 70 years of age. Congenital bicuspid aortic valve anatomy is found in 0.5%–2.0% of the population although it is relatively uncommon compared to calcific aortic stenosis. A precise evaluation of the severity of aortic valve stenosis is crucial for patient management and risk stratification, and to allocate symptoms legitimately to the valvular disease. Echocardiography is the main method to assess aortic valve stenosis severity. It relies on three parameters, such as the peak velocity (PVel), the mean pressure gradient (MPG), and the aortic valve area (AVA), as shown in Table 1 (Messika-Zeitoun & Lloyd, 2018).

The development of left ventricular hypertrophy in aortic stenosis is accompanied by coronary microcirculatory dysfunction (Rajappan et al., 2002) that may gradually affect systolic and diastolic function (Tzivoni, 1993). Left ventricular ejection fraction is used routinely to assess left ventricular systolic function and is an important parameter for prognosis stratification (Taniguchi et al., 2018). However, left ventricular ejection fraction depends not only on the contractility of the left ventricle but also on loading conditions. In fact, ejection fraction may appear to be preserved despite underlying reduced contractility. The characterization of myocardial dysfunction is of primary importance to identify patients with reduced contractility. STE assessment of myocardial strain usually provides a better quantification of systolic function than global left ventricle ejection fraction (Delgado et al., 2009). Although strain echocardiography can provide prognostic information in patients with aortic stenosis (Kearney et al., 2012), the shortening indices, calculated from cardiac strains, do not reflect myocardial work or oxygen demand. As opposed to the normal left ventricle, where all segments contract almost synchronously and myocardial energy is used effectively, regional dysfunction that could be induced by myocardial fibrosis (Hein et al., 2003), could bring a significant loss of efficient work. For instance, the impairment of myocardial diastolic and systolic function, due to fibrosis (Villari et al., 1995), has shown to induce significant mechanical dispersion in patients with severe aortic stenosis (Klaeboe et al., 2017).

A noninvasive method for left ventricle work analysis, which is based on an estimated left ventricle pressure curve, has been proposed by Russell et al. (2013, 2012). As strain is largely influenced by left ventricle afterload (Donal et al., 2009), model-based myocardial work might be a robust complementary tool, taking into account aortic stenosis severity and arterial pressures values. In previous works of our team, we have shown that the noninvasive estimation of global myocardial work, when using a left ventricle pressure curve estimation as proposed by Russell et al. (2012), is correlated with that obtained when using the observed invasive left ventricle pressure curve, in the context of cardiac resynchronization therapy (Hubert et al., 2018). However, the accuracy of estimated left ventricle pressure has never been evaluated in the case of aortic stenosis, where high-pressure gradients could be observed between the left ventricle and the aorta (Eleid et al., 2013). The experimental observation of left

TABLE 1 Aortic stenosis grades of severity as assessed using echocardiography and computed tomography (Messika-Zeitoun & Lloyd, 2018).

Echo parameters	Sclerosis	Mild aortic stenosis	Moderate aortic stenosis	Severe aortic stenosis
PVel—Peak velocity (m/s)	<2.5	2.5–3	3–4	>4
MPG—Mean pressure gradient (mmHg)	Normal	<20	20–40	40
AVA—Aortic valve area (cm^2)	Normal	≥ 1.5	1–1.5	$<1\,cm^2$

ventricle pressure is notably difficult to perform clinically because it requires an invasive, intraventricular measurement. As a consequence, it is necessary to propose novel tools to assess noninvasive left ventricle pressure and to calculate myocardial work in the case of aortic stenosis.

3 Myocardial work

Myocardial work is a novel technique used in the advanced assessment of left ventricular function. It includes left ventricle pressure and provides incremental information to left ventricle ejection fraction and strain which are sensitive to left ventricle afterload. Noninvasive myocardial work measures the amount of work done by the left ventricle during isovolumic contraction, mechanical systole, and isovolumic relaxation (IVR), and is essentially the force multiplied by distance. In addition, since the pressure-strain loop area reflects myocardial metabolic demand and oxygen consumption, the work method provides insight into myocardial energetics (Russell et al., 2012). The original experimental study was published by Suga H. in 1979, in which the concept was validated by invasive measures by including instantaneous left ventricle pressure recordings during cardiac catheterization. This research showed that the work takes into account the dynamic relationship between intracardiac pressure and myocardial contraction. This is computed into the invasively derived PV loop which is correlated to that of myocardial oxygen consumption (Suga, 1979).

In the past few years, a more easily obtainable method has been achieved in recent research pioneered by Russell et al. (2013, 2012) that introduced a noninvasive method to measure myocardial work with the incorporation of left ventricle strain by speckle-tracking echocardiograph (STE). The estimated pressure curve component of this noninvasive measure is generated by incorporating the peripheral systolic blood pressure with cardiac event times (including isovolumic contraction, systolic ejection, and isovolumic relaxation [IVR]) that are derived from the timing of echocardiographic valvular events. Russell et al. (2012) compared absolute left ventricle systolic pressure, which was set to be equivalent to arterial pressure measured invasively, with noninvasive left ventricle systolic pressure estimation. The left ventricle pressure-strain loop is developed from the noninvasive left ventricle pressure measurement by brachial artery and strain analysis by two-dimensional (2D) STE. This novel method demonstrated a means to quantify myocardial work by noninvasive measures that have been validated with invasive methods and demonstrated a strong correlation with oxygen consumption as well as regional myocardial glucose metabolism by positron emission tomography (Russell et al., 2013, 2012).

The calculation of myocardial force is difficult, so the pressure is used as a substitute. During this phase, pressure is defined as the force applied to the left ventricle wall surface divided among different wall segments. Noninvasive myocardial work is derived from the estimated measurement of left ventricular systolic pressure via brachial cuff blood pressure and measurement of strain (Edwards et al., 2020). Chan et al. (2019) applied myocardial work by left ventricle pressure-strain analysis in a cohort of patients referred to coronary angiography. The patients were subdivided into three main groups: controls, hypertensives, and patients with cardiomyopathy. Although the groups are unbalanced, their study shows interesting results, particularly in hypertensive patients with a systolic blood pressure > 160 mmHg. In this group, the global work index was significantly higher when compared with controls despite global longitudinal strain and left ventricular ejection fraction being normal and relatively unchanged. The findings confirm the inability of the traditional methods to detect increased performance and load imposed on the myocardium.

The steps of myocardial work and its components, namely global work index (GWI), global constructed work (GCW), global wasted work (GWW), and global work efficiency (GWE), are as follows. Global work index (GWI) represents the area of the loop from mitral valve closure (MVC) to mitral valve opening (MVO). Global constructive work (GCW) reflects the work that contributes to pump from aortic valve opening (AVO) to MVO. There is shortening during the ejection period and lengthening during isovolumic relaxation (IVRT). Global wasted work (GWW) represents the work that does not contribute to ejection. The main components of myocardial work are given in Table 2. All MW components apart from GWE have the same units and are expressed in mmHg% as they reflect the power over the cardiac cycle longitudinal strain. GWE is expressed in % as it is the fraction of GCW divided by the sum of GCW + GWW (Papadopoulos et al., 2021).

4 Computational simulation of myocardial work in aortic stenosis

This section presents the basis of the FE numerical procedure for loose coupling solid-fluid interaction in order to simulate the heartbeat cycle in aortic stenosis. The parametric left ventricle model with specific parts: base part, valves (aortic and mitral), and connecting part (the connection between base and valves) has been implemented (Anić et al., 2020). In the case of modeling the solid mechanics of the heart, two sources of stress should be accounted—the passive and the active. Passive mechanical stresses are calculated using an orthotropic material model based on the experimental investigation of passive material properties of the myocardium (Sommer et al., 2015). Active stresses are calculated using the Hunter excitation model (Hunter et al., 1998). The passive and active models are implemented in the FE program PAK (Kojic et al., 2010) according to the procedure presented by (Kojic et al., 2008). The incompressibility of the material is imposed by using a standard penalty procedure.

TABLE 2 Components of myocardial work (Papadopoulos et al., 2021).

Myocardial work component	Explanation
Global work index (GWI)	Total work—represents the area of the loop from mitral valve closure (MVC) to mitral valve opening (MVO)
Global constructive work (GCW)	Segmental shortening during the systole, i.e. effective energy for blood ejection. Total work contributing to pump from aortic valve opening (AVO) to MVO
Global wasted work (GWW)	Work that does not contribute to ejection; elongation of myocytes during systole and shortening against a closed aortic valve
Global work efficiency (GWE)	Fraction of GCW divided by the sum of GCW + GWW

Fluid flow is based on the continuity equation and differential Navier-Stokes equations, represented as (Kojic et al., 2008)

$$-\mu\nabla^2 v_l + {}^l\rho(v_l \cdot \nabla)v_l + \nabla p_l = 0 \tag{1}$$

$$\nabla v_l = 0 \tag{2}$$

where v_l is the blood flow velocity, p_l is the pressure, μ is the coefficient of dynamic viscosity of blood, and ρ is the density of blood. In the loose coupling procedure, the fluid field is solved first, then forces of fluid acting on the surrounding solid domain are calculated, and the solid deformation is determined. Finally, when solving for the solid domain, the velocities of the solid are transferred to the corresponding fluid nodes. A usual iterative procedure (Kojic et al., 2008) is used for both solid and fluid models.

Muscles are activated by electrical signals transmitted from the neural system to muscle cells. The signals produce a change of the cell membranes potentials and a change of calcium concentration within the cell, which is directly related to the generation of the active stress within muscle fibers. This stress acts as a distributed external loading on the muscle tissue causing muscle deformation and motion. Active stress—calcium concentration relation for the heart muscle is formulated by Hunter et al. (1998) and previously used by (Kojic et al., 2019). In order to simulate realistic mechanical conditions for the left ventricle, blood velocities are prescribed at the inlet mitral valve and outlet aortic valve with mild stenosis (Messika-Zeitoun & Lloyd, 2018).

As result, the PV diagram is presented, with plotted volume along the X-axis and pressure on the Y-axis (Fig. 1). The area of the loop is equal to the stroke volume, which refers to the amount of blood pumped out of the left ventricle in one cardiac cycle. The effects of isolated

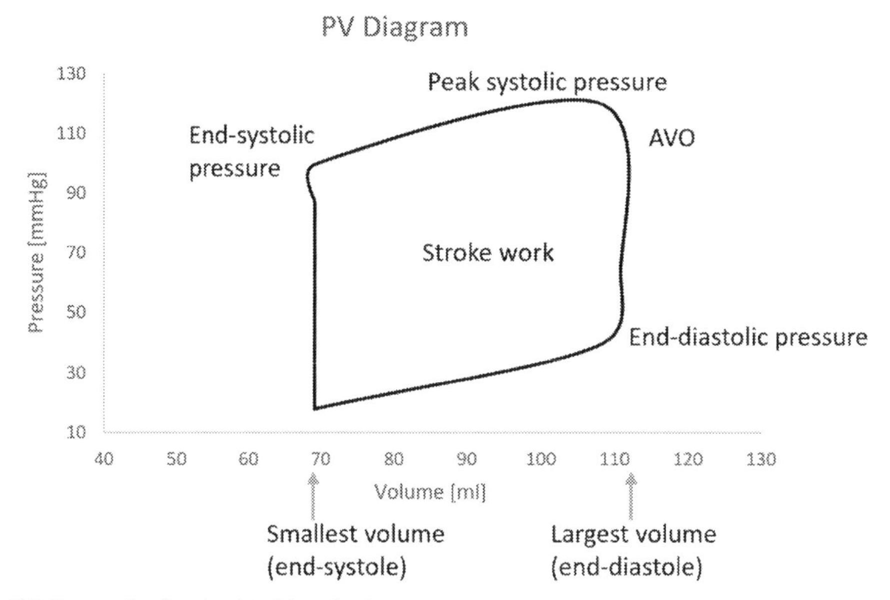

FIG. 1 PV diagram for the simulated heartbeat.

changes in preload are best demonstrated on the PV diagram, which relates ventricular volume to the pressure inside the ventricle throughout the cardiac cycle. The maximum right point on the diagram is denoted as the end-diastolic volume (EDV), while the minimum left point as the end-systolic volume (ESV). Also, as EDV increases, the proportion of blood ejected by the heart increases slightly; this is the ejection fraction (EF) calculated by the equation:

$$EF = \frac{EDV - ESV}{EDV} \tag{3}$$

The simulation resulted in EF of 52.38%. A decrease in preload will result in a leftward shift down the end-diastolic PV line, decreasing EDV, stroke volume, and a slight decrease in ejection fraction (Villars et al., 2004).

The GWE resulted in 62.4%, and it is calculated as

$$GWE = \frac{GCW}{GCW + GWW} \tag{4}$$

Estimation of left ventricle systolic function is essential in all cardiac diseases, and EF continues to be the most commonly used parameter. Although speckle-tracking echocardiography with global longitudinal strain (GLS) provides more detailed information about the global and segmental systolic function, giving the opportunity to detect subclinical dysfunction, it remains load-dependent (Stanton et al., 2009). Myocardial work, on the other hand, is a less load-dependent tool for left ventricle function evaluation as it incorporates the left ventricle afterload. Myocardial work gives a rough estimation of the work that every segment of the left ventricle produces during the cardiac cycle. This work is influenced by the power of the contraction of myocardial fibers, left ventricle loading conditions, and the wall stress applied on the left ventricle segments.

In the case of aortic valve malformations such as aortic stenosis, the reduced flow across the valve, a large pressure gradient results in a significantly elevated peak systolic pressure in the left ventricle (Chambers, 2006; Roemer et al., 2021). According to Laplace's law, an increase in the intracavitary pressure leads to an increase in wall stress as no geometrical changes occurred (i.e., remodeling). In the recent study of aortic valve stenosis, a "corrected" method for determining the amount of work taking place is proposed (Jain et al., 2021). The study demonstrated this method that uses the sum of the cuff systolic blood pressure and aortic valve mean gradient to give a more accurate account of the increased myocardial consumption from the increased afterload. In addition, myocardial work is studied in the nonobstructive hypertrophic cardiomyopathy population and great value in myocardial work analysis is found, but patients with obstruction had been excluded (Hiemstra et al., 2020).

5 Computational simulation of patient-specific aortic root

In attrition to myocardial work, this section describes the computational simulation of the patient-specific aortic root with aortic stenosis. In the field of computational simulations, the analysis of aortic root with stenosis gives better insight into biomechanical characteristics and functioning. Moreover, aortic valve evaluation with assumptions of an ideal fluid, geometry, and governing equations could increase the uncertainty of the degree of stenosis and valve area. With the effect of aortic stenosis altering blood flow in the left ventricle and ascending aorta, it is essential to analyze the orifice area and leaflet stress distribution (Zakikhani et al., 2019).

FE analysis has been used to obtain the complex motion of the leaflets during a cardiac cycle and to identify the regions of stress concentrations in the valvular structure (Gnyaneshwar et al., 2002). A similar dynamic study simulated the opening phase of the normal TAV by not including fluid-structure interaction, and with idealized geometry (Howard et al., 2009). As a result, the leaflets of the compliant root opened smoothly and symmetrically. On the other hand, Conti et al. (2010) reported on a comparative dynamic FE analysis of a tricuspid aortic valve and bicuspid aortic valve to demonstrate that the bicuspid valve opened asymmetrically with an elliptic orifice.

The tricuspid aortic valve analyzed in this study includes applied boundary conditions for mild aortic stenosis ($AVA \approx 2\,cm^2$). In addition, the computational simulation employs structural analysis. The creation of a 3D patient-specific model is based on computed tomography scans, whereas the segmentation procedure is based on a previous study (Djorovic et al., 2019). Since the patient-specific wall thickness was unknown, we assumed a thickness for aortic root and leaflets. The average thickness of the aortic substructures was adopted from literature which is a common practice due to the lack of patient-specific information. We assumed a uniform thickness of 2 mm for the aortic root substructures, while leaflets had variable thickness going from the commissures to the free margins (Grande et al., 1998). Also, we adopted existing values of material properties (Gnyaneshwar et al., 2002), due to a lack of experimental data.

Dynamic analysis during the cardiac cycle has been employed. As the first boundary condition, the time-dependent pressure is applied on the leaflet surfaces and aortic root substructures based on published transvalvular pressure patterns in the dynamic FE analysis (Gnyaneshwar et al., 2002). The base of the annulus and the sinotubular junction was fixed from motion, as the second boundary condition. After the applied appropriate boundary conditions and material characteristics, the PAK software for structural analysis was used for this computational simulation, which calculated Von Mises stress distribution (Fig. 2). Fig. 2 shows the isometric and outlet view of open leaflets at the systolic moment of cardiac cycle. The complex interplay between the aortic annulus, sinuses, and leaflets leads to increased stresses in the coaptation area. Also, slightly increased Von Mises stresses are present at interleaflet triangles. It should be noted that the calculated Von Mises stress values can be applied only for this study, due to the proposed material model and analysis simplification. Also, the linear material model does not fully describe biological aortic tissue, considering that aortic root has highly nonlinear and anisotropic properties. Consequently, the model or aortic valve did not replicate the realistic opening.

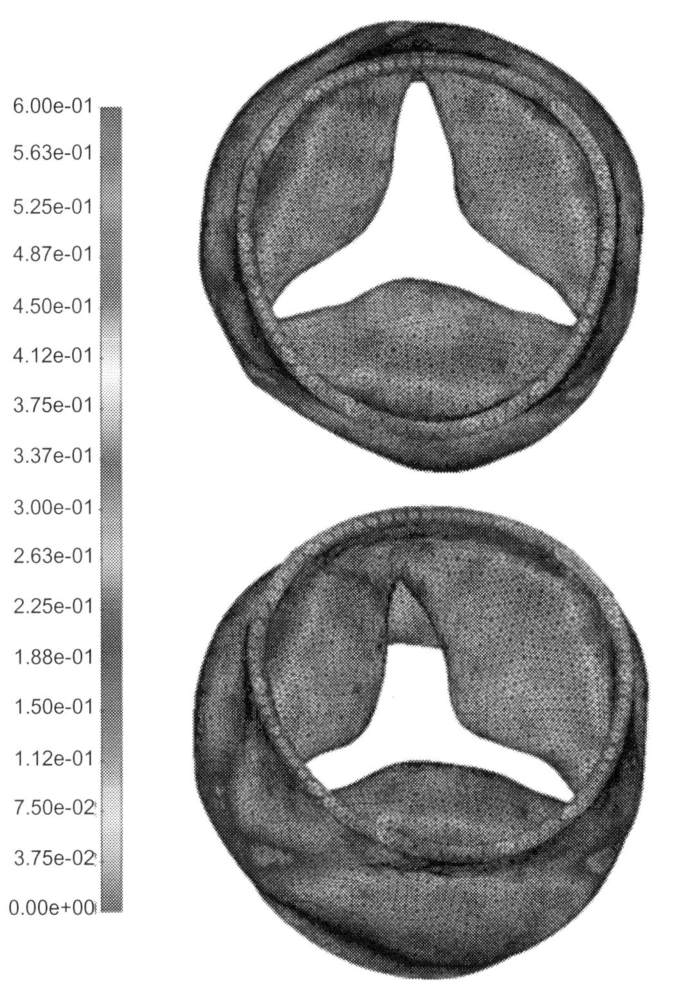

FIG. 2 Von Mises stress distribution (outlet and isometric view respectively; units: MPa).

6 Conclusions

This study discussed the basis of myocardial work and aortic stenosis, presenting also computer-based methods and models as the additive value of myocardial function assessment. The presented results consist of (i) computational simulation of myocardial work in aortic stenosis employing a parametric left ventricle model and (ii) structural analysis of patient-specific aortic root with aortic stenosis. This study has given insight into the model-based simulation of myocardial work and aortic stenosis, employing the in-house FE PAK software, and aimed to propose an advanced approach for the assessment of cardiac work indices and biomechanical characteristics based on computational modeling. The results have shown the possibility to visualize and analyze PV loops and further biomechanical characteristics (i.e. Von Mises stress distribution), by combining different modeling approaches into one unique solver.

Myocardial work provides incremental information of left ventricle ejection fraction and strain, which are sensitive to left ventricle afterload and related to noninvasive assessment of different types of cardiomyopathies. Altered myocardial work, associated with aortic stenosis is commonly related to significant morbidity, mortality, and healthcare costs. Computational modeling offer solutions to some of the problems in current practice but its use in myocardial work and aortic stenosis, thus far, has been limited, primarily due to the data available and the difficulty in personalizing the physiology in these models. With further improvement and increased robustness, the presented computational modeling can be used to produce noninvasive PV loops in daily clinical practice, as PV loops are the gold standard in assessing cardiac hemodynamics. Moreover, computational models can be a useful clinical tool for decision-making. Studies have shown that modeling holds promise in the health-care setting and its ability to run simulations and be predictive could help identify the likely outcomes for patients. The presented results are the first step in further investigation and development of more advanced and complex models. In future studies, numerical simulations will be improved that will eliminate current modeling limitations so that new models can more closely replicate the physiological problem, which is our main goal.

Acknowledgments

This study was supported by the European Union's Horizon 2020 research and innovation programme under grant agreements No 777204 and No 952603. This article reflects only the author's view. The Commission is not responsible for any use that may be made of the information it contains. The study is also funded by Serbian Ministry of Education, Science, and Technological Development [451-03-68/2020-14/200107 (Faculty of Engineering, University of Kragujevac)].

References

Anić, M., Savić, S., Milovanović, A., Milošević, M., Milićević, B., Simić, V., & Filipović, N. (2020). Solution of fluid flow through left heart ventricle. *Applied Engineering Letters: Journal of Engineering and Applied Sciences*, *5*, 120–125.

Bastos, M., Burkhoff, D., Maly, J., Daemen, J., den Uil, C., Ameloot, K., et al. (2020). Invasive left ventricle pressure-volume analysis: Overview and practical clinical implications. *European Heart Journal*, *41*, 1286–1297.

Bluemke, D., Kronmal, R., Lima, J., Liu, K., Olson, J., Burke, G., et al. (2008). The relationship of left ventricular mass and geometry to incident cardiovascular events: The MESA (multi-ethnic study of atherosclerosis) study. *Journal of the American College of Cardiology*, *52*, 2148–2155.

Chambers, J. (2006). The left ventricle in aortic stenosis: Evidence for the use of ACE inhibitors. *Heart*, *92*(3), 420–423.

Chan, J., Edwards, N., Khandheria, B., Shiino, K., Sabapathy, S., Anderson, B., et al. (2019). A new approach to assess myocardial work by non-invasive left ventricular pressure–strain relations in hypertension and dilated cardiomyopathy. *European Heart Journal Cardiovascular Imaging*, *20*, 31–39.

Conti, C., Della Corte, A., Votta, E., Del Viscovo, L., Bancone, C., De Santo, L., & Redaelli, A. (2010). Biomechanical implications of the congenital bicuspid aortic valve: A finite element study of aortic root function from in vivo data. *The Journal of Thoracic and Cardiovascular Surgery*, 890–896.

Cuspidi, C., Facchetti, R., Bombelli, M., Sala, C., Tadic, M., Grassi, G., et al. (2016). Prevalence and correlates of new-onset left ventricular geometric abnormalities in a general population: The PAMELA study. *Journal of Hypertension, 34*, 1423–1431.

de Simone, G., Gottdiener, J., Chinali, M., & Maurer, M. (2008). Left ventricular mass predicts heart failure not related to previous myocardial infarction: The cardiovascular health study. *European Heart Journal, 29*, 741–747.

Delgado, V., Tops, L., van Bommel, R., van der Kley, F., Marsan, N., Klautz, R., et al. (2009). Strain analysis in patients with severe aortic stenosis and preserved left ventricular ejection fraction undergoing surgical valve replacement. *European Heart Journal, 30*, 3037–3047.

Djorovic, S., Saveljic, I., & Filipovic, N. (2019). Computational simulation of carotid artery: from patient-specific images to finite element analysis. *Journal of the Serbian society for Computational Mechanics, 13*(1), 120–129.

Donal, E., Bergerot, C., Thibault, H., Ernande, L., Loufoua, J., Augeul, L., et al. (2009). Influence of afterload on left ventricular radial and longitudinal systolic functions: A two-dimensional strain imaging study. *European Journal of Echocardiography, 10*, 914–921.

Dweck, M., Boon, N., & Newby, D. (2012). Calcific aortic stenosis: A disease of the valve and the myocardium. *Journal of the American College of Cardiology, 60*, 1854–1863.

Edwards, N., Scalia, G., Shiino, K., Sabapathy, S., Anderson, B., Chamberlain, R., et al. (2020). Global myocardial work is superior to global longitudinal strain to predict significant coronary artery disease in patients with normal left ventricular function and wall motion (published correction appears). *Journal of the American Society of Echocardiography, 33*, 947–957.

Eleid, F., Sorajja, P., Michelena, I., Malouf, F., Scott, G., & Pellikka, P. (2013). Flow-gradient patterns in severe aortic stenosis with preserved ejection fraction: Clinical characteristics and predictors of survival. *Circulation, 128*, 1781–1789.

Garcia, D., Barenbrug, P., Pibarot, P., Dekker, A., van der Veen, F., Maessen, J., et al. (2005). A ventricular-vascular coupling model in presence of aortic stenosis. *American Journal of Physiology. Heart and Circulatory Physiology, 288*(4), H1874–H1884.

Gnyaneshwar, R., Kumar, R., & Balakrishnan, K. (2002). Dynamic analysis of the aortic valve using a finite element model. *The Annals of Thoracic Surgery, 73*, 1122–1129.

Grande, K., Cochran, R., Reinhall, P., & Kunzelman, K. (1998). Stress variations in the human aortic root and valve: The role of anatomic asymmetry. *Annals of Biomedical Engineering*, 534–545.

Hein, S., Arnon, E., Kostin, S., Schonburg, M., Elsasser, A., Polyakova, V., et al. (2003). Progression from compensated hypertrophy to failure in the pressure overloaded human heart: Structural deterioration and compensatory mechanisms. *Circulation, 107*, 984–991.

Hiemstra, Y., van der Bijl, P., El Mahdiui, M., Bax, J., Delgado, V., & Marsan, N. (2020). Myocardial work in nonobstructive hypertrophic cardiomyopathy: Implications for outcome. *Journal of the American Society of Echocardiography*, 1201–1208.

Howard, I., Patterson, E., & Yoxall, A. (2009). On the opening mechanism of the aortic valve: Some observations from simulations. *Journal of Medical Engineering & Technology*, 259–266.

Hubert, A., Le Rolle, V., Leclercq, C., Galli, E., Samset, E., Casset, C., et al. (2018). Estimation of myocardial work from pressure-strain loops analysis: An experimental evaluation. *European Heart Journal Cardiovascular Imaging, 19*(12), 1372–1379.

Hunter, P., McCulloch, A., & ter Keurs, H. (1998). Modelling the mechanical properties of cardiac muscle. *Progress in Biophysics & Molecular biology, 69*, 289–331.

Inoko, M., Kihara, Y., Morii, I., Fujiwara, H., & Sasayama, S. (1994). Transition from compensatory hypertrophy to dilated, failing left ventricles in dahl salt-sensitive rats. *The American Journal of Physiology, 267*, H2471–H2482.

Jain, R., Bajwa, T., Roemer, S., Huisheree, H., Allaqaband, S., Kroboth, S., et al. (2021). Myocardial work assessment in severe aortic stenosis undergoing transcatheter aortic valve replacement. *European Heart Journal Cardiovascular Imaging, 22*, 715–721.

Kearney, L., Lu, K., Ord, M., Patel, S., Profitis, K., et al. (2012). Global longitudinal strain is a strong independent predictor of all-cause mortality in patients with aortic stenosis. *European Heart Journal Cardiovascular Imaging, 13*, 827–833.

Klaeboe, L., Haland, T., Leren, I., Ter Bekke, R., Brekke, P., Rosjo, H., et al. (2017). Prognostic value of left ventricular deformation parameters in patients with severe aortic stenosis: A pilot study of the usefulness of strain echocardiography. *Journal of the American Society of Echocardiography, 30*(8), 727–735.

Kojic, M., Filipovic, N., Stojanovic, B., & Kojic, N. (2008). *Computer modelling in bioengineering theory, examples and software*. J Wiley and Sons.

Kojic, M., Milosevic, M., Simic, V., Milicevic, B., Geroski, V., Nizzero, S., ... Ferrari, M. (2019). Smeared multiscale finite element models for mass transport and electrophysiology coupled to muscle mechanics. *Frontiers in Bioengineering and Biotechnology, 7*, 381.

Kojic, M., Slavkovic, R., Zivkovic, M., Grujovic, N., Filipovic, N., & Milosevic, M. (2010). *PAK—finite element program for linear and nonlinear analysis*. Kragujevac, Serbia: Univ Kragujevac and R&D Center for Bioengineering.

Korurek, M., Yildiz, M., & Yuksel, A. (2010). Simulation of normal cardiovascular system and severe aortic stenosis. *Anadolu Kardiyoloji Dergisi, 10*(6), 471–478.

Lavie, C., Patel, D., Milani, R., Ventura, H., Shah, S., & Gilliland, Y. (2014). Impact of echocardiographic left ventricular geometry on clinical prognosis. *Progress in Cardiovascular Diseases, 57*, 3–9.

Lieb, W., Gona, P., Larson, M., Aragam, J., Zile, M., Cheng, S., et al. (2014). The natural history of left ventricular geometry in the community: Clinical correlates and prognostic significance of change in LV geometric pattern. *JACC: Cardiovascular Imaging, 7*, 870–878.

Lind, L., & Sundstrom, J. (2019). Change in left ventricular geometry over 10 years in the elderly and risk of incident cardiovascular disease. *Journal of Hypertension, 37*, 325–330.

Marwick, T., Gillebert, T., Aurigemma, G., Chirinos, J., Derumeaux, G., Galderisi, M., et al. (2015). Recommendations on the use of echocardiography in adult hypertension: A report from the European Association of Cardiovascular Imaging (EACVI) and the American Society of Echocardiography (ASE)dagger. *European Heart Journal Cardiovascular Imaging, 16*, 577–605.

Messika-Zeitoun, D., & Lloyd, G. (2018). Aortic valve stenosis: Evaluation and management of patients with discordant grading. *e-Journal of Cardiology Practice, 15*, 26.

Milani, R., Lavie, C., Mehra, M., Ventura, H., Kurtz, J., & Messerli, F. (2006). Left ventricular geometry and survival in patients with normal left ventricular ejection fraction. *The American Journal of Cardiology, 97*, 959–963.

Papadopoulos, K., Özden Tok, Ö., Mitrousi, K., & Ikonomidis, I. (2021). Myocardial work: Methodology and clinical applications. *Diagnostics (Basel), 11*(3), 573.

Pibarot, P., & Dumesnil, J. (2012). Improving assessment of aortic stenosis. *Journal of the American College of Cardiology, 60*, 169–180.

Rajappan, K., Rimoldi, O., Dutka, D., Ariff, B., Pennell, D., Sheridan, D., et al. (2002). Mechanisms of coronary microcirculatory dysfunction in patients with aortic stenosis and angiographically normal coronary arteries. *Circulation, 105*, 470–476.

Robnik-Šikonja, M., Radović, M., Đorović, S., Anđelković-Ćirković, B., & Filipović, N. (2018). Modeling ischemia with finite elements and automated machine learning. *Journal of Computational Science, 29*, 99–106.

Roemer, S., Jaglan, A., Santos, D., Umland, M., Jain, R., Tajik, A., & Khandheria, B. (2021). The utility of myocardial work in clinical practice. *Journal of the American Society of Echocardiography, 34*, 807–818.

Russell, K., Eriksen, M., Aaberge, L., Wilhelmsen, N., Skulstad, H., Gjesdal, O., et al. (2013). Assessment of wasted myocardial work: A novel method to quantify energy loss due to uncoordinated left ventricular contractions. *American Journal of Physiology. Heart and Circulatory Physiology, 305*, H996–1003.

Russell, K., Eriksen, M., Aaberge, L., Wilhelmsen, N., Skulstad, H., Remme, E., et al. (2012). A novel clinical method for quantification of regional left ventricular pressure-strain loop area: A non-invasive index of myocardial work. *European Heart Journal, 33*, 724–733.

Schramm, W. (2010). The units of measurement of the ventricular stroke work: A review study. *Journal of Clinical Monitoring and Computing, 24*, 213–217.

Sommer, G., Schriefl, A., Andrä, M., Sacherer, M., Viertler, C., Wolinski, H., & Holzapfel, G. (2015). Biomechanical properties and microstructure of human ventricular myocardium. *Acta Biomaterialia, 24*, 172–192.

Stanton, T., Leano, R., & Marwick, T. (2009). Prediction of all-cause mortality from global longitudinal speckle strain: Comparison with ejection fraction and wall motion scoring. *Circulation. Cardiovascular Imaging, 2*, 356–364.

Suga, H. (1979). Total mechanical energy of a ventricle model and cardiac oxygen consumption. *The American Journal of Physiology, 236*, H498–H505.

Taniguchi, T., Morimoto, T., Shiomi, H., et al. (2018). Prognostic impact of left ventricular ejection fraction in patients with severe aortic stenosis. *JACC. Cardiovascular Interventions, 11*, 145–157.

Tzivoni, D. (1993). Effect of transient ischaemia on left ventricular function and prognosis. *European Heart Journal, 14*, 2–7.

Velagaleti, R., Gona, P., Pencina, M., Aragam, J., Wang, T., Levy, D., et al. (2014). Left ventricular hypertrophy patterns and incidence of heart failure with preserved versus reduced ejection fraction. *The American Journal of Cardiology, 113*, 117–122.

Villari, B., Vassalli, G., Monrad, E., Chiariello, M., Turina, M., & Hess, O. (1995). Normalization of diastolic dysfunction in aortic stenosis late after valve replacement. *European Heart Journal Cardiovascular Imaging, 91*, 2353–2358.

Villars, P., Hamlin, S., Shaw, A., & Kanusky, J. (2004). Role of diastole in left ventricular function, I: Biochemical and biomechanical events. *American Journal of Critical Care, 13*(5), 394–403.

Zakikhani, P., Ho, R., Wang, W., & Li, Z. (2019). Biomechanical assessment of aortic valve stenosis: Advantages and limitations. *Medicine in Novel Technology and Devices, 2*, 1000091-9.

CHAPTER

7

Lab-on-a-chip for lung tissue from in silico perspective

Milica Nikolic[a,b]

[a]Steinbeis Advanced Risk Technologies Institute doo Kragujevac, SARTIK, Kragujevac, Serbia
[b]Institute for Information Technologies Kragujevac, Kragujevac, Serbia

1 Introduction

Research outcomes from cardiovascular and respiratory bioengineering have significantly improved in the last decade with the improvement in technology and in vitro approach, development of microfluidic, and organ-on-a-chip systems. The traditional in vitro approach is still used as a less invasive technique, where cells are extracted from a living bodies and examined under laboratory conditions to understand the mechanisms of behavior and interactions between cells. Many processes occurring between cells can be captured with in vitro experiments, but the main issue with in vitro testing is the translation of the research results to in vivo testing. The reason lies in the fact that cells are isolated from the system, and therefore have reduced interactions and functionality. In the first place, the vascular components are missing. That is why scientists came up with idea to create a better environment for the examined cells. Since the cells miss the interaction with vascular cells and blood, most microfluidic chips have been developed to create interactions with the cells and improve environmental conditions. They are small in size and usually contain several channels where the cells are seeded and blood can circulate, thus simulating more realistic physiological behavior of the organ and the biomechanical response of the cells. The cells of the investigated organ are lined in one channel and can interact with endothelial cells in the concept of artificial vasculature, thus the cells can obtain nutrients from blood supplies to the tissue. External forces can be applied in accordance with biomechanics felt by specific organs as improvement for in vivo like conditions. Therefore, the organ-on-a-chip concept is a more complex system that can better mimic cells' environment and provide results that are more in alliance with in vivo results and behavior of real organs.

https://doi.org/10.1016/B978-0-12-823956-8.00007-9

So far, many platforms for simulating organ-on-a-chip have been developed. Different organs are tested in an improved environment—the heart, lungs, liver, kidney, bone, brain, etc. (Hao, 2018; Huh, 2015; Kitsara, 2019; Knowlton, 2016; Lee, 2018; Wheeler, 2008). One of the pioneers and founders of lung-on-a-chip design is Huh, who replicated the functional unit of the human lung by creating the alveolar-capillary interface (Huh, 2015). Under applied external dynamic conditions mimicking the breathing, the cells can show their physiological functions and be used to simulate complex disease processes of the lungs. Different cell lines are used for investigation in lung-on-a-chip concept. Among the most used are A549 cell line, epithelial cells derived from adenocarcinoma, Calu-3 cell line derived from lung cancer, and primary cells NHBE (human bronchial epithelial cells).

Wyss Institute for Biologically Inspired Engineering at Harvard University[a] developed the first lung-on-a-chip system, containing two chambers divided by membrane. The blood circulates in the lower chamber, while the air circulates in the upper chamber. Endothelial cells are seeded below the membrane and epithelial cells above the membrane. With this design, they created airways in the upper chamber, and artificial blood vessels in the lower chamber that supplied the lung tissue. External cyclic mechanical stretching is applied to simulate breathing. This and later similar systems developed later can successfully simulate the behavior of the healthy lungs and with changed conditions, it can be used to simulate different diseases and malfunctioning of the lungs. For instance, lung-on-a-chip systems have been employed to analyze the inflammatory effect of the nanoparticles, originating from the air and/or water pollution, or to test novel/existing drugs for the treatment of specific lung diseases.

Huh adapted the lung-on-a-chip system to the alveolus-on-a-chip model to simulate pulmonary thrombosis, by seeding alveolar epithelial cells in the upper chamber. Thrombosis was initiated with the addition of tumor necrosis factor (TNF-α) to the upper chamber to provoke inflammation. Inflammation causes an increase of cytokines, and further leukocytes and platelet blood cells, leading to the formation of thrombi. Then he applied drug parmodulins to reduce inflammation and obtained reduction of the process. Huh (2015) also analyzed pulmonary edema, caused by heart failure or as a side effect of cancer drugs, manifesting in the generation of blood clots and fluid in the lung. The state of disease is established with the injection of the cancer drug to the artificial blood vessel and as a result migration of the fluid and protein into the upper chamber was noticed. Many researches in this area focused on the analysis of asthma (Benam, 2015) and chronic obstructive pulmonary disease, COPD. The focus of the research is tumors-on-a-chip systems as well (Yang, 2018). More detailed state of the art in lung-on-a-chip systems can be found in review papers (Nikolic, 2018a).

Computational models, commonly referred to in the bioengineering area as in silico models, can be very useful in predicting the outcomes from the system. They can reduce the number of experiments and more easily test the effects of changes in boundary/loading conditions than experiments can do. A rapid increase in technology has enhanced the

[a]Wyss Institute. (November 7, 2012). Available online at: https://wyss.harvard.edu/wyss- institute-models-a-human-disease-in-an-organ-on-a-chip.

development of the in silico models as well. However, there are not many models developed for the lung-on-a-chip system, especially mathematical models. In this chapter, a review of the current in silico models developed for the lung-on-a-chip system is given, together with author's research in that area highlighting the current advantages and limitations of in silico models.

2 In silico lung-on-a-chip models

In silico models can be developed to reduce the number of experimental trials and also to predict optimized parameters for in vitro or organ-on-a-chip models. Further, they can be used to translate outputs to the macromodel, for in vivo like conditions, by observing biological phenomena at different scales. From an ethical point of view, in silico models have higher approval, since they can reduce the number of experiments performed on animals and avoid the usage of human cell lines.

Usually in silico models are presented at more scales, creating a multiscale approach. The macroscale can be used to simulate behavior of the whole system at the macroscopic level, for instance, with the finite element method (FEM). If the behavior of the system needs to be captured at the molecular, cell level, then in silico models that are at the microscale can be solved with molecular dynamics. Behavior of the system in between, like molecular clusters, a specific region on the system, can be presented as mesoscale in silico model (methods such as dissipative particle dynamics, DPD). In the creation of a multiscale approach, special attention should be paid to the translation of the findings at different scales. Regardless of the scale, all in silico models have to be verified and validated. Verification is related to the confirmation of the solver and developed mathematical model, while validation assumes that the outputs are in line with experimental measurements and valid for the biological system of interest.

When it comes to the lungs and lung-on-a-chip systems there are not many models in the literature. The majority of the models are related to the imaging processing methods for cell biology (Konar, 2016). The model developed by Konar (2016) can mimic the environment of the lung tissue and present air-liquid interface, but the results are obtained upon image reconstruction and analysis and therefore they can be used only as a support to the experimental work, by providing more information on the output of the experiments. A numerical solution should be proposed in terms of possible reduction of the experiments and higher usage of computational models.

More applicable numerical model was developed by Hancock (2018) in COMSOL commercial software with FEM based on Huh's model (Huh, 2015). They utilized built-in possibilities of COMSOL software to apply Neo-Hookean, Mooney-Rivlin, or Ogden nonlinear material models with a thickness of 10 μm for polydimethylsiloxane (PDMS) membrane between the chambers. To simulate cyclic stretching, also known as breathing, they applied a vacuum pressure waveform. For simulation of air and blood flow, computational fluid dynamics (CFD) was used. Nonstationary analysis is conducted to identify deformation of the membrane and chambers. Another feature from COMSOL, particle tracing, can be used to predict nanoparticle uptake, bacteria, drug, or nutrient transport within the culture media and through the membrane. Although the model is capable of replicating the model developed by Hun, information on the mathematical model are not extracted from this paper.

Mathematical model that describes the airway epithelial model wound and its closure were developed by Savla (2004). The model takes into account the spreading, migration, and proliferation of epithelial cells. Mathematical representation of the wound healing process is given with an extended diffusion equation. Based on their work the model to simulate behavior of epithelial cells was developed and is presented below.

3 Mathematical model for epithelial cell behavior and fluid flow

In the previous work, the model for simulation of epithelial cell behavior after the seeding as well as the model for simulation of blood flow were established. Derivations of the models have been already disseminated (Filipovic, 2020; Nikolic, 2019a) and are, therefore, explained briefly here. The components of the lung-on-a-chip system have been analyzed separately, to combine them in the future work in the proper representation of the lung-on-a-chip system.

3.1 Model of the bioreactor with monocytes

The three-dimensional (3D) geometry of the model was created on geometrical parameters provided by Sharifi (2019). To mathematically present the fluid flow inside the bioreactor, Navier-Stokes equation together with the continuity equation for incompressible fluid is used (Eqs. 1, 2).

$$\rho\frac{\partial V}{\partial t}-\mu\nabla^2 V+\rho(V\cdot\nabla)V+\nabla p=0 \tag{1}$$

$$\nabla\cdot V=0 \tag{2}$$

where fluid velocity is represented by variable V, dynamic viscosity of the fluid by μ, ρ is the density of the fluid, p is the fluid pressure, t is the simulation time, and ∇ is the operator nabla.

The fluid flow is assumed to be laminar, so the expected fluid velocity profile is parabolic. Boundary conditions of the model are constant inlet velocity at the entrance of the bioreactor and zero pressure at the end of the bioreactor, meaning open end.

Besides fluid flow, traveling of the monocyte cells with the fluid was analyzed. In the computational model, this is included as mass transfer in fluid, since their movement is driven by fluid flow (Eq. 3)

$$\frac{\partial N}{\partial t}+\nabla(-D\nabla N+NV)=0 \tag{3}$$

D represents the diffusion coefficient of the monocytes, N is the concentration (number of monocytes in solution). Monocyte cells are small in size, but since their movement is dependent on the fluid velocity and behavior of the whole system, the system of the Eqs. (1)–(3) can be solved at the macroscopic level with FEM (Kojic, 1998a). In house built solver PAK (Kojic, 1998b, 2001, 2008) was used to perform the calculations. The output of the bioreactor model with monocytes predicted properly fluid flow velocity and distribution of the monocytes through the domain. The results were compared with the one obtained in commercial software ANSYS (Nikolic, 2018b) and COMSOL (Nikolic, 2019b). The velocity of the monocytes was calculated upon drag force and Stoke's law.

3.2 Model of epithelial cell behavior

In this model, the behavior of epithelial cells upon seeding was observed. The mathematical description of the model takes into account several processes of the cells—diffusion and sedimentation, spreading and proliferation. The behavior of the cells was observed under static conditions, but with a goal to be later connected with the bioreactor model to achieve predictions of lung-on-a-chip systems.

For the mathematical representation of the system, diffusion equation was used (Eq. 4).

$$\frac{\partial n}{\partial t} = D\left(\frac{\partial^2 n}{\partial x^2} + \frac{\partial^2 n}{\partial y^2} + \frac{\partial^2 n}{\partial z^2}\right) \tag{4}$$

where n represents the epithelial cell density (cells/mm^2), t is time (h), and D is diffusion coefficient (mm^2/h).

The model aims to capture cell spreading and proliferation processes so these have to be included in the definition of the equation. Based on the work of Savla (2004), the diffusion equation has been extended to include additional terms, taking spreading and proliferation of the cells into account (Eq. 5). Experimentally, it has been determined that proliferation of the cells starts a day or two after the seeding, depending on the cell type, and this also needs to be included in the computational model.

$$\frac{\partial n}{\partial t} = D\left(\frac{\partial^2 n}{\partial x^2} + \frac{\partial^2 n}{\partial y^2} + \frac{\partial^2 n}{\partial z^2}\right) + (k_s + k_p)\left(1 - \frac{n}{n_0}\right)n,\ k_p = \begin{cases} 0, & t < 24\,\text{h}\,(48\,\text{h}) \\ \text{value}, & t \geq 24\,\text{h}\,(48\,\text{h}) \end{cases} \tag{5}$$

Newly introduced parameters are spreading coefficient, k_s, (1/h), proliferation coefficient, k_p, (1/h), and the initial number of cells, n_0 (cells/mm^2).

The remaining process to be included is the sedimentation of the epithelial cells upon seeding, so the term from the Mason-Weaver equation is used. The final equation of the model is presented in Eq. (6).

$$\frac{\partial n}{\partial t} = D\left(\frac{\partial^2 n}{\partial x^2} + \frac{\partial^2 n}{\partial y^2} + \frac{\partial^2 n}{\partial z^2}\right) + (k_s + k_p)\left(1 - \frac{n}{n_0}\right)n + sg\frac{\partial n}{\partial z} \tag{6}$$

The last added term of the equation contains s for coefficient of sedimentation (h), g for gravitational acceleration ($g = 9.81\,\text{m/s}^2 = 127.14 \times 10^9\,\text{mm/h}^2$), and the product is referred to as terminal velocity V_t.

Initially, used boundary conditions of the extended diffusion equation were cell density at the top of the model, where the cells are seeded, but different conditions are lately tested and their effect on sedimentation of the cells was analyzed. Besides that, there is no leakage through any boundary. The equation has been written in the form required for FEM analysis and solved with PAK solver. The values of spreading and proliferation coefficient were calculated from experimental images in the work conducted within the European project[b]. After

[b]PANBioRA—Personalised and generalised integrated biomaterial risk assessment.

seeding, cells move toward the bottom glass side. The quantity of interest from the model is time needed for cells to reach the bottom surface, and form a monolayer, which can be used as a component in the lung-on-a-chip system.

The proposed model showed strong dependence on the values of the parameters, leading to a convergence issue. Therefore, additional effort in defining the input parameters is required together with later validation of the obtained in silico results. The sedimentation coefficient was marked as the most sensitive parameter and because of that, the sedimentation process was simulated separately using simplified geometry and finite difference method, FDM, in Matlab software. Derivation of the Mason-Weaver equation and finite difference approximation together with different schemes were tested and the obtained results are explained in the next section.

4 Mason-Weaver equation, finite difference method

To simulate the sedimentation process of the cells, Mason-Weaver equation was used (Eq. 7).

$$\frac{\partial n}{\partial t} = D\frac{\partial^2 n}{\partial z^2} + sg\frac{\partial n}{\partial z} \tag{7}$$

The variables from Eq. (7) have been already defined in the previous section. If the sedimentation process is driven by gravitational force, then only the z-direction can be taken into account, leading to 1D problem.

Boundary condition for the Mason-Weaver equation is that there is no flux change through the top and bottom surfaces of the domain (Eq. 8).

$$D\frac{\partial n}{\partial z} + \mathrm{sgn} = 0 \tag{8}$$

Since Eqs. (7), (8) are partial differential equations (PDEs) finite difference method seems to be the most suitable for testing and analysis of the obtained results and impact of different initial conditions on the solution. Different FDM schemes were applied—ordinary scheme (central, forward, backward differences), Crank-Nicolson scheme; both dimensionless and with the units. The simulated height of the domain was 10 μm and sedimentation time was taken to be around 7 min.

4.1 Crank-Nicolson scheme, dimensionless

Crank-Nicolson scheme takes in approximation values of neighboring nodes in space domain for the current and the next time step. Eqs. (7) and (8) are first transferred in dimensionless by introducing two new variables, τ and ζ, to make a connection with real-time and space.

By considering

$$t_0 = \frac{D}{(sg)^2} \bigwedge \tau = \frac{t}{t_0} \tag{9}$$

$$z_0 = \frac{D}{sg} \bigwedge \zeta = \frac{z}{z_0} \tag{10}$$

Mason-Weaver equation becomes dimensionless:

$$\frac{\partial n}{\partial \tau} = \frac{\partial^2 n}{\partial \zeta^2} + \frac{\partial n}{\partial \zeta} \tag{11}$$

with boundary conditions

$$\frac{\partial n}{\partial \zeta} + n = 0 \tag{12}$$

Members of dimensionless Mason-Weaver equation (Eq. 11) and boundary condition (Eq. 12) are then written with Crank-Nicolson scheme. In the following equations, indices i and j stand for the position of the node in space domain and time step, respectively.

$$\frac{\partial n}{\partial \tau} = \frac{n_i^{j+1} - n_i^j}{\delta \tau} \tag{13}$$

$$\frac{\partial^2 n}{\partial \zeta^2} = \frac{n_{i+1}^{j+1} - 2n_i^{j+1} + n_{i-1}^{j+1} + n_{i+1}^j - 2n_i^j + n_{i-1}^j}{2\delta\zeta^2} \tag{14}$$

$$\frac{\partial n}{\partial \zeta} = \frac{n_{i+1}^{j+1} - n_{i-1}^{j+1} + n_{i+1}^j - n_{i-1}^j}{4\delta\zeta} \tag{15}$$

By introducing the approximated members from Eqs. (13)–(15) into Eq. (11), Crank-Nicolson scheme is achieved.

$$\frac{n_i^{j+1} - n_i^j}{\delta \tau} = \frac{n_{i+1}^{j+1} - 2n_i^{j+1} + n_{i-1}^{j+1} + n_{i+1}^j - 2n_i^j + n_{i-1}^j}{2\delta\zeta^2} + \frac{n_{i+1}^{j+1} - n_{i-1}^{j+1} + n_{i+1}^j - n_{i-1}^j}{4\delta\zeta} \tag{16}$$

Or with the rearrangement by time steps:

$$\begin{aligned} &n_{i+1}^{j+1}\left(-\frac{1}{2\delta\zeta^2} - \frac{1}{4\delta\zeta}\right) + n_i^{j+1}\left(\frac{1}{\delta\tau} + \frac{1}{\delta\zeta^2}\right) + n_{i-1}^{j+1}\left(-\frac{1}{2\delta\zeta^2} + \frac{1}{4\delta\zeta}\right) \\ &= n_{i+1}^{j}\left(\frac{1}{2\delta\zeta^2} + \frac{1}{4\delta\zeta}\right) + n_i^{j}\left(\frac{1}{\delta\tau} - \frac{1}{\delta\zeta^2}\right) + n_{i-1}^{j+1}\left(\frac{1}{2\delta\zeta^2} - \frac{1}{4\delta\zeta}\right) \end{aligned} \tag{17}$$

The parameters from Eq. (17) can be presented as the members of tridiagonal matrix, a, b, and c.

The same approach is applied for boundary conditions. For the bottom boundary ($\zeta = 0$) forward approximation is used.

$$\frac{\partial n}{\partial \zeta} = \frac{n_2^{j+1} - n_1^{j+1} + n_2^j - n_1^j}{2\delta\zeta} \tag{18}$$

$$n = \frac{n_2^{j+1} + n_1^{j+1} + n_2^j + n_1^j}{4} \tag{19}$$

Approximation in Eqs. (18), (19) are slightly different from the usual and they were selected on work of Midelet (2017), where they claimed to have achieved better mass conservation. Mass conservation is actually one of the main issues in numerical solving of the Mason-Weaver equation.

Boundary equation now can be rearranged as

$$n_2^{j+1}\left(\frac{1}{2\delta\zeta}+\frac{1}{4}\right)+n_1^{j+1}\left(-\frac{1}{2\delta\zeta}+\frac{1}{4}\right)=n_2^j\left(-\frac{1}{2\delta\zeta}-\frac{1}{4}\right)+n_1^j\left(\frac{1}{2\delta\zeta}-\frac{1}{4}\right) \tag{20}$$

Similarly, boundary condition is established for the top boundary using backward approximation.

$$n_{\zeta n}^{j+1}\left(\frac{1}{2\delta\zeta}+\frac{1}{4}\right)+n_{\zeta n-1}^{j+1}\left(-\frac{1}{2\delta\zeta}+\frac{1}{4}\right)=n_{\zeta n}^j\left(-\frac{1}{2\delta\zeta}-\frac{1}{4}\right)+n_{\zeta n-1}^j\left(\frac{1}{2\delta\zeta}-\frac{1}{4}\right) \tag{21}$$

Eqs. (17), (20), (21) are implemented in Matlab to predict the sedimentation process. The total number of diffusions by time is 200 and the total number of divisions by height is 400. τ and ζ are set to change linearly. Exponential function for τ was also analyzed, like Midelet (2017) selected, but there has not been recorded improvement in response, in comparison to the linear function. Therefore, a linear function has been selected as more simplified for implementation.

In simulation analysis, different initial conditions and their impact on numerical solutions were tested. The first simulation was conducted with cell concentration homogeneously distributed through the domain, as this was the condition in the derivation of the equation. As expected, the model was able to simulate the sedimentation process from the initially homogeneous solution to the equilibrium state. The obtained results are presented in Fig. 1.

After that, it was interesting to find whether initially cell-free boundaries will affect the equilibrium state. As is presented in Fig. 2, simulated sedimentation in time steps does not change significantly in comparison to the whole domain covered.

In the next test, the goal was to observe whether sedimentation can reach steady-state if the cells are seeded only on top of the domain. Literature survey showed that this initial condition has not been used before for the Mason-Weaver equation. This test applied initially the cells in one-quarter of the fluid height at the upper side. The interesting finding from the simulation was that an equilibrium state can be achieved. True, it differs from previous tests, but finally diagram of steady-state shape is reached. Yet, it is not completely clear why disturbance at the border zone between cell populated fluid height and cell-free fluid height persists in time (Fig. 3).

Finally, random distribution of the cells as an initial condition was simulated and analyzed (Fig. 4). In all, 50% of the nodes got the concentration of the cells as an input value and the other 50% were cell-free. The model, like the others presented above, simulates sedimentation of the cells with a tendency to reach steady-state, but with higher oscillations. Seems like randomly seeded cells will affect the transition of concentration between the nodes.

4.2 Crank-Nicolson scheme, with the units

The code developed for solving dimensionless Mason-Weaver equation with Crank-Nicolson scheme was corrected to give solution in time and with the meaning of physical

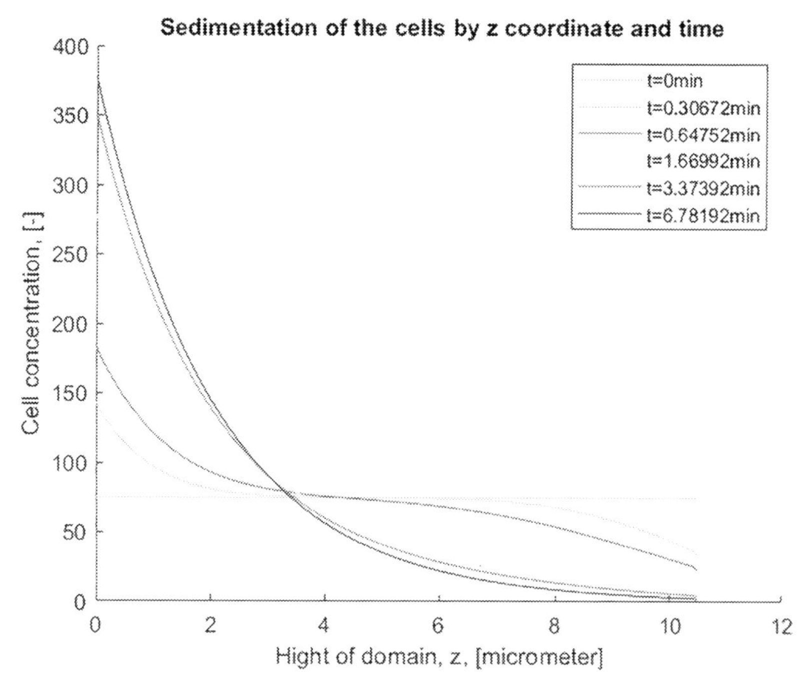

FIG. 1 Sedimentation representation of the cells with the initially homogeneous solution for Crank-Nicolson approximation and dimensionless Mason-Weaver equation.

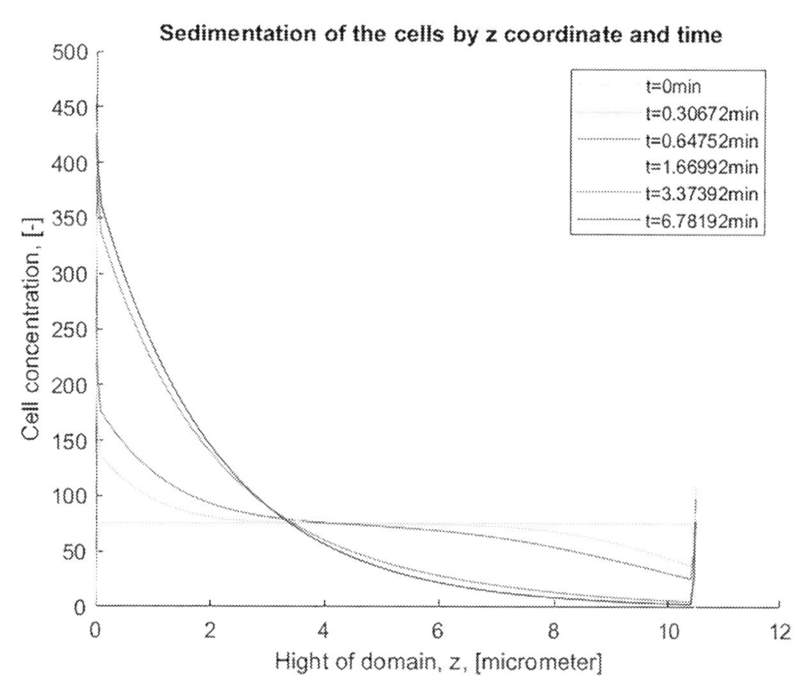

FIG. 2 Sedimentation of the cells with initially cell-free boundaries for Crank-Nicolson approximation and dimensionless Mason-Weaver equation.

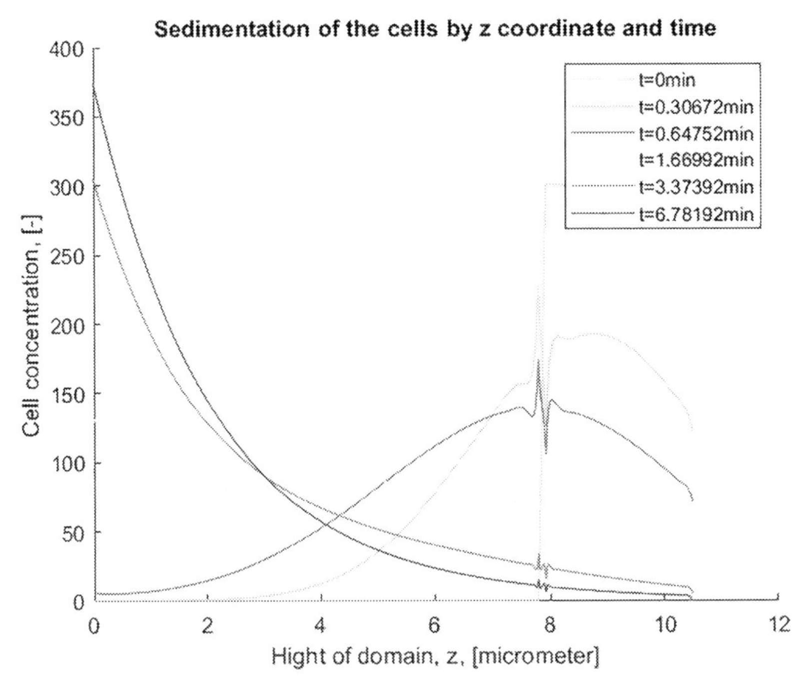

FIG. 3 Sedimentation of the cells with initially seeded cells in one-quarter of the fluid height for Crank-Nicolson approximation and dimensionless Mason-Weaver equation.

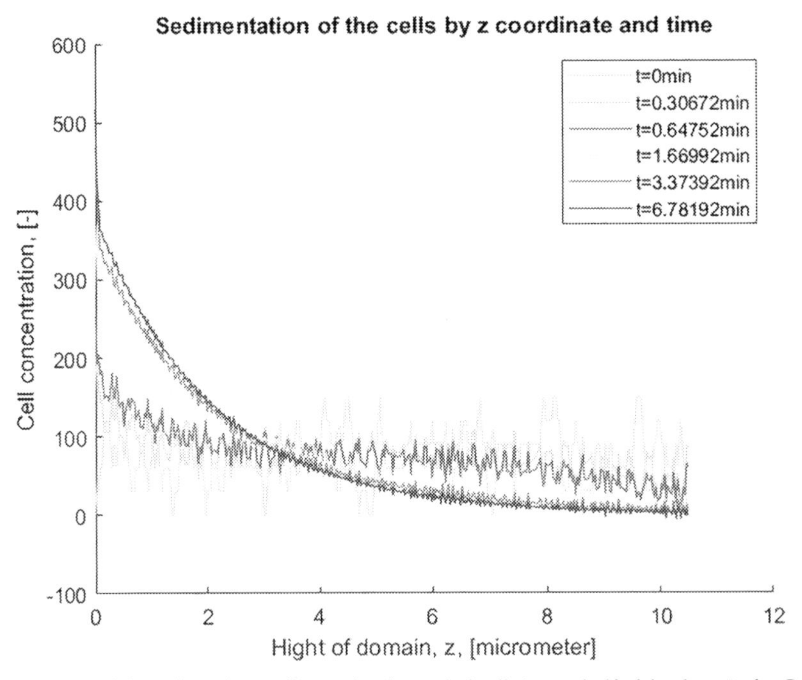

FIG. 4 Sedimentation of the cells with initially randomly seeded cells in one half of the domain for Crank-Nicolson approximation and dimensionless Mason-Weaver equation

units, instead of time and space ratios. The FDM approximations are the same as that for the first model, but only as a starting point in derivation, regular Mason-Weaver equation was used, instead of dimensionless. Therefore, in the following text, only figures from the simulations are presented and briefly explained. A number of time steps and divisions in the z-direction are the same as for the dimensionless model.

No differences have been observed in solving the Mason-Weaver dimensionless and regular model for the initially homogeneous distribution of the cells (compare Figs. 5 and 1). The same output has been recorded for the cell-free boundaries initial condition (Fig. 6) as well as cells seeded on the top of the fluid (25% of the fluid height), which can be observed from Fig. 7.

However, oscillations from the randomly distributed cells in the domain (one-half in total) seem to be higher (Fig. 8).

Some researchers managed to obtain a proper solution for the sedimentation process with the Mason-Weaver equation and Crank-Nicolson scheme. However, they could not obtain mass conversion with ordinary approximation (Midelet, 2017), which was also the issue in our previously modeled behavior of epithelial cells. To resolve this issue, FDM model with ordinary differences was developed and solved with Matlab.

4.3 Ordinary differences, with the units

In deriving FDM ordinary approximations—central, forward, backward, starting points are Eqs. (7), (8). Ordinary approximations are given below—forward for the first member and central for the remaining two.

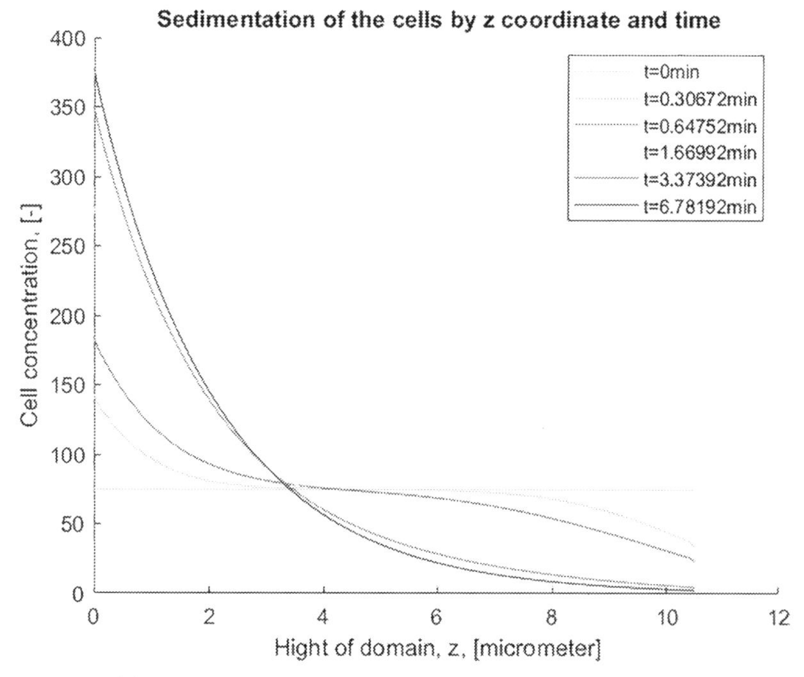

FIG. 5 Sedimentation of the cells with the initially homogeneous solution for Crank-Nicolson approximation and regular Mason-Weaver equation.

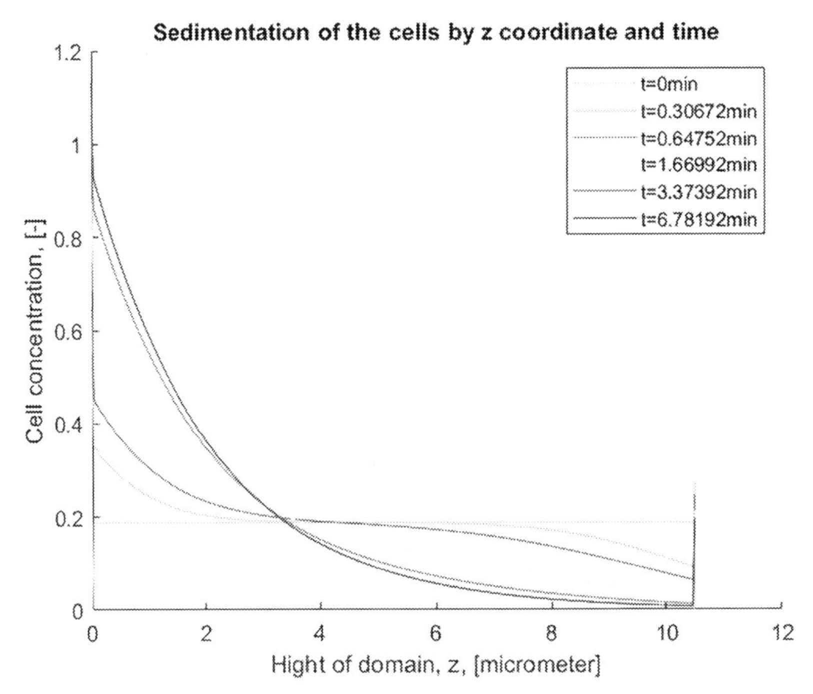

FIG. 6 Sedimentation of the cells with initially cell-free boundaries for Crank-Nicolson approximation and regular Mason-Weaver equation.

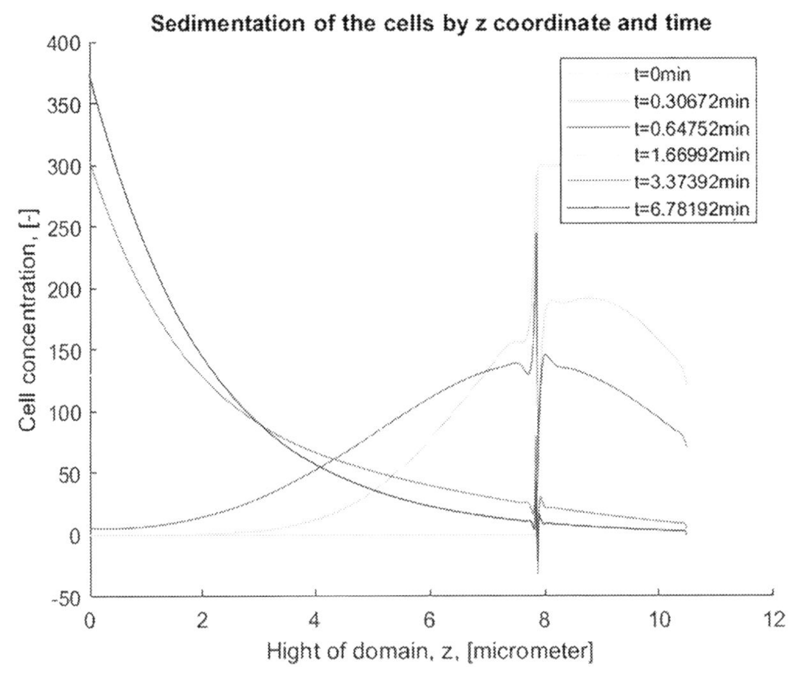

FIG. 7 Sedimentation of the cells with initially seeded cells in one-quarter of the fluid height for Crank-Nicolson approximation and regular Mason-Weaver equation.

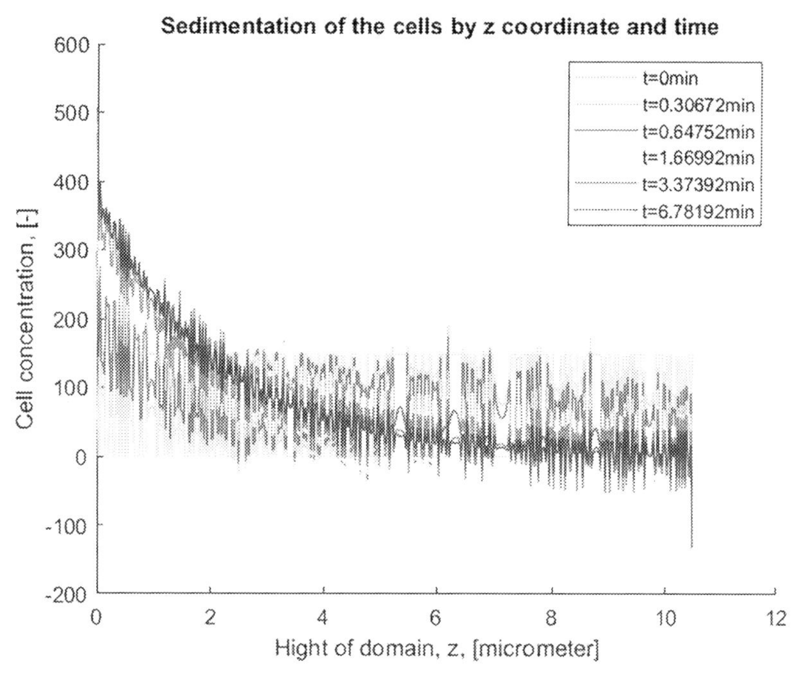

FIG. 8 Sedimentation of the cells with initially randomly seeded cells in one-half of the domain for Crank-Nicolson approximation and regular Mason-Weaver equation.

$$\frac{\partial n}{\partial t} = \frac{n_i^{j+1} - n_i^j}{\delta t} \tag{22}$$

$$D\frac{\partial^2 n}{\partial z^2} = D\frac{n_{i+1}^j - 2n_i^j + n_{i-1}^j}{\delta z^2} \tag{23}$$

$$sg\frac{\partial n}{\partial z} = sg\frac{n_{i+1}^j - n_{i-1}^j}{2\delta z} \tag{24}$$

On rearrangement, the Mason-Weaver equation has a form:

$$n_i^{j+1}\left(\frac{1}{\delta t}\right) = n_{i+1}^j\left(\frac{D}{\delta z^2} + \frac{sg}{2\delta z}\right) + n_i^j\left(\frac{1}{\delta t} - \frac{2D}{\delta z^2}\right)D + n_{i-1}^j\left(\frac{D}{\delta z^2} - \frac{sg}{2\delta z}\right) \tag{25}$$

Boundary conditions at the top and the bottom of the domain are approximated as well. The bottom of the domain ($z = 0$) used central difference for the first derivative. The same approximation was applied for the top of the domain, leading to the equation of the same shape.

$$D\frac{\partial n}{\partial z} = D\frac{n_{i+1}^j - n_{i-1}^j}{2\delta z} \tag{26}$$

$$sgn = sgn_i^j \tag{27}$$

On rearrangement, boundary condition at the top and bottom of the domain has the shape:

$$n_{i+1}^{j}\left(\frac{D}{2\delta z}\right)+n_{i}^{j}(sg)+n_{i-1}^{j}\left(-\frac{D}{2\delta z}\right)=0 \tag{28}$$

Initially, the model showed an issue with mass conservation, but it was determined that it requires a fine definition of time step to successfully simulate the sedimentation process. Therefore, the number of divisions in space is kept to be 400, but the number of time steps has been increased up to 150,000. With these numerical parameters, ordinary FDM model properly simulated the sedimentation process of homogenously seeded cells (Fig. 9), initially cell-free boundaries (Fig. 10) and initially seeded cells at the top of fluid (Fig. 11) and randomly distributed cells in half of the domain in total (Fig. 12).

The model with ordinary differences not only predicted the sedimentation process but also showed improvement in analysis when cells were seeded at top of the domain or randomly. As it can be seen from Figs. 11 and 12, disturbances and oscillations are eliminated from the numerical output.

Finally, three developed FDM models were tested for mass conservation (cell concentration). Crank-Nicolson models showed up to 0.8% increase in total cell concentration, while the ordinary FDM scheme showed 0.35% (Fig. 13).

Both model types are capable to capture the sedimentation process. Errors in numerical simulations are always present, but it is important to be as small as possible. From that point

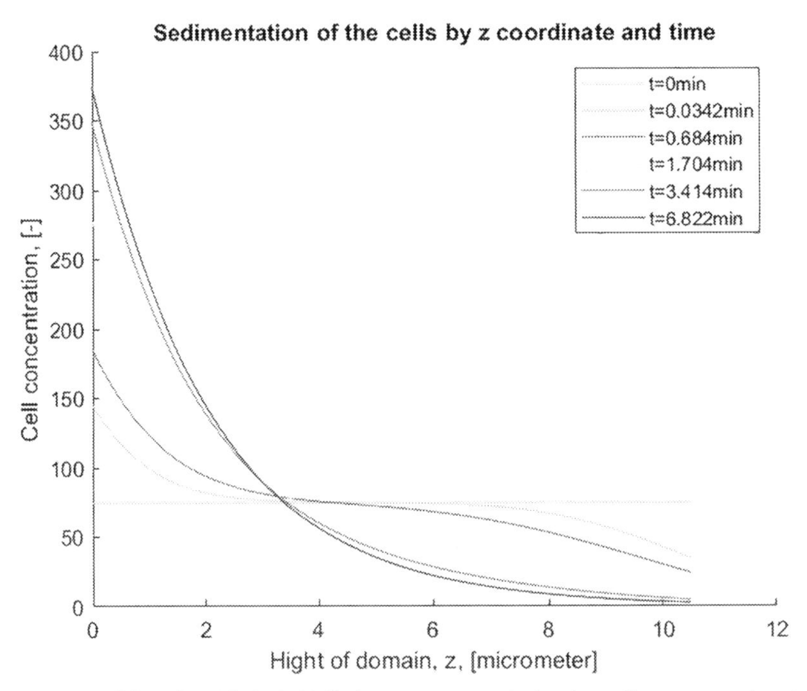

FIG. 9 Sedimentation of the cells with the initially homogeneous solution for ordinary approximation and regular Mason-Weaver equation.

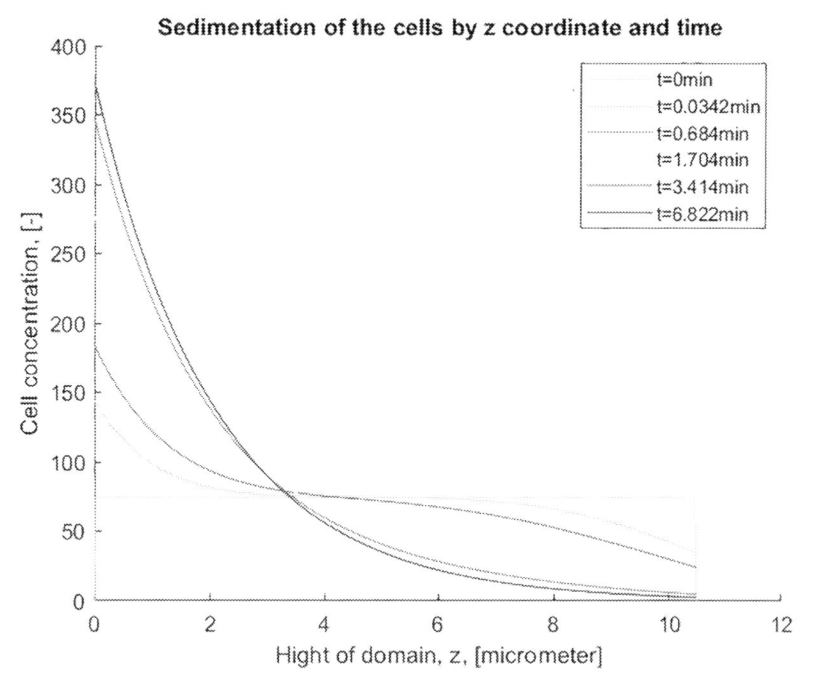

FIG. 10 Sedimentation of the cells with initially cell-free boundaries for ordinary approximation and regular Mason-Weaver equation.

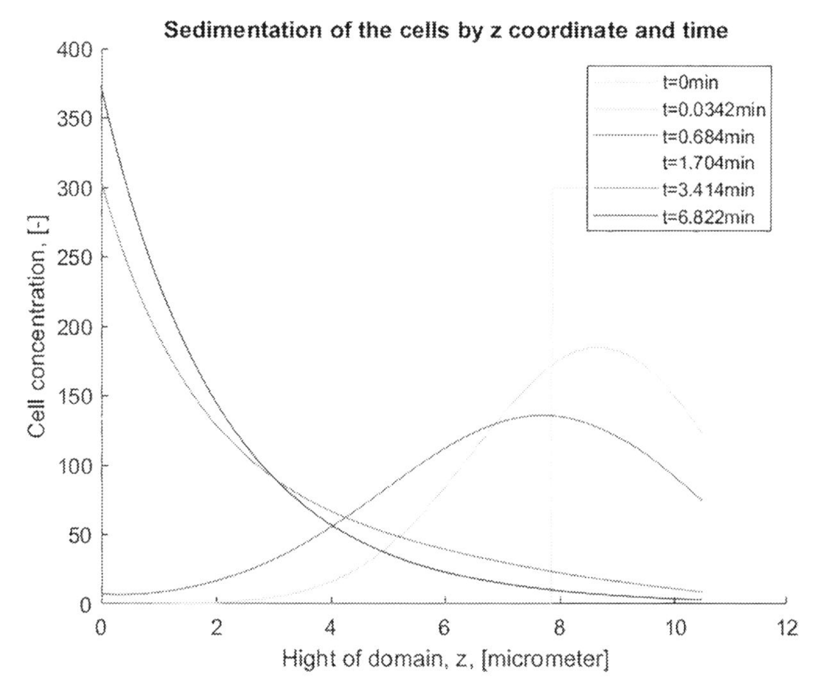

FIG. 11 Sedimentation of the cells with initially seeded cells in one-quarter of the fluid height for ordinary approximation and regular Mason-Weaver equation.

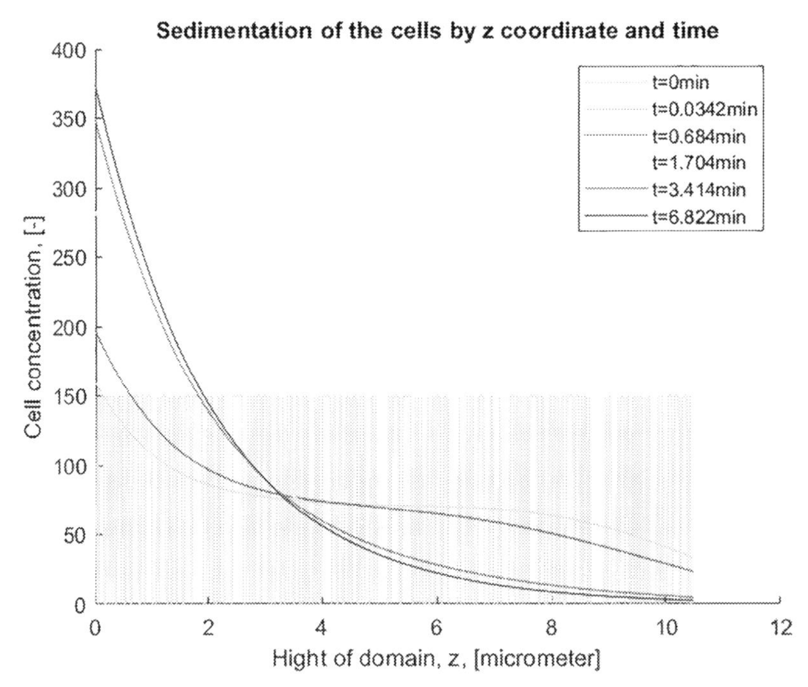

FIG. 12 Sedimentation of the cells with initially randomly seeded cells in one-half of the domain for ordinary approximation and regular Mason-Weaver equation.

of view, an ordinary FDM scheme provides better output. Also, a more smooth response on the ordinary FDM scheme was recorded. The disadvantage of the ordinary approximation is that it requires fine selection on time increment. These findings were used to fit the data from the experiment, performed by the partner in PANBioRA project. Experimental and numerical data showed good matching, leading to the calculation of sedimentation coefficient to be improved in the epithelial cell behavior model and FEM. Further effort will be done in the validation of the epithelial cell behavior model and connection with vascular, bioreactor, component to provide in silico model for more like lung-on-a-chip system.

5 Conclusions

In this chapter, the novelty and importance of the ongoing development of lung-on-a-chip systems have been highlighted together with pioneer and current lung-on-a-chip models. Their advantage in comparison to the conventional in vitro models has been explained. A small number of in silico models can be seen as a lack in further progress in lung research. For instance, in the cardiovascular area, significant progress has been made in recent years with in silico models. They were used for prediction of health/diseased stages, for testing of existing/novel drug therapies, for improvement of regenerative strategies. On the other side, most in silico models in the lung area are based on imaging reconstruction and analysis.

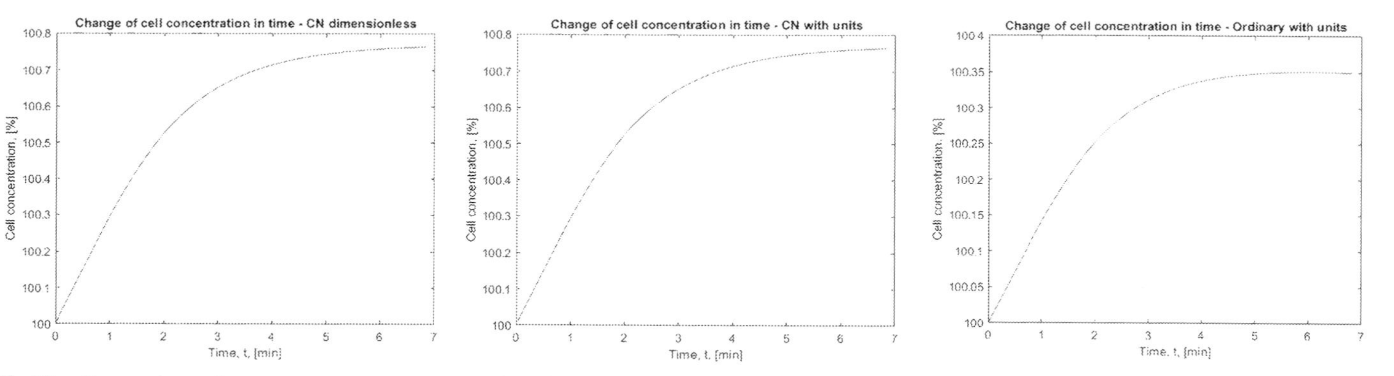

FIG. 13 Comparison of concentration conversion within developed models.

These models certainly provide benefit to the research, but they cannot reduce the number of experiments or give more predictions because they are not simulating the processes within the cells or tissue. To create a more relevant computational model, the behavior of epithelial and endothelial cells should be included, together with the behavior of the blood flow and interaction between blood and the cells. In our previous work, the bioreactor model was established as well as an initial model for simulating the behavior of epithelial cells. From this research, the sedimentation process has been detected as important for model stability and successful convergence. In this chapter, the author has provided more information on numerical modeling of the sedimentation process. Further integration of collected knowledge and developed computational models would lead to more detailed models with better replication of lung-on-a-chip systems. Proper numerical replication would lead to better predictions of healthy lung tissue behavior or corresponding disease.

Acknowledgments

This study was funded by the European Project H2020 PANBioRA (grant number 760921) and Serbian Ministry of Education, Science, and Technological Development (451-03-9/2021-14/200107 (Faculty of Engineering, University of Kragujevac) and 451-03-9/2021-14/200378 (University of Kragujevac, Institute for Information Technologies, Kragujevac)). This article reflects only the author's view. The Commission is not responsible for any use that may be made of the information it contains.

References

Benam, K. V. (2015). Small airway-on-a-chip enables analysis of human lung inflammatory and drug responses in vitro. *Nature Methods*, 151–157.

Filipovic, N. N. (2020). Simulation of organ-on-a-chip systems. In *Biomaterials for organ and tissue regeneration, new technologies and future prospects* (pp. 753–790). ScienceDirect (chapter 28).

Hancock, M. (2018). *Modelling a lung-on-a-chip microdevice*. COMSOL Software.

Hao, S. H. (2018). A spontaneous 3D bone-on-a-chip for bone metastasis study of breast cancer cells. *Small*, *14*(12).

Huh, D. (2015). A human breathing lung-on-a-chip. *Annals of the American Thoracic Society*, S42–S44.

Kitsara, M. K. (2019). Heart on a chip: Micro- nanofabrication and microfluidic steering the future of cardiac tissue engineering. *Microelectronic Engineering*, 44–62.

Knowlton, S. (2016). A bioprinted liver-on-a-chip for drug screening applications. *Trends in Biotechnology*, 681–682.

Kojic, M. F. (1998a). *PAK-F finite element program for laminar flow of incompressible fluid and heat transfer*. Serbia: University of Kragujevac.

Kojic, M. S. (1998b). *Metod konacnih elemenata I—Linearna analiza*. Faculty of Mechanical Engineering.

Kojic, M. F. (2001). *PAK-FS finite elmenet program for fluid-solid interaction*. Serbia: University of Kragujevac.

Kojic, M. F. (2008). *Computer modeling in bioengineering—Theoretical backgroung, examples and software*. Chichester, England: John Wiley and Sons.

Konar, D. D. (2016). Lung-on-a-chip technologies for disease modelling and drug development. *Biomedical Engineering and Computational Biology*, 17–27.

Lee, J. K. (2018). Kidney-on-a-chip: A new technology for predicting drug efficacy, interactions, and drug-induced nephrotoxicity. *Current Drug Metabolism*, 577–583.

Midelet, J. (2017). The sedimentation of colloidal nanoparticles in solution and its study using quantitative digital photography. *Particle and Particle Systems Characterization*, 1700095. Wiley-VCH Verlag.

Nikolic, M. S. (2018a). In vitro models and on-chip systems: Biomaterial interaction studies with tissues generated using lung epithelial and liver metabolic cell lines. *Frontiers in Bioengineering and Biotechnology*, 1–13.

Nikolic, M. (2018b). Modelling of monocytes behaviour inside the bioreactor. In *Belgrade bioinformatics conference, BelBi. Belgrade, Serbia: Biologia Serbica.*

Nikolic, M. (2019a). Lung on a chip and epithelial lung cells modelling. In *Computational modelling in bioengineering and bioinformatics* (pp. 105–135). ScienceDirect (chapter 4).

Nikolic, M. S. (2019b). Monocytes behaviour under perfusion conditions for development of granuloma on-a-chip. In *Rhodes, Greece: TERMIS EU 2019 conference book. (p. 1356).*

Savla, U. O. (2004). Mathematical modelling of airway epithelial wound closure during cyclic mechanical strain. *Journal of Applied Physiology*, 566–574.

Sharifi, F., et al. (2019). A foreign body response-on-a-chip platform. *Advanced Healthcare Materials*, *8*(4). https://doi.org/10.1002/adhm.201801425.

Wheeler, B. (2008). Building a brain on a chip. In *Conference proceedings: Annual international conference of the IEEE engineering in medicine and biology society*.

Yang, X. L. (2018). Nanofiber membrane supported lung-on-a-chip microdevices for anti-cancer drug testing. *Lab on a Chip*, 486–495.

CHAPTER

8

Chaotic mixing and its role in enhancing particle deposition in the pulmonary acinus: A review

Akira Tsuda[a,*] *and Frank S. Henry*[b]

[a]Tsuda Lung Research, Shrewsbury, MA, United States [b]Department of Mechanical Engineering, Manhattan College, Riverdale, NY, United States

1 Introduction

Oxygen is a key component in the production of energy within our cells, and the primary purpose of the lung is to ensure a constant supply of the gas (Butler & Tsuda, 2011; Tsuda, Henry, & Butler, 2008). Oxygen-rich air is drawn into the lung by the diaphragm and intercostal muscles, and carbon dioxide and other gasses are exhaled with the outgoing air when the muscles relax. The inhalation/exhalation rhythmic, motion occurs approximately 12 times a minute (Weibel, 1984).

The anatomy of the lung may be divided into three regions: upper airways (nasal pharynx area), conducting airways (trachea to terminal bronchioles), and the pulmonary acinus (respiratory bronchioles to terminal alveolar ducts). Each region has its own unique anatomy and flow regime (West, 2012). Despite these differences in anatomy and flow type, fine particles entering the mouth or nose can travel to, and deposit in, the lung periphery.

The deposition of particles onto the alveolar surface can be considered a two-step process. First, the particle has to travel with the ambient air through a network of ducts and end up close to the alveolar surface. Second, if the particle is close enough, short-distance forces acting on the particle (e.g., the van der Waals force, electrostatic force, Brownian force, etc. [Friedlander, 1977]) will be sufficient to bring it to the surface.

Not all particles reach the alveolar surface. In the lung physiology literature (e.g., Oberdörster et al., 2007; Tsuda, 1998a, 1998b; West, 2012), three mechanisms are typically

*Retired from Harvard University, Boston, MA, United States.

https://doi.org/10.1016/B978-0-12-823956-8.00008-0

defined as contributing to particle deposition. These are inertial impaction, gravitational sedimentation, and Brownian motion. This first two, impaction and sedimentation, tend to prevent larger particles reaching the alveolar region.

Inertial impaction: describes the situation in which a particle with relatively large mass (typically a relatively large particle, since mass is proportional to the cube of the particle's diameter) cannot follow the curvilinear airflow patterns faithfully; and as a result, it deviates from the airflow streamlines and the particle's own inertia carries it to the surface (Friedlander, 1977; Tsuda et al., 1994a). This phenomenon is significant when airflow velocity (U) is large, and thus it occurs predominantly in the upper/large airways. A particle's inertia is considered significant when the Stokes number $Stk > 1$. The Stokes number is expressed as $Stk = \rho_p d^2 U / 18\ \eta L$, where ρ_p is the particle density, d is the particle diameter, η is the air viscosity, and L is the characteristic length scale.

Sedimentation: particles with large mass are also subject to the external gravitational force, which makes the particles deposit in the direction of gravity (Haber et al., 2003; Kojic & Tsuda, 2004; Tsuda et al., 1994a, 2013). This phenomenon becomes significant when the particle sedimentation velocity (expressed in terms of the terminal velocity $v_s = \rho_p d^2 g / 18\ \eta$, where g = gravitational acceleration) becomes comparable to, or greater than, the airflow velocity. Deposition by sedimentation occurs primarily in the large airways and at the beginning of the acinus.

Brownian motion: the potential for particles to cross flow streamlines and deposit due to Brownian motion is characterized by the Péclet number, $Pe = UL/D$. A balance between thermal effects and viscous drag exerted on the particle determines the magnitude of the diffusivity, D. Particles with small Péclet numbers are more likely to deposit due to Brownian motion. Particles of very small size (diameters $<0.005\,\mu m$) may deposit in the upper/large airways because those small particles have an extremely high diffusivity and low Péclet numbers. Particles with small mass, which can follow the curvilinear airflow patterns with little inertia/gravitational effects, can enter the pulmonary acinus with the airflow. In this review, we focus on the deposition of particles in the pulmonary acinus, which occupies more than 95% of the lungs in volume (Weibel, 1984). We concentrate on particles in the diameter range 0.005–0.5 μm because particles in this range have been found to deposit preferentially in the acinus (Tsuda et al., 2013). Such particles have diffusivities that are small enough to prevent them from depositing before reaching the acinus but are also light enough to exclude deposition through inertial impaction or sedimentation (Tsuda et al., 1994b).

2 Classical view

Small particles ($<2.5\,\mu m$) are transported (convected) by the inhaled air to the alveolar region, and the particle-laden air mixes with the residual gas in the acinus (Tsuda et al., 2013). Thus, acinar airflow patterns and the mixing of fresh air with residual gas are very important for particle deposition. The momentum equation for flow of a Newtonian fluid with constant properties and negligible body forces can be written in Cartesian tensor notation (Cengel & Cimbala, 2018) as

$$\frac{\partial u_i}{\partial t} + \frac{\partial u_j u_i}{\partial x_j} = -\frac{1}{\rho}\frac{\partial p}{\partial x_i} + \nu \frac{\partial^2 u_i}{\partial {x_j}^2} \tag{1}$$

If we define dimensionless variables as $t' = t/T$, $u_i' = u_i/U$, $x_i' = x_i/L$, and $p' = p/\rho U^2$, where T, U, and L are characteristic time, velocity, and length, respectively, then Eq. (1) can be written

$$St\frac{\partial u_i'}{\partial t'} + \frac{\partial u_j' u_i'}{\partial x_j'} = -\frac{\partial p'}{\partial x_i'} + \frac{1}{\mathrm{Re}}\frac{\partial^2 u_i'}{\partial x_j'^2} \tag{2}$$

where $St = L/UT$ is the Strouhal number, and $Re = UL/\nu$ is the Reynolds number.

An alternate form of Eq. (2), for oscillatory flow, is gained by multiplying Eq. (2) by Re and introducing the Womersley number, $\alpha = \sqrt{2\pi L^2/T\nu}$; i.e.,

$$\frac{\alpha^2}{2\pi}\frac{\partial u_i'}{\partial t'} + \mathrm{Re}\frac{\partial u_j' u_i'}{\partial x_j'} = -\mathrm{Re}\frac{\partial p'}{\partial x_i'} + \frac{\partial^2 u_i'}{\partial x_j'^2} \tag{3}$$

We note that $\alpha^2 = 2\pi St Re$. Also, now, T is the period of oscillation.

In a flow where the Reynolds number is much less than one, the convective terms of Eq. (3) are negligible compared to the other terms. While, the pressure gradient in Eq. (2) is also multiplied by the Reynolds number, it has to be assumed that this term is not negligible as it is the pressure gradient that is driving the flow. In a flow where the Reynolds number is much less than one, the Womersley number is also likely to be less than one (as $\alpha \propto Re$), and hence Eq. (3) can be reduced to

$$\mathrm{Re}\frac{\partial p'}{\partial x_i'} = \frac{\partial^2 u_i'}{\partial x_j' 2} \tag{4}$$

In flows where Eq. (4) applies, the direction of the flow is not discernable from inspection of the streamlines; e.g., the upstream streamline pattern for flow over a cylinder would be a mirror image of that of the downstream streamline pattern. This is called a kinematically reversible flow (Taylor, 1960). In terms of our cylinder example, it would not be possible to tell if the flow was traveling from left to right or vice versa.

It used to be considered that flow in the lung periphery was perfectly kinematically reversible (Davies, 1972). From this assumption, it followed that particles traveling along streamlines would not deposit (Davies, 1972): they would enter alveoli over inspiration and leave over expiration. For particles to deviate from streamlines in a kinematically reversible flow, they must exhibit some intrinsic motion resulting from; for instance, gravitational or diffusional forces. However, in 1988, Heyder et al. showed that the experimental data did not conform to the idea of perfectly kinematically reversible flow in the lung periphery; they found that the half-width of the expired bolus was a linear function of the volume to which the bolus penetrated at volumetric penetrations of 100–800 cm^3 (Fig. 1). Their results suggest that convective mixing is not confined to central airways but can also occur in the lung periphery.

If the flow is not kinematically reversible, the convection terms in Eq. (3) must have some effect on the resulting flow. The Reynolds number at the entrance to the respiratory region of the lung is of the order of 1.0. The Reynolds number is a measure of the importance of inertial

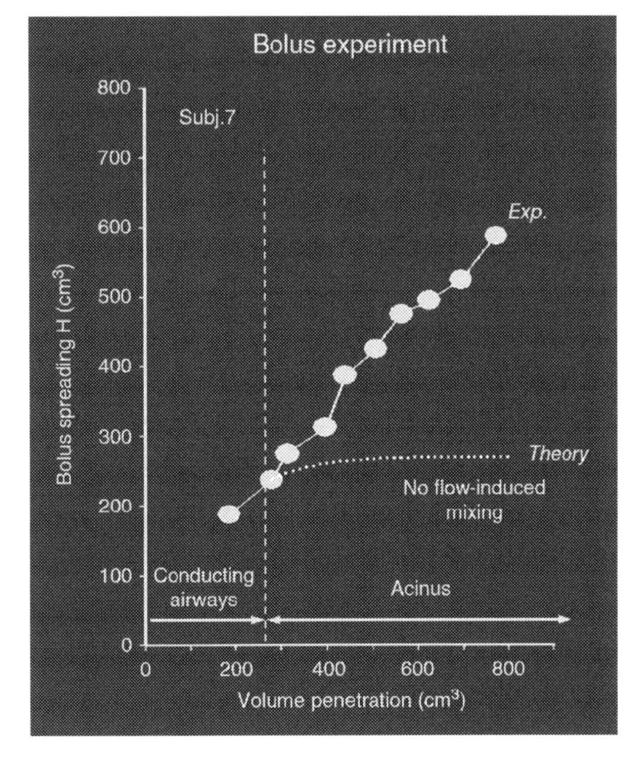

FIG. 1 Spreading of a bolus of monodisperse aerosols vs depth of volume to which the bolus penetrated (Heyder et al., 1988). The figure demonstrates, experimentally, the presence of kinematical irreversibility in the acinus. *Adapted from Tsuda, A., & Henry, F. S., & Butler, J. P. (2013). Particle transport and deposition: Basic physics of particle kinetics.* Comprehensive Physiology, 3(4), *1437–1471.*

transport (convection) over diffusional transport. Thus, in the entrance region of the lung, the flow inertia, while small, cannot be ignored. We note that the Womersley number of the flow in the lung periphery is of the order of 0.1; and hence, the flow in this region of the lung can be considered quasi steady.

In 1995, Tsuda et al. showed that the nonzero inertia in the flow in the respiratory region of the lung probably produces chaotic, or convective mixing, and it is believed that such mixing is the mechanism responsible for the experimental findings of Heyder et al. Tsuda et al. (1995) showed that the path lines of fluid particles, expressed as $x_i = \int u_i dt$, can be highly complex in alveolar flow. They intentionally built a very simple alveolated duct model; simple but possessing the basic features of an acinar duct; namely, a thoroughfare central channel surrounded by dead-end side pockets. The model was intentionally designed to be simple to highlight the fact that, contrary to classical theory, kinematically irreversible flow could be generated by a geometric model that expands and contracts in a self-similar, kinematically reversible, fashion (Fig. 2).

3 Computational simulation and Hamiltonian chaos

The origin of kinematical irreversibility in the alveoli can be explained by the concept of Hamiltonian chaos (Henry et al., 2009; Tabor, 1989; Tsuda et al., 2011). Basically, this is

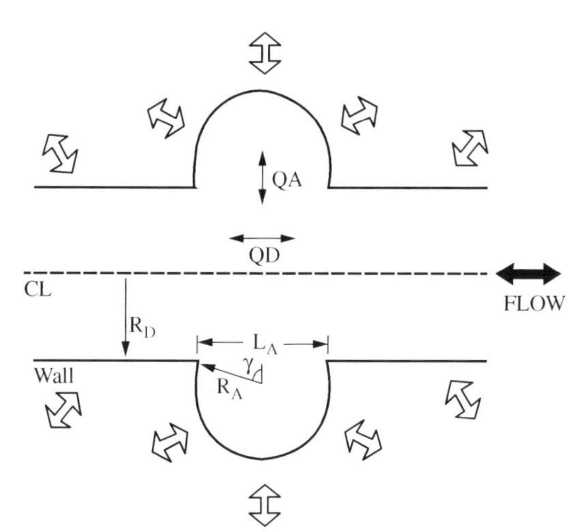

FIG. 2 Rhythmically expanding and contracting alveolated duct model consisting of circular channel surrounded by torus. Q_A, alveolar flow; Q_D, ductal flow; CL, center line; RD, channel radius; RA, alveolar radius; LA, alveolar opening size; γ, opening half angle. *Adapted from Tsuda, A., Henry, F. S., & Butler, J. P. (1995). Chaotic mixing of alveolated duct flow in rhythmically expanding pulmonary acinus.* Journal of Applied Physiology, 79(3), *1055–1063.*

due to an interplay between two mechanisms operating in the system; the central channel flow and rotation flows inside of the alveolus. Both are cyclic (Fig. 3). The frequency of the central channel flow is externally determined by intercostal muscles and diaphragm cyclic motion (controlled by the central nervous system). This frequency represents the breathing frequency (f_{br}). Rotational flows inside of the alveolus are intrinsic to the lung. Fluid in the alveolus (dead-end side pocket) is rotated by the viscous shear force imposed on the alveolar fluid by central channel flow as it passes along an alveolar opening. How strongly and how quickly the alveolar fluid rotates is determined by a balance between shear force exerted by the central channel flow along the alveolar opening and the alveolar wall drag force, which is acting in the opposite directions to the rotation of the alveolar flow. As a result, fluid near the alveolar wall moves more slowly due to the drag force exerted by the walls, while the fluid at the center of the alveolus moves faster due to being further from the effect of the wall (Fig. 4). In other words, the frequency of alveolar recirculation flows (f_{alv}) is a family of frequencies—not one fixed number—depending on geometric factors such as the alveolar cavity shape and the size of the alveolar opening.

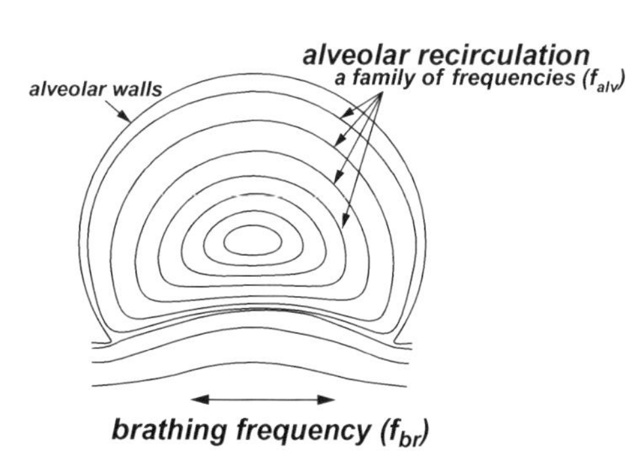

FIG. 3 Representative alveolated duct flow. The frequency of the cyclic central channel flow represents breathing frequency (f_{br}), while the frequency of alveolar recirculation flows (f_{alv}) is a family of frequencies. *Adapted from Tsuda, A., Laine-Pearson, F. E., & Hydon, P. E. (2011). Why chaotic mixing of particles is inevitable in the deep lung.* Journal of Theoretical Biology, 286, 57–66.

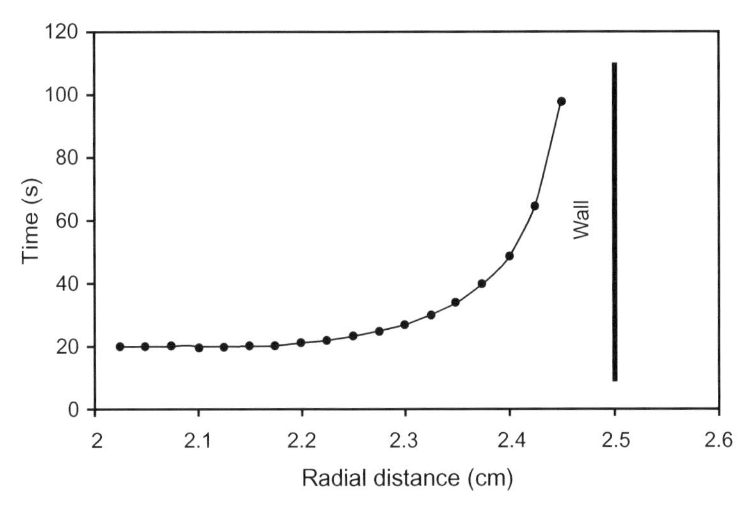

FIG. 4 Time taken by a fluid element to make one rotation in a model alveolus. The fluid near the alveolar wall moves more slowly due to the drag force exerted by the walls, while the fluid far from the walls (at the center of the alveolus) moves faster. *Adapted from Karl, A., Henry, F. S., & Tsuda, A. (2004). Low Reynolds number viscous flow in an alveolated duct.* ASME Journal of Biomechanical Engineering, 126(1), *13–19.*

Particles in the alveolus are carried passively with the rotating flow around the alveolar cavity and these paths are called as recirculation orbits. The frequency of particle rotation depends on which orbit the particle resides. The motion of a particle depends on the ratio f_{alv}/f_{br}; and this ratio dictates whether or not a particle will follow a chaotic path. When the frequencies f_{alv} and f_{br} resonate (i.e., when the ratio f_{alv}/f_{br} is a rational number), the interaction between the frequencies produces a net drift if the system is perturbed. This drift results in Hamiltonian chaos, which produces qualitative changes in those trajectories (Tsuda et al., 2011). Phrased differently, if the system is not perturbed (e.g., the Reynold number is strictly zero), passive particles in the alveoli simply recirculate. In this case, resonance does not produce a qualitative change in particle trajectories: each particle moves back and forth on a single (closed) path forever.

In the lung, the unperturbed situation cannot occur for a number of reasons. For one, *Re* is never exactly zero, as long as air moves (Henry et al., 2002, 2009; Tsuda et al., 1995); for another, the alveolar walls move, as long as we live (i.e., blood has to be perfused to avoid necrosis) (Henry et al., 2002; Laine-Pearson & Hydon, 2006; Tsuda et al., 1995); and lastly, small but nonzero geometric hysteresis may exist (Butler & Tsuda, 2005; Haber et al., 2000; Haber & Tsuda, 2006; Tsuda et al., 1999).

The explanation of the mathematical theory of 'perturbed Hamiltonian dynamical systems' is rather involved, and is beyond the scope of this review. Interested readers are encouraged to read the paper by Tsuda et al. (2011).

The important point here is that f_{alv} is a family of frequencies, the magnitude of which increases as the distance from the wall increases (Fig. 4). Hence, in areas of the acinus where the alveolar flow rotates, it is theoretically guaranteed that there are series of orbits with f_{alv} that resonate with f_{br}. Indeed, Henry et al. (2009) showed that chaotic mixing occurs inside the

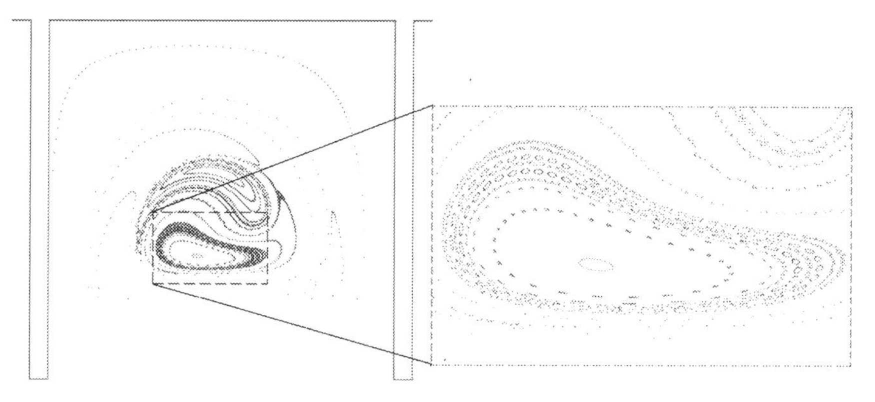

FIG. 5 Poincaré sections for a flow with $Re = 1.0$ and $a = 0.096$ ($T = 3\,s$). Notice chains of islands surrounded by a sea of chaos. Colors have no significance other than to differentiate between individual orbits. *Adapted from Henry, F. S., Laine-Pearson, F. E., & Tsuda, A. (2009). Hamiltonian chaos in a model alveolus.* ASME Journal of Biomechanical Engineering 131, *011006(1)–011006(7).*

alveolar space as long as the flow exhibits recirculation, even if the walls do not move. That is, a small but nonzero Reynolds number was enough to perturb the flow and produce chaos (Fig. 5).

The strength of chaotic mixing is dependent on the degree of perturbation. For instance, alveoli near the entrance of the acinus create stronger chaotic mixing than those deeper in the acinus due to larger Reynolds number. If alveolar expansion is more or less constant regardless of location along the acinar tree (Weibel, 1984), then the ratio between a flow entering into the alveolus (Q_A) and a flow passing by the same alveolus (Q_D) is inversely proportional to *Re*. Since this ratio Q_A/Q_D is considered as one of the important fluid mechanics parameters to uniquely describe alveolar flow, the ratio Q_A/Q_D, instead of *Re* (Tsuda et al., 1995), is often used in the discussion of chaotic mixing in the pulmonary acinus.

4 Proof of the occurrence of chaotic mixing in vivo

The presence of chaotic mixing in the pulmonary acinus was demonstrated experimentally using excised rat lungs (Tsuda et al., 2002). Briefly, using polymerizable viscous fluids of two colors (white and blue), Tsuda et al. (2002) studied the mixing phenomena deep in the lungs. They first filled the lungs with white fluid representing alveolar residual air, then they ventilated the lungs with blue fluid as a tidal fluid with physiologically relevant ventilatory conditions (e.g., Reynolds number and Womersley parameter were matched to physiological conditions). After letting the two-color fluids polymerize, they studied blue-white mixing patterns. First, at the end of the first inspiration (Fig. 6), most of the large, medium size, and alveolar airways were entirely filled with tidal (blue) fluid with no sign of significant mixing in the acini, but many alveoli (i.e., dead-end pockets) exhibit circular patterns, showing that the fluid was indeed recirculating inside the alveoli. After only one breathing cycle (Fig. 7), remarkably complex stirring patterns emerged on transverse cross sections of the

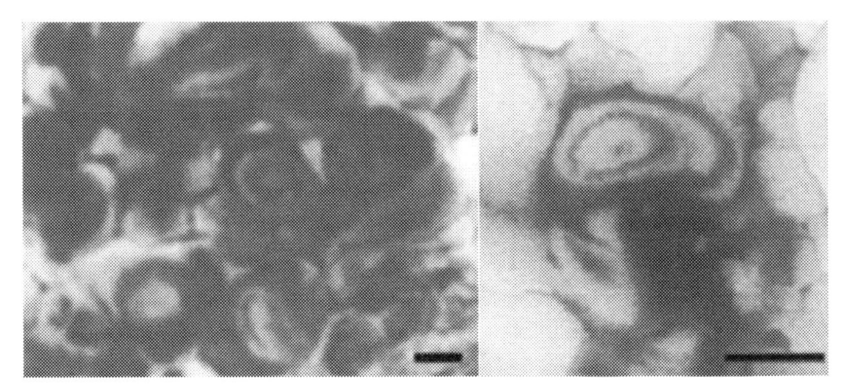

FIG. 6 Alveolar recirculation. In many alveoli, circular blue/white color (gray/white in print version) patterns were observed at $N = ½$. (Bar = 100 μm). *Adapted from Tsuda, A., Rogers, R. A., Hydon, P. E., & Butler, J. P. (2002). Chaotic mixing deep in the lung.* Proceedings of the National Academy of Science of the United States of America, 99, *10173–10178.*

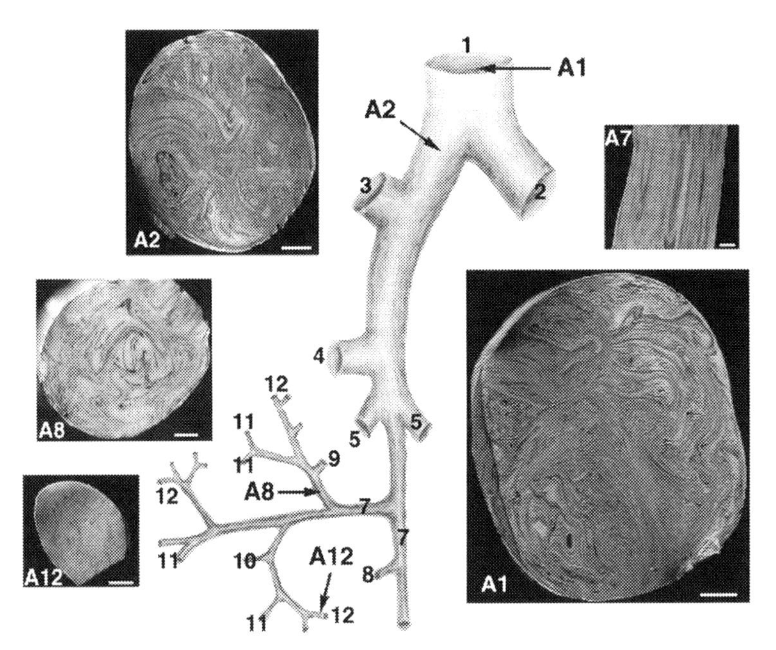

FIG. 7 Typical mixing patterns observed on airway cross sections at different locations in the tracheobronchial tree after one ventilatory cycle. The entire transverse cross section of the trachea (A1) was filled with myriad extremely fine blue-white (gray/white in print version) striations, which formed convoluted swirling patterns. Similar patterns were seen on the transverse cross sections of the main stem bronchi (A2). This pattern was consistently seen through the eighth (A8), and even up to the 12th-generation (A12) airways (counted from the trachea). In contrast, the longitudinal airway sections showed much simpler patterns, displaying fine laminae of blue and white striations (see A7 for example; also tested at several other locations). The fact that there were very few complex patterns observed on the longitudinal sections suggests that inertia-based secondary flows such as turbulent eddies were not generated in this experiment. Images bar = 500 μm (A1), 500 μm (A2), 200 μm (A8), 100 μm (A12), 100 μm (A7). Representative of five rat lungs analyzed. *Adapted from Tsuda, A., Rogers, R. A., Hydon, P. E., & Butler, J. P. (2002). Chaotic mixing deep in the lung.* Proceedings of the National Academy of Science of the United States of America, 99, *10173–10178.*

airways. We performed a box-counting analysis (Bassingthwaighte et al., 1994) on transverse cross sections of the airways and found a nearly linear relationship between the overall mean intensity and box size with a slope of about −0.1, showing that the color pattern is indeed fractal with a fractal dimension $D = 1.1$. The facts that (1) the fractal dimension $D = 1.1$ is invariant throughout the airways from the trachea down to the 12–13th generation (the most distal airways we examined for this analysis) and (2) the front of the tidal fluid represented by the interface of the blue and white colors is enormously stretched by the end of the inspiration sampling millions of alveolar spaces indicate that the observed fractal pattern is a result of chaotic mixing occurring deeper than the 12–13th generation, most likely in the pulmonary acinus.

To quantify the extent of mixing in the acini, we examined the time evolution of mixing patterns on the transverse cross sections of acinar airways (approximately 200 μm in diameter) at the end of each breathing cycle for four breathing cycles (Fig. 8). After the first cycle ($N = 1$, Fig. 8), most of the acinar airways appeared predominantly white, with microscopic

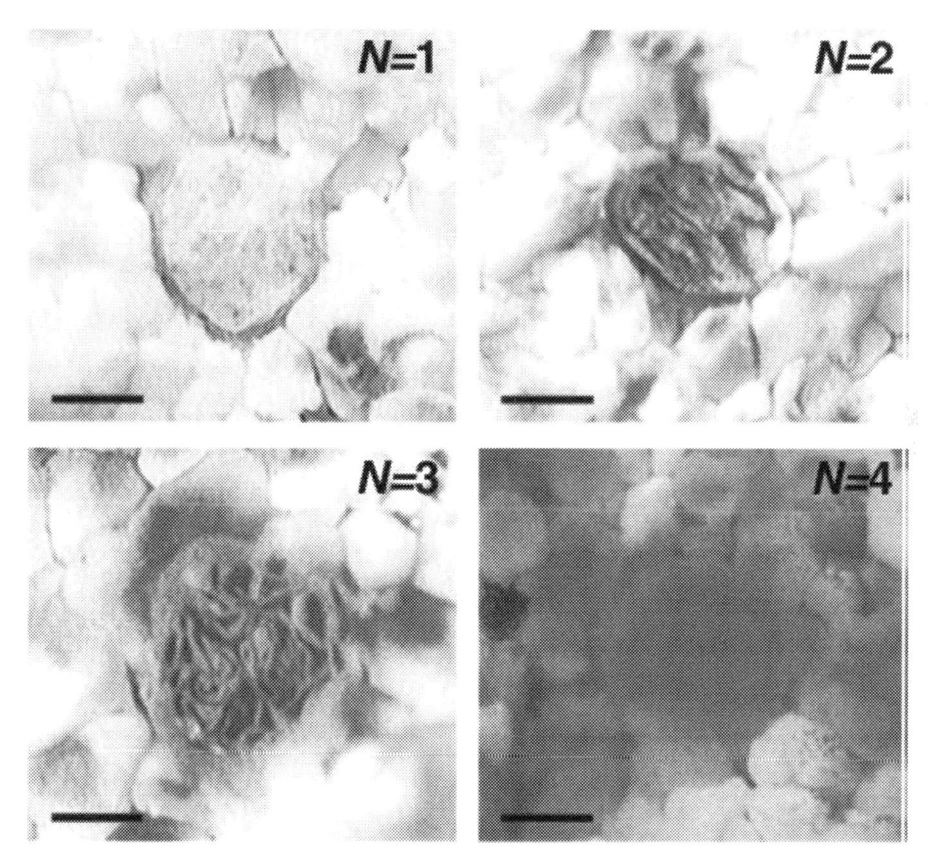

FIG. 8 Typical mixing pattern of two colors observed in approximately 200 μm acinar airways of adult rats after ventilatory cycles of $N = 1$, 2, 3, and 4 (Bars = 100 μm). *Adapted, with permission, from Tsuda, A., Rogers, R. A., Hydon, P. E., & Butler, J. P. (2002). Chaotic mixing deep in the lung.* Proceedings of the National Academy of Science of the United States of America, 99, *10173–10178.*

traces of blue. After the second or the third cycle ($N = 2, 3$, Fig. 8), however, a large amount of tidal (blue) fluid appeared on the cross-sectional images, indicating that substantial net axial transport had occurred along the bronchial-acinar tree. The cross-sectional images of acinar airways showed clearly delineated interface patterns with both blue and white fluids being stretched and folded (discussed below). After the fourth cycle ($N = 4$, Fig. 8), the clarity of the interface patterns had largely disappeared, and the blue and white patterns changed into smeared (mixed) bluish-white uniformity.

These findings suggest that there may be a specific cycle-by-cycle folding factor f, characteristic of the acinar duct structure. To determine the cycle-by-cycle folding factor f, we analyzed the cross-sectional images of randomly selected acinar airways (approximately 200 μm diameter) for each of $N = 1, 2, 3$, and 4 (from 12 animals; 3 animals for each cycle number). For each image, we computed the characteristic distance between neighboring blue and white interfaces defined as the harmonic mean wavelength, l_w, obtained by two-dimensional (2D) spectral analysis. The time evolution of l_w, normalized by airway diameter, d, showed a sharp decrease from $N = 1$ to $N = 3$, reflecting the decreasing lateral length scales associated with folding. This was followed by a sudden increase in l_w/d at $N = 4$, caused by the diffusive loss of high-frequency components in the pattern. To determine the folding factor $f$in rat acini, we developed a simple convection-diffusion mathematical model with a parameterized folding factor and fit this to the data. The model represents a simple evolving sine-wave convective pattern, whose wavelength is divided by a folding factor f for each breath cycle, and whose amplitude is allowed to smooth out by diffusion. We found that a best fit to the data was obtained with a folding factor $f = 2.3$. The significance of this finding is discussed below.

The temporal evolution of convection patterns observed in our study fundamentally differs from the one predicted by the classical theory based on kinematically reversible fluid flow. To illustrate this difference, consider the following two systems. One is kinematically reversible (Fig. 9, left) and the other has irreversible stretch and fold convection (Fig. 9, right). We introduce a Brownian tracer into both systems and track the evolution of diffusive and convective length scales in both systems. In the kinematically reversible system (Fig. 9, left), there is no net convective transport; mixing is therefore characterized by a diffusion distance δ, which increases slowly with time t, that is, $\delta \propto \sqrt{Dt}$, where D is the tracer diffusivity. Significant mixing only occurs when δ becomes of the order of L, the fixed system size, which in our case is a typical alveolar dimension of a few hundred microns. For fine aerosol particles, this process would be very slow because such particles have low diffusivities. By contrast, in the system with stretch and fold convection, diffusion and convection interact.

The diffusion length scale δ initially increases as $\sqrt{Dt}$ but asymptotically approaches $\sqrt{D/\alpha}$, where α is the stretching rate. Importantly, however, the length scale over which diffusion must operate to effect mixing is no longer fixed at the system size L, but, because of convective folding, decreases exponentially with cycle number N. This can be expressed as f^{-N}, where f is the characteristic cycle-by-cycle folding factor. In this interaction, mixing is initially very slow, but suddenly increases after a few cycles when the rapidly decreasing folded scales Lf^{-N} become comparable to the asymptotically constant value of diffusion length scale δ (Fig. 9, bottom right). This phenomenon—a sudden increase in mixing that can be described by an equivalent entropy burst (Butler & Tsuda, 1997)—is a characteristic feature of chaotic mixing (Metcalfe, 2010; Ottino, 1989; Ottino et al., 1988) and is quantified by the folding factor f. We showed above that in rat acinar airways with diameter of

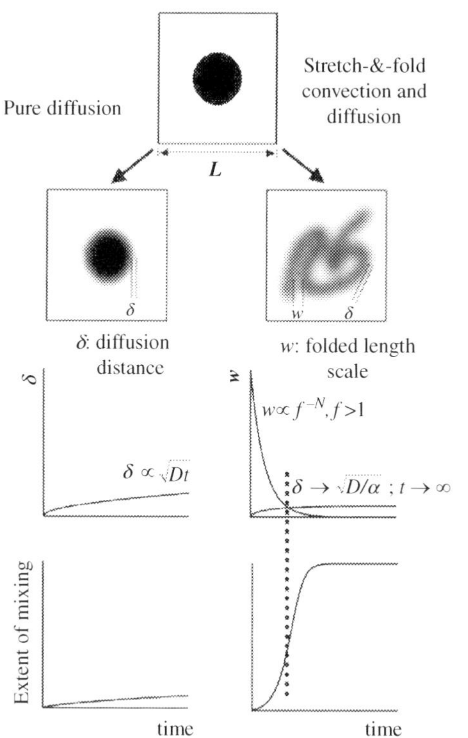

FIG. 9 Comparison of Brownian tracer mixing between a system with pure diffusion and a system with stretch and fold convection and diffusion. (Top) Schematic view of these two systems. (Middle left) The slowly increasing length scale for mixing (δ) in pure diffusion. (Middle right) With stretch and fold convection, δ also increases slowly (but approaches an asymptotic value); by contrast, the folding length scale, w, decreases exponentially rapidly. (Bottom) Representation of the evolving extent of mixing corresponding to diffusion alone and to diffusion coupled with stretch and fold convection. At the time when the two length scales are comparable (vertical dotted line), there is sharp jump in mixing (entropy burst). *Adapted from Tsuda, A., Rogers, R. A., Hydon, P. E., & Butler, J. P. (2002). Chaotic mixing deep in the lung.* Proceedings of the National Academy of Science of the United States of America, 99, *10173–10178.*

approximately 200 μm, the folding factor f was about 2.3, which means that the lateral length scales over which the complexity of the convective flow patterns is evolving are decreased by more than half at every breath. Equivalently, the complexity of the pattern itself more than doubles each breath. This is an exponentiating phenomenon, which in consequence implies that only a modest number of breaths are required to ensure that these mixing lengths become sufficiently small that true diffusive and irreversible mixing can take place, even for aerosols with very low diffusivities.

5 Effects of acinar chaotic mixing on inhaled particle deposition

One way to test whether chaotic mixing is indeed crucial for the deposition of submicron particles in the pulmonary acinus is to compare the deposition in the case of the presence of acinar chaotic mixing against a case without such mixing (Semmler-Behnke et al., 2012). As shown in Fig. 10, there are striking differences in the geometry of the lung parenchyma in the developing lungs. Rats at postnatal day 4 (P4) have largely saccular airways with no visible alveoli (Fig. 10A and D), but at postnatal day 21 (P21) the acini show marked alveolation (Fig. 10B and E) with fully shaped, although somewhat smaller, alveoli compared with the adult animal (Fig. 10C and F). The size of the airspace is minimal at ~21 days in the rat, when bulk alveolation and septal thinning have just completed.

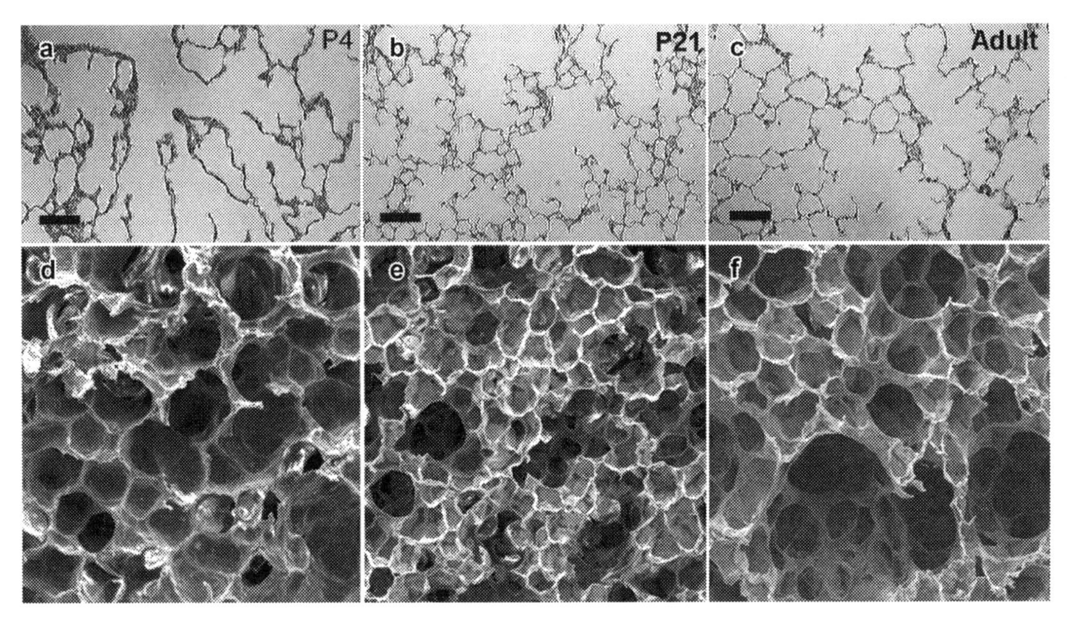

FIG. 10 The lung parenchyma of the developing lungs in rats at various time points. At 4 days of age in rats, the airways are mostly saccular with smooth-walled primary septa with no visible alveoli (A and D). Acinar structure after the bulk alveolation phase (at 21 days of age in rats), the acinus becomes markedly alveolated, filled with numerous alveoli with a shape similar to, but smaller than, fully developed matured alveoli (B and E). Acinar structure of fully developed adult lungs (at ≥90 days of age in rats) (C and F). (Upper) H&E staining. (Scale bar, 100 μm.) (Lower) SEM imaging, 400 ×. [Burri noted that there is no noticeable difference in appearance between human immature parenchyma vs rodent parenchyma, except their size (Burri, 1985)]. *Adapted from Semmler-Behnke, M., Kreyling, W. G., Schulz, H., Takenaka, S., Butler, J. P., Henry, F. S., Tsuda, A. (2012). Nanoparticle delivery in infant lungs.* Proceedings of the National Academy of Science of the United States of America, 109(13), *5092–5097.*

Based on this morphological information, we built an alveolar duct model (Fig. 11). In the case of very immature lungs (e.g., 0–~½ year old human infants, postnatal 7 days old or earlier rodent babies), alveoli were too shallow (Fig. 11, left) for alveolar recirculation to occur (Fig. 12, left); but in the case of slightly older but still immature lungs (e.g., ~½–~2 years

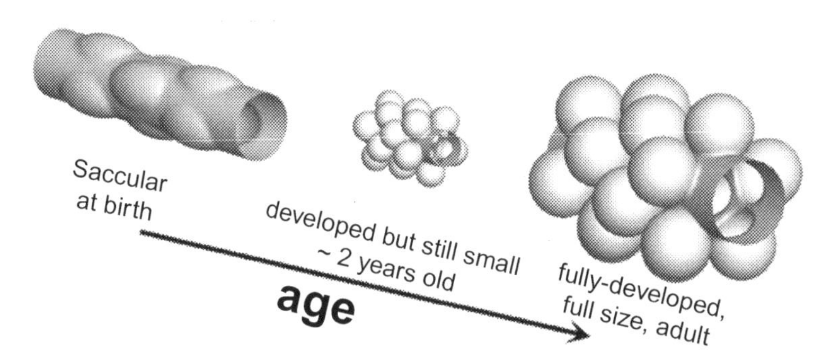

FIG. 11 An alveolar duct model of the developing lungs at various time points. *Adapted from Semmler-Behnke, M., Kreyling, W. G., Schulz, H., Takenaka, S., Butler, J. P., Henry, F. S., Tsuda, A. (2012). Nanoparticle delivery in infant lungs.* Proceedings of the National Academy of Science of the United States of America, 109(13), *5092–5097.*

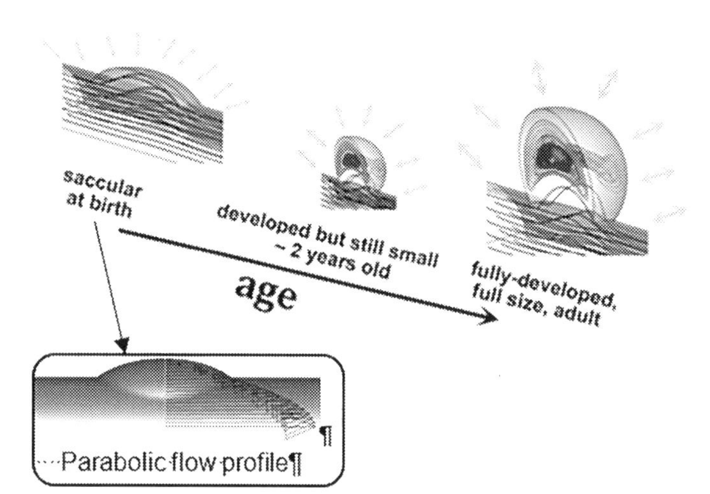

FIG. 12 Flow patterns in the alveolar model of the developing lungs. Breathing patterns at various ages were also measured and use for the flow simulation. In awake, spontaneously breathing infant rats the tidal volume (VT), breathing frequency (f), and minute ventilation (MV = VT f) were found to scale allometrically with body weight (BW) to exponents of 1.06, −0.12, and 0.91, respectively. *Adapted from Semmler-Behnke, M., Kreyling, W. G., Schulz, H., Takenaka, S., Butler, J. P., Henry, F. S., Tsuda, A. (2012). Nanoparticle delivery in infant lungs.* Proceedings of the National Academy of Science of the United States of America, 109(13), *5092–5097.*

old human infants, postnatal from 1 week to 3 weeks rodent babies), alveoli were sufficiently deep (Fig. 11, middle) to promote recirculation flow inside to the alveoli (Fig. 12, middle). In the final stage of lung development (from human toddler, ~2 years old, to adult human, or postnatal rats older than 3 weeks to fully matured adult rats), the shape of acinus does not change but the size increases. Similar to the 2 years old case, alveoli are deep enough (Fig. 11, right) to allow recirculation flow to occur inside to the alveolus (Fig. 12, right).

Exposing P7, P14, P21, P35 immature Wistar-Kyoto rats and fully matured adult (P90<) rats to insoluble, radioactively labeled iridium (^{192}Ir) particles of 20 and 80 nm, we measured total deposition (Fig. 12). We found that total deposition strongly depended on the age of the rats. Though the deposition was generally higher for 20-nm particles compared with 80-nm particles, due to their relatively higher intrinsic diffusivity, the deposition of both 20- and 80-nm particles peaked in 21-days-old rats. At this age, the acini have just completed structural alveolation and septal thinning, and their shapes approximate those of the adult animal, although smaller in size (Fig. 10B and E vs Fig. 10C and F). Thus, the alveoli at this age may be sufficiently deep to promote rotational flows, and consequent chaotic mixing, as was suggested by the corresponding alveolar shapes in our computational studies (Fig. 12, middle). This finding supports the idea that age-dependent changes in acinar fluid mechanics do indeed play a critical role in determining the fate of inhaled particles (Fig. 13). In animals younger than 21 days, the acini have few, relatively shallow, alveoli. Airflows in such acini are shown in computational studies to be simpler and reversible, with a tongue-like Poiseuille flow pattern (Fig. 12, left; and inset). In addition to less-effective diffusional deposition due to the relatively large airspaces in animals younger than 21 days, the absence of chaotic mixing results in less flow-induced mixing. This finding is consistent with our observation of lower deposition fraction in animals younger than 21 days compared with older animals (Fig. 13).

In rats of 21 days and older, the acini are fully alveolated, and therefore their airflow patterns are likely rotational and chaotic (Fig. 12, right). These chaotic flow patterns, as noted above, are important in enhancing deposition, but the size of each airspace, through which a particle is transported for deposition, also plays a role in determining the magnitude of deposition. Because alveolar size tends to increase after age 21 days (Burri, 1985), the deposition efficiency

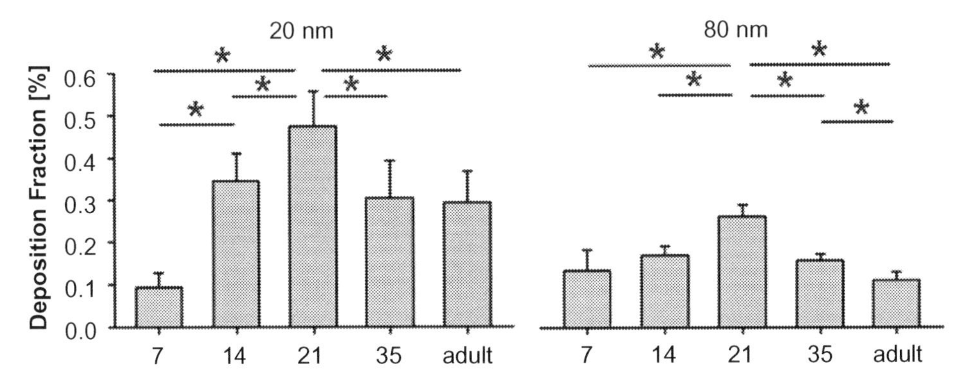

FIG. 13 Total deposition fraction vs rat age measured immediately after exposure for inhaled 20- and 80-nm particles. Deposition of (left) 20- and (right) 80-nm particles peaks in 21-day-old rats, whose acini are largely alveolated already but are still small in size ($n = 8$ for each age group). $n = 16$ for age group; $^*P < 0.05$; error bars indicate ± SD. Note that the five chosen age groups represent distinct stages of postnatal acinar structural development (Burri, 1985). *Adapted from Semmler-Behnke, M., Kreyling, W. G., Schulz, H., Takenaka, S., Butler, J. P., Henry, F. S., Tsuda, A. (2012. Nanoparticle delivery in infant lungs.* Proceedings of the National Academy of Science of the United States of America, 109(13), *5092–5097.*

is expected to decrease after that age. Overall, therefore, it appears that the geometry of the 21-days-old rat's acinar structure, with its small but fully formed alveoli, is an optimal combination for efficient deposition of submicron particles, and results in a peak deposition (Fig. 13).

In a numerical study, Henry and Tsuda (2016) considered the effect of chaotic mixing on nanoparticle deposition in the pulmonary acinus by comparing the deposition based on the full solution of the flow in a model alveolus to that of a case in which the alveolar flow is set to zero. Henry and Tsuda (2016) showed (Fig. 14) that deposition rates for all three particle sizes considered (20, 50, and 80nm) were much higher in the presence of alveolar recirculation

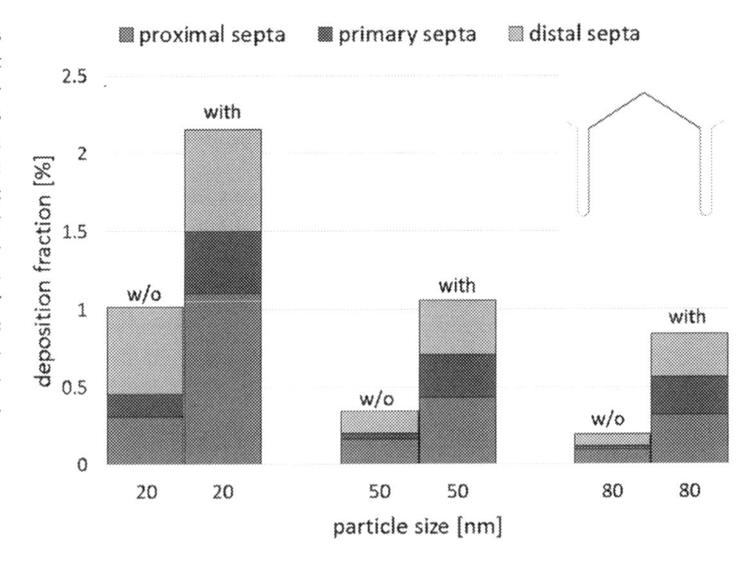

FIG. 14 Percentage of particles entering the alveolus that deposit on various regions of the septal surface in a fully developed alveolus (day 21) at generation 0 with and without (w/o) alveolar flow. Blue: deposition on the proximal septa; red: deposition on the primary septa; green: deposition on the distal septa. *Adapted from Henry, F. S., Tsuda, A. (2016). Onset of alveolar recirculation in the developing lungs and its consequence on nanoparticle deposition in the pulmonary acinus.* Journal of Applied Physiology, 120, *38–54.*

compared to the case in which the alveolar flow was set to zero. Further, the difference was shown to be more significant for the less diffusive particles. In the case of no alveolar flow, deposition in the alveolus is solely due to diffusion; and thus, the differences in the deposition rates in three different size particles roughly follow the differences in their diffusivity. Conversely, in the case with alveolar recirculation, deposition in the alveolus appeared to be greatly enhanced by convective mixing, since the differences in the deposition rates between the three particle sizes did not correspond to the differences in their diffusivity.

6 Physiological significance

Prior to 1988, the possibility of convective mixing in the pulmonary acinus was theoretically excluded (Davies, 1972). However, this misconception was exposed by the experiment of Heyder et al. (1988), suggesting that convective mixing is not confined to central airways but can also occur in the lung periphery. Their work was seminal because it demonstrates convective mixing in the pulmonary acinus experimentally. However, the mechanism behind this experimental observation was not known at that time; and an explanation had to await until our proposal of chaotic mixing (Tsuda et al., 1995). The essence of chaotic mixing is as follows. If the system has two (or more) independent oscillating subsystems interplaying with each other, the resonant orbits have a possibility of breaking into chaos when a perturbation is introduced (Henry et al., 2009; Tabor, 1989; Tsuda et al., 2011). The air in an alveolus, whose geometry is sufficiently deep, may rotate when the drag force exerted by the cyclic breathing ductal flow (i.e., one oscillating subsystem), which passes by the alveolar opening, is sufficiently strong. In other words, an alveolus in which the air rotates and the direction of rotation changes cyclically (i.e., another oscillation subsystem) (Tippe & Tsuda, 2000) provides an environment for the production of chaotic mixing.

A small (but nonzero) amount of inertia or a small (but nonzero) geometric hysteresis can be considered as a perturbation; thus, as the pulmonary acinus expands and contracts, perturbations are intrinsic to alveolar flow. A stagnation saddle point, recently confirmed experimentally by Lv et al. (2020), must exist when alveolar fluid rotates in a rhythmically expanding and contracting alveolus (Tsuda et al., 1995). Therefore, finding the saddle point in the alveolar space is another way of identifying chaotic mixing (i.e., one of subsystems).

Three-dimensional (3D) simulations (e.g., Henry et al., 2012) of rotating flow in model alveoli show that instantaneous streamlines spiral away from the symmetry plane. During inspiration, when the alveolar walls are expanding, streamlines that start on one side of the symmetry plane end at the wall on the same side of the symmetry plane; that is, the walls appear to act as sinks (Henry et al., 2012). The fact that streamlines move away from the symmetry plane, implies that on the symmetry plane itself there will be rotating flow with closed streamlines (as seen in 2D simulations; e.g., Tsuda et al., 1995). While we have extensively studied the fluid mechanics (Haber et al., 2000, 2003; Haber & Tsuda, 2006; Henry et al., 2002, 2009; Henry & Tsuda, 2010, 2016; Tsuda et al., 1995, 2011) and explained Hamiltonian chaos of the symmetry plane (Tsuda et al., 2011), the same concept can be extended to the off-symmetry area. Indeed, we have already observed kinematic irreversibility on the off-symmetry area (Fig. 7C and D of Haber et al., 2000; Fig. 3 of Semmler-Behnke et al., 2012).

7 Summary

The rate of particle mixing measured experimentally by Heyder et al. (1988) could not be explained by any of the mechanisms known at the time. We have demonstrated that the gas exchange region of the lung, which is necessarily alveolated to increase the surface area for efficient gas exchange (Tsuda, Filipovic, et al., 2008), affords the required additional mixing mechanism; namely, chaotic mixing, which occurs and in turn increases deposition inside the alveoli. While this mechanism may be small on an individual alveolus basis, the respiratory region of the lung, which accounts for more than 95% of lung volume, contains hundreds of millions of alveoli. Thus, the deposition caused by chaotic mixing in the gas exchange region of the lungs must play a crucial role in the deposition of submicron particles, which represent the size range of many important particles; such as, cigarette smoke particles, inhaled bacterial particles, and possible therapeutic drug delivery particles.

Acknowledgment

None.

References

Bassingthwaighte, J. J., Liebovitch, L. S., & West, B. J. (1994). *Fractal physiology*. Oxford: Oxford University Press.

Burri, P. H. (1985). Development and growth of the human lung. In *Vol. I. Handbook of physiology: The respiratory system* (pp. 1–46). Bethesda, MD: American Physiological Society. Chapter 1.

Butler, J. P., & Tsuda, A. (1997). Effect of convective "stretching and folding" to aerosol mixing deep in the lung, assessed by approximate entropy. *Journal of Applied Physiology*, *83*(3), 800–809.

Butler, J. P., & Tsuda, A. (2005). Logistic trajectory maps and aerosol mixing due to asynchronous flow at airway bifurcation. *Respiratory Physiology & Neurobiology*, *148*, 195–206.

Butler, J. P., & Tsuda, A. (2011). Transport of gases between the environment and alveoli—Theoretical considerations. *Comprehensive Physiology*, *1*(3), 1301–1316.

Cengel, Y. A., & Cimbala, J. M. (2018). *Fluid mechanics, fundamentals and applications* (4th ed.). McGraw Hill Education.

Davies, C. N. (1972). Breathing of half-micron aerosols. II. Interpretation of experimental results. *Journal of Applied Physiology*, *35*, 605–611.

Friedlander, S. K. (1977). *Smoke, dust, and haze*. New York: Wiley.

Haber, S., Butler, J. P., Brenner, H., Emanuel, I., & Tsuda, A. (2000). Flow field in selfsimilar expansion on a pulmonary alveolus during rhythmical breathing. *Journal of Fluid Mechanics*, *405*, 243–268.

Haber, S., & Tsuda, A. (2006). Cyclic model for particle motion in the pulmonary acinus. *Journal of Fluid Mechanics*, *567*, 157–184.

Haber, S., Yitzhak, D., & Tsuda, A. (2003). Gravitational deposition in a rhythmically expanding and contracting alveolus. *Journal of Applied Physiology*, *95*, 657–671.

Henry, F. S., Butler, J. P., & Tsuda, A. (2002). Kinematically irreversible flow and aerosol transport in the pulmonary acinus: A departure from classical dispersive transport. *Journal of Applied Physiology*, *92*, 835–845.

Henry, F. S., Haber, S., Haberthür, D., Filipovic, N., Milasinovic, D., Schittny, J. C., & Tsuda, A. (2012). The simultaneous role of an alveolus as flow mixer and flow feeder for the deposition of inhaled submicron particles. *ASME Journal of Biomechanical Engineering*, *134*, 121001-1–121001-11.

Henry, F. S., Laine-Pearson, F. E., & Tsuda, A. (2009). Hamiltonian chaos in a model alveolus. *ASME Journal of Biomechanical Engineering*, *131*, 011006(1)–011006(7).

Henry, F. S., & Tsuda, A. (2010). Radial transport along the human acinar tree. *ASME Journal of Biomechanical Engineering*, *132*(10), 101001–101011.

Henry, F. S., & Tsuda, A. (2016). Onset of alveolar recirculation in the developing lungs and its consequence on nanoparticle deposition in the pulmonary acinus. *Journal of Applied Physiology*, *120*, 38–54.

Heyder, J., Blanchard, J. D., Feldman, H. A., & Brain, J. D. (1988). Convective mixing in human respiratory tract: Estimates with aerosol boli. *Journal of Applied Physiology, 64*, 1273–1278.

Kojic, M., & Tsuda, A. (2004). Gravitational deposition of aerosols from oscillatory pipe flows. *Journal of Aerosol Science, 35*(2), 245–261.

Laine-Pearson, F. E., & Hydon, P. E. (2006). Particle transport in a moving corner. *Journal of Fluid Mechanics, 559*, 379–390.

Lv, H., Dong, J., Qiu, Y., Yang, Y., & Zhu, Y. (2020). Microflow in a rhythmically expanding alveolar chip with dynamic similarity†. *Lab on a Chip, 20*, 2394–2402.

Metcalfe, G. (2010). Applied fluid chaos: Designing advection with periodically reoriented flows for micro to geophysical mixing and transport enhancement. In R. L. Dewar, & F. Detering (Eds.), *World scientific lecture notes in complex systems: Vol. 9. Complex physical, biophysical, and econophysical systems* World Scientific Publishing.

Oberdörster, G., Stone, V., & Donaldson, K. (2007). Toxicology of nanoparticles: A historical perspective. *Nanotoxicology, 1*(1), 2–25.

Ottino, J. M. (1989). *The kinematics of mixing: Stretching, chaos, and transport*. Cambridge University Press.

Ottino, J. M., Leong, C. W., Rising, H., & Swanson, P. D. (1988). Morphological structures produced by mixing in chaotic flows. *Nature (London), 333*, 419–425.

Semmler-Behnke, M., Kreyling, W. G., Schulz, H., Takenaka, S., Butler, J. P., Henry, F. S., & Tsuda, A. (2012). Nanoparticle delivery in infant lungs. *Proceedings of the National Academy of Sciences of the United States of America, 109*(13), 5092–5097.

Tabor, M. (1989). *Chaos and integrability in nonlinear dynamics*. Wiley.

Taylor, G. I. (1960). *Low Reynolds number flow (16 mm film)*. Newton, MA: Educational Services Inc.

Tippe, A., & Tsuda, A. (2000). Recirculating flow in an expanding alveolar model: Experimental evidence of flow-induced mixing of aerosols in the pulmonary acinus. *Journal of Aerosol Science, 31*(8), 979–986.

Tsuda, A. (1998a). Airway fluid mechanics and aerosol deposition. *Respiration & Circulation, 46*(9), 849–859 (in Japanese).

Tsuda, A. (1998b). Aerosol deposition in the lung. *Respiratory Research, 17*(11), 1221–1229 (*in Japanese*).

Tsuda, A., Butler, J. P., & Fredberg, J. J. (1994a). Effects of alveolated duct structure on aerosol kinetics: Part II gravitational deposition and inertial impaction. *Journal of Applied Physiology, 76*(6), 2510–2516.

Tsuda, A., Butler, J. P., & Fredberg, J. J. (1994b). Effects of alveolated duct structure on aerosol kinetics: Part I diffusion in the absence of gravity. *Journal of Applied Physiology, 76*(6), 2497–2509.

Tsuda, A., Filipovic, N. D., Haberthür, D., Dickie, R., Stampanoni, M., Matsui, Y., & Schittny, J. C. (2008). The finite element 3D reconstruction of the pulmonary acinus imaged by synchrotron X-ray tomography. *Journal of Applied Physiology, 105*, 964–976.

Tsuda, A., Henry, F. S., & Butler, J. P. (1995). Chaotic mixing of alveolated duct flow in rhythmically expanding pulmonary acinus. *Journal of Applied Physiology, 79*(3), 1055–1063.

Tsuda, A., Henry, F. S., & Butler, J. P. (2008). Gas and aerosol mixing in the acinus. *Respiratory Physiology & Neurobiology, 163*(1–3), 139–149.

Tsuda, A., Henry, F. S., & Butler, J. P. (2013). Particle transport and deposition: Basic physics of particle kinetics. *Comprehensive Physiology, 3*(4), 1437–1471.

Tsuda, A., Laine-Pearson, F. E., & Hydon, P. E. (2011). Why chaotic mixing of particles is inevitable in the deep lung. *Journal of Theoretical Biology, 286*, 57–66.

Tsuda, A., Otani, Y., & Butler, J. P. (1999). Acinar flow irreversibility caused by boundary perturbation of reversible alveolar wall motion. *Journal of Applied Physiology, 86*(3), 977–984.

Tsuda, A., Rogers, R. A., Hydon, P. E., & Butler, J. P. (2002). Chaotic mixing deep in the lung. *Proceedings of the National Academy of Sciences of the United States of America, 99*, 10173–10178.

Weibel, E. R. (1984). *The pathway for oxygen-structure and function in the mammalian respiratory system*. Cambridge, MA: Harvard University Press.

West, J. B. (2012). *Respiratory physiology: The essentials*. Baltimore: William & Wilkins.

CHAPTER

9

Three-dimensional reconstruction and modeling of the respiratory airways, particle deposition, and drug delivery efficacy

Nenad Filipovic[a] *and Akira Tsuda*[b,*]

[a]Bioengineering Research and Development Center (BioIRC), Kragujevac, Serbia [b]Tsuda Lung Research, Shrewsbury, MA, United States

1 Introduction

The respiratory system can be divided into two anatomical areas: the extrathoracic airways consisting of the mouth, nose, pharynx, larynx, and trachea; and the intrathoracic airways, which begin at the level of the intrathoracic trachea and extend all the way down to the alveoli. The intrathoracic airways can be further subdivided into conducting zone (generations 0–16) and respiratory zone where gas exchange takes place (generations 17–23). Here, we focus on the extrathoracic and upper conducting airways, which we group together in the term 'upper airways' for simplicity.

Many studies of the upper airways have adopted simplified representations of the airway geometry. Idealized models allow us to elucidate the flow and particle dynamics in the airways without the added complexity of the realistic geometry, and can provide representative estimations of global deposition. For the extrathoracic airways, a number of idealized geometries have been developed, such as the University of Alberta replica, which is representative of a physiologically averaged adult airway (Stapleton et al., 2000). Built from simple geometric shapes, it nonetheless captures all the basic anatomical features of the real

*Retired from Harvard University, Boston, MA, United States.

https://doi.org/10.1016/B978-0-12-823956-8.00011-0

extrathoracic airways and has widely been used in the literature, both in numerical (Ball et al., 2008; Debhi, 2011; Matida et al., 2004; Nicolaou & Zaki, 2016) and in vitro studies (DeHaan & Finlay, 2001; Grgic et al., 2004; Heenan et al., 2003; Johnstone et al., 2004). Similarly, Jayaraju et al. (2008) adopted a smoothed geometry based on a representative CT-image chosen among a set of healthy subjects. Others have adopted an idealized geometry with variable circular cross sections based on the hydraulic diameters of a human cast replica (Cui & Gutheil, 2011; Kleinstreuer & Zhang, 2003; Radhakrishnan & Kassinos, 2009; Zhang et al., 2002).

As such, the majority of more recent studies have focused on patient-specific models of the airways (Choi et al., 2009; Jayaraju et al., 2007; Koullapis et al., 2016; Lambert et al., 2011; Lin et al., 2007; Nicolaou & Zaki, 2013). We obtained the realistic models from medical images of the airways, typically via computed tomography (CT) scans or magnetic resonance imaging (MRI), which were digitally reconstructed via an image registration method. The resulting 3D geometry models, typically generated as stereolithography (STL) files, can then be meshed for CFPD simulations and used to manufacture casts for in vitro studies. Anatomically accurate models are, however, generally, still limited to the first 6 or 7 generations due to imaging resolution Fig. 1.

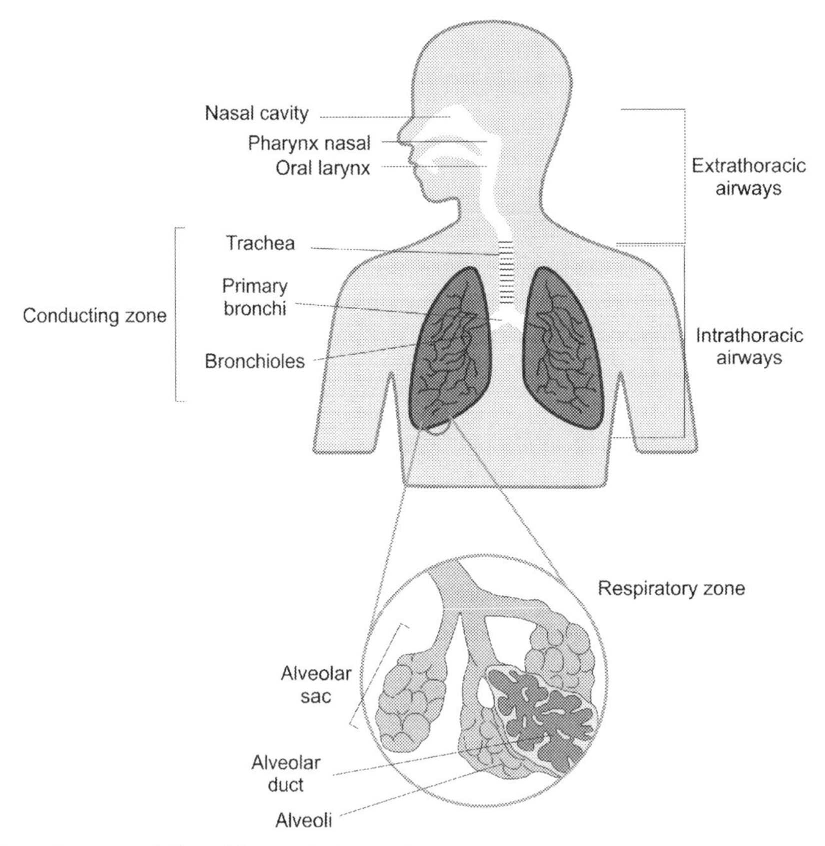

FIG. 1 Schematic representation of the respiratory system.

2 Methods

2.1 Geometric model: 3D reconstruction of lung using CT images

As input image data, unlabeled CT scans are used to extract bronchial tree and airways structural features. The lung volume is defined as CT-based lung segmentation and annotation is required. For lung segmentation, the FAST heterogeneous medical image computing and visualization framework will be applied (Smistad et al., 2015). The result of the lung segmentation process is a binary mask. The algorithm could be described as:

1. A second region growing takes place starting from a single random point inside any of the segmented regions only if all its immediate neighbors bear the same label.
2. To advance the region growing front, all points neighboring a candidate voxel must not include background voxel. This region is given a new label.
3. Steps 1 and 2 are repeated for the other lung volume. The result is an image with three labels (background and two lung volumes).
4. To distinguish left or right, we employ the directed graph extracted from the main airways and follow the generic rule according to which the topological distance between the bifurcations of the first and the second generation is longer in the left lung Fig. 2.

The next step involves the segmentation of the first generations of the airways that are identifiable in the patient's CT image, but any available airway tree segmentation method can be also applied. For this purpose, we investigated two algorithms. The first algorithm is the gradient vector flow (Bauer, Bischof, & Beichel, 2009; Bauer, Pock, et al., 2009), which achieved high accuracy with a low false-positive rate (only 1.44%) in a comparative study (Lo et al., 2012) in the context of the EXACT09 airway segmentation challenge. The second is a standard and stable approach based on seeded region growing (Adams & Bischof, 1994). Fig. 3 presents the up to four generation centerline of the airways.

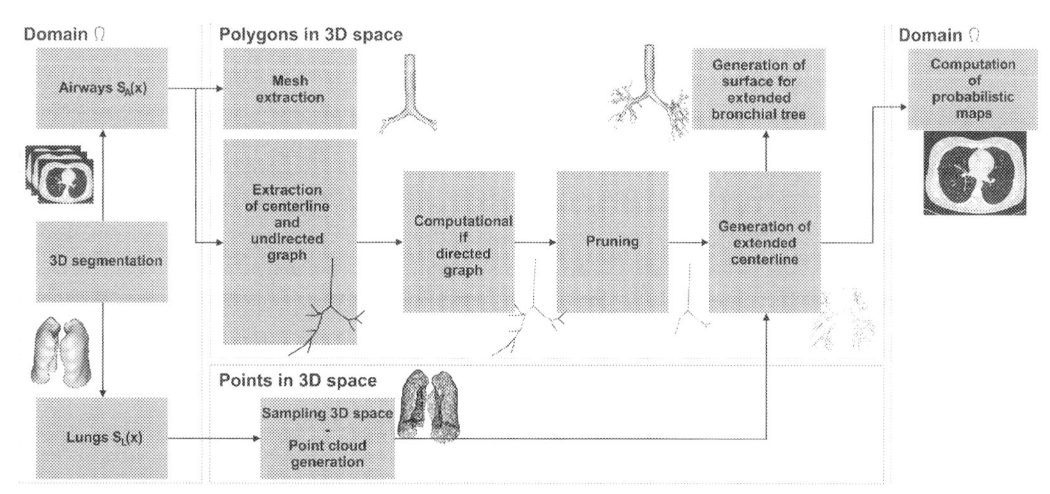

FIG. 2 3D reconstruction processing.

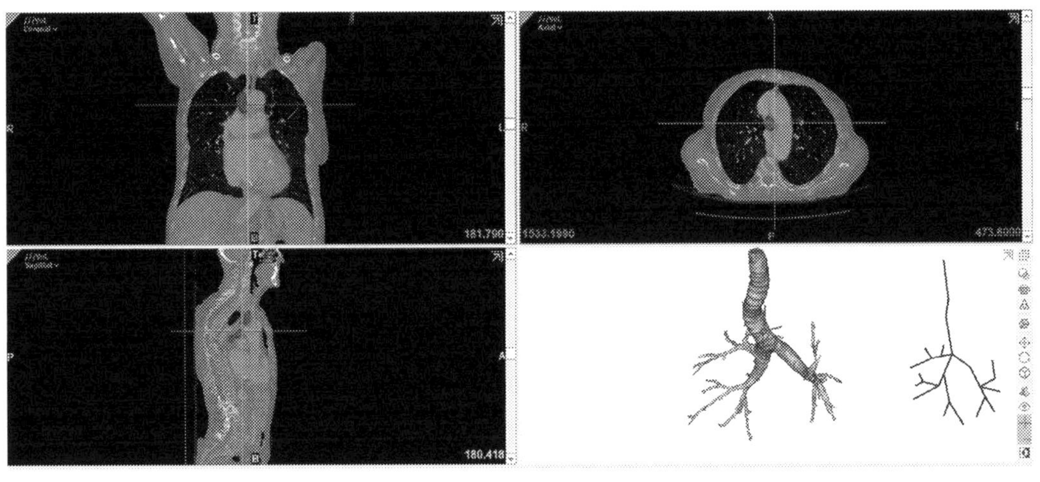

FIG. 3 Extraction of airway surface and centerline.

2.2 Skeletonization and detection of confirmed airway branches (Nadeem, Hoffman, Comellas, & Saha, 2020)

Centerline representation of the airway lumen tree is computed using a minimum-cost path-based skeletonization algorithm. This method generates the curve skeleton of the airway tree by iteratively growing new skeletal branches computed as a minimum-cost path. The meaningfulness of a skeletal branch is defined by its global context and scales. It is necessary to remove spurious skeletal branches from the airway tree centerline, since they introduce critical challenges for computerized airway labeling due to erroneous branching patterns and branch features. A local scale-based method is developed and applied to prune spurious centerline branches while preserving the true ones. The basic idea of this step is that for the valid airway tree centerline branches, the ratio of skeletal depth to local scale will be higher.

The new method relies on leveraging the knowledge of the overall airway branching structure to compartmentalize the labeling process. We determined from the airway tree structure the 32 anatomical segments of interest (Fig. 4). Starting at the trachea, the left and right main bronchi (LMB and RMB) lead to the left and right lungs. Beneath the LMB and RMB, bronchi lead to the left and right upper lobes, right middle lobe, and left and right lower lobes. In the right lung, the right upper lobe (RUL) branch leads to the upper lobe while the bronchus intermedius (BronchInt) feeds the middle and lower lobes. For the first three generations, the airway branching pattern is standard, and branching variabilities are mostly observed beyond this depth. Moreover, such variabilities of bronchopulmonary segments are only local phenomena. An example of such variability is trifurcations of RB1, RB2, and RB3 versus one bifurcation leading to RB1 shortly followed by another bifurcation leading to RB2 and RB3 branches.

After pruning, the airway tree centerline is rotated and translated such that the trachea is aligned with the image z-axis with its root at the origin. The first step of the ML method identifies the seven anatomical airway centerline branches (trachea, RMB, LMB, RUL, LUL, BronchInt, and LLB6 in Fig. 5) in the first three generations including the trachea using their

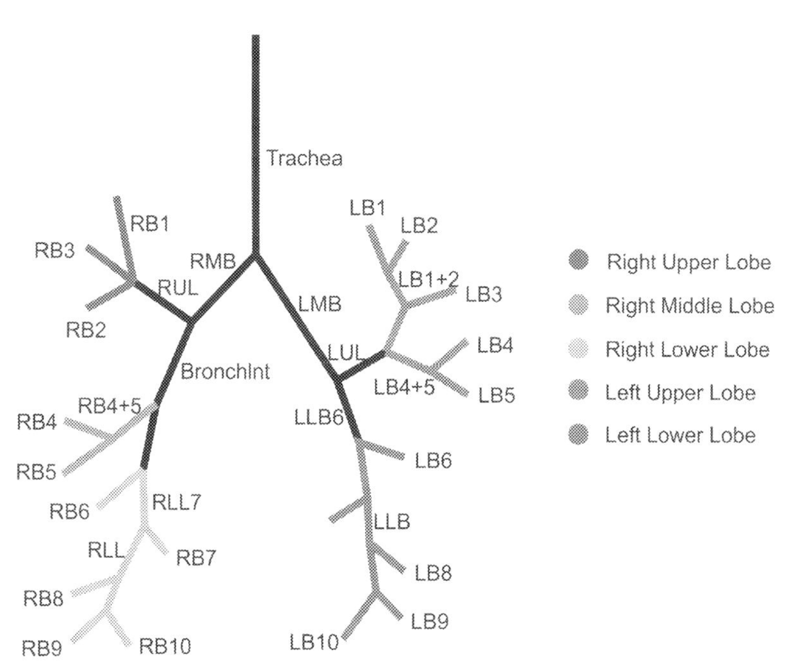

FIG. 4 Graphical description of topological and spatial relationship among 32 anatomical human bronchi up to the segmental level. Segmental branches for different lobes are color-coded.

tree generation and relative coordinate positions. These branches are used to identify the entry points into each of the five lung lobes.

Results of intermediate steps of the airway labeling algorithm are presented in Fig. 5. As observed in Fig. 5C, the airway tree centerline includes no apparent spurious branches, while visually valid branches are preserved after our pruning step. As shown in Fig. 5D, the method successfully identifies the first seven anatomical bronchi in the first three generations. Results of anatomical labeling of segmental bronchi are presented in Fig. 5E and F. In these figures, red, orange, yellow, green, and blue are used to denote segmental bronchi in the right upper, right middle, right lower, left upper, and left lower lung lobes. Results of anatomical labeling of segmental bronchi in right and left lungs, shown in Fig. 5E and F, respectively, are correct as visually confirmed by an expert using an ITK-based graphical user interface with reference and computed labeled airway masks.

2.3 Using AI for 3D reconstruction of airway

ML method uses geometric and topologic features of the current as well as ancestral and descendant generations through a series of neural network (NN)-based machine learning classifiers. We will define the ML algorithm with two different steps. During the first ML step, candidate anatomical branches will be differentiated from insignificant topological branches, often, responsible for variations in airway branching patterns. The second ML step will be designed for lung lobe-based classification of anatomical labels for valid candidate branches.

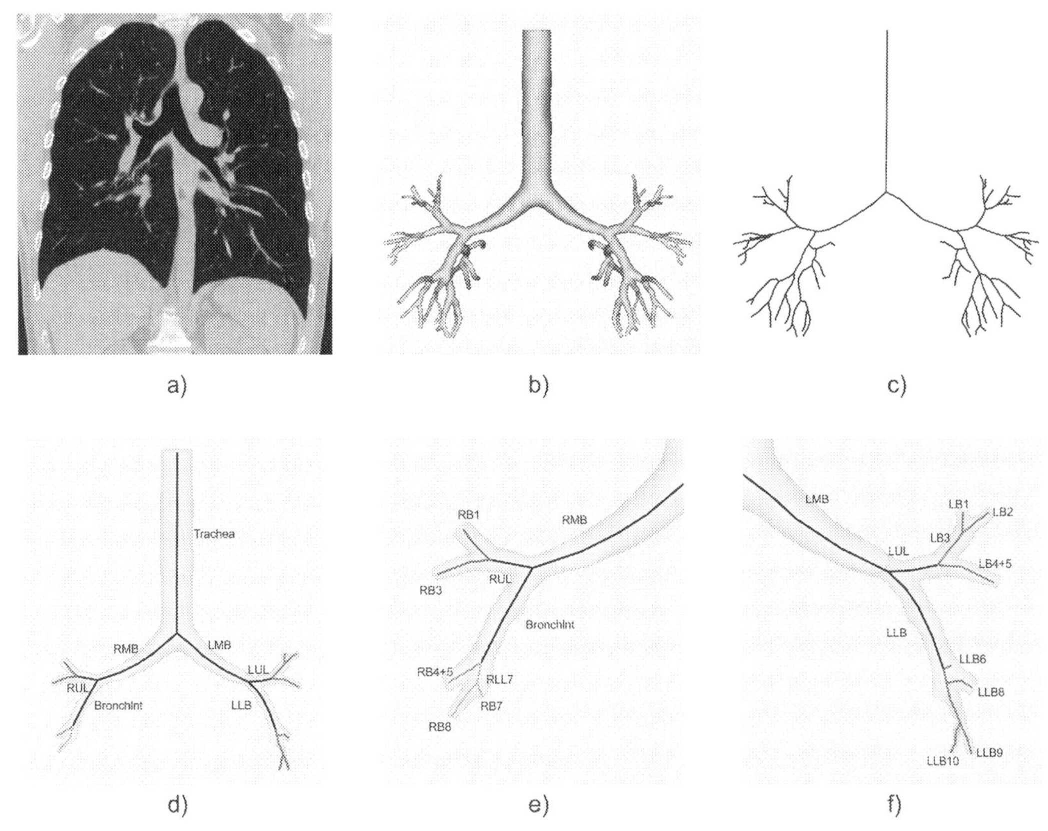

FIG. 5 Results of intermediate steps within our anatomical human bronchial labeling. (A) Input chest CT image at total lung capacity, (B) segmented airway tree volume computed from (A), (C) airway tree centerline after skeletal pruning computed from (B), (D) anatomical labeling results for the seven branches in the first three generations of the airway tree, results of complete anatomical bronchial labeling at the segmental level with right and left lung results enlarged in (E) and (F), respectively.

2.4 3D U-Net + freeze-and-grow algorithm

Region growing methods rely on CT attenuation differences between airway lumen, airway wall, and surrounding lung parenchyma. However, due to limitations of spatial resolution and various imaging artifacts, CT attenuation differences become less pronounced at smaller bronchi causing segmentation leakages into the lung parenchyma, especially in the presence of parenchymal pathologies associated with decreased parenchymal density.

2.5 Freeze-and-grow algorithm

For CT intensity-based airway segmentation, the FG algorithm starts with the CT image as input and maintains two volume markings:

(a) Confident airway volume (CAV) representing validated segmentation of airway volume.

(b) Forbidden volume (FV) around leakage-roots prohibiting connectivity paths from entering into leakage regions. FV is essential for iterative threshold relaxation. The algorithm consists of three major steps—(1) initialization, (2) iterative parameter relaxation and segmentation volume updating, and (3) termination.

(1) *Initial lumen segmentation and centerline computation*

During iteration, at a threshold *t*, the airway tree volume *Vt* is computed using region growing from CAV over Z3-FV and applying a fuzzy distance transform (FDT)-based dilation for smoothing and extending the lumen volume to partially occupied voxels.

(2) *Leakage detection*

Leakage detection is completed in two steps:

(a) preliminary detection of potential leakages—potential leakages are identified by locating voxels in the skeleton SCAV with a large change in associated airway volume.

(b) secondary screening for likely leakages—potential leakages are subjected to further scrutiny for secondary screening of likely leakages, which are passed into a correction phase.

(3) *Leakage correction*

(a) This step confirms a leakage, locates its root and deletes the subtree volume following the leakage root. It maximally recovers valid airway branches beyond the initial skeletal voxel location of a likely leakage. The confirmed leakage root is located using a tree traversal approach starting from the likely leakage voxel and checking the validity of individual branches.

(4) *Segmentation and forbidden volume augmentation*

New forbidden regions are added around individual leakage roots to arrest future growth through the same leakage sites.

(5) *Final leakage removal*

A final leakage removal step is applied to remove small leakages mostly occurring near terminal branches. Such leakages are missed by the rapid growth criterion used for initial leakage screening, and such leakages slowly accumulate over iterations. To remove such leakages, a classifier-based approach using SVM was applied using topologic and geometric features at skeletal voxels.

2.6 Deep learning-based freeze-and-grow algorithm

L is applied as a preprocessing step computing a voxel-level airway lumen likelihood map from a CT image, which is fed to an FG algorithm as an input (Fig. 6). A modified 3D U-Net classifier was developed with three pooling and three deconvolutional layers. The proposed network extends the previous U-Net architecture from Ronneberger et al. by replacing all 2D operations with their 3D counterparts. High memory and computational complexity restrict the use of an entire CT image through the network; so smaller subregions of $64 \times 64 \times 64$ voxels were used as samples.

- Data augmentation was avoided since it involves image transformation, introducing additional image blur and potential artifacts related to systematic feature perturbation.

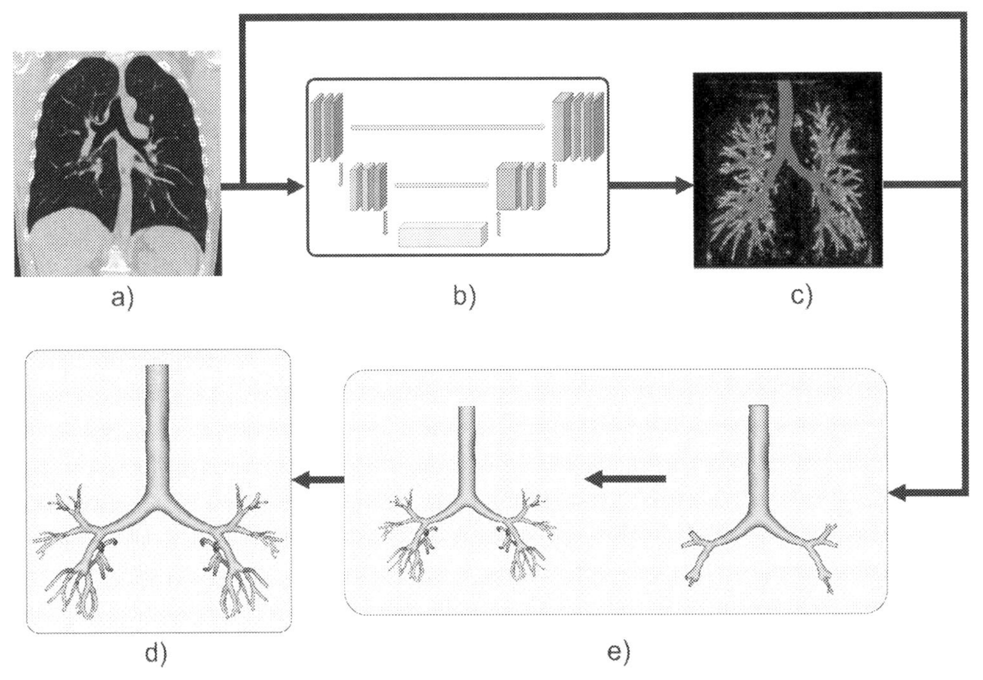

FIG. 6 Centerline extraction on two airway tree volumes (left-middle); Orthogonal planes computed on an airway tree (right).

Kernels of size $3 \times 3 \times 3$ were used at every convolutional layer except at the last layer, where a $1 \times 1 \times 1$ kernel was used.

- At the highest resolution, 64 feature maps were used, and along the contracting path, the number of feature maps was doubled at each downsampling step leading to 128, 256, and 512 feature maps at three lower resolutions. At each upsampling step along the expansive path, the number of feature maps was halved after a concatenation with the feature map from the contracting path at the same resolution.

2.7 Deep learning-based final leakage removal

A 3D U-Net classifier was developed to remove small terminal leakages in DL and FG-based airway tree segmentations. The classifier was posed as a three-class segmentation problem-valid airway lumen, leakages, and background. CT intensity and initial segmentation label are fed as separate channels resulting in a $64 \times 64 \times 64 \times 2$ input sample. Similar network architecture and data normalization, described in the previous section, were adopted. Weighted categorical cross-entropy loss function was used for training and validation, where the weight for a class was computed over all training samples as one minus the average volume fraction of that class. Training samples were generated by randomly sampling $64 \times 64 \times 64$ subregions along with the terminal, and two-generation ancestral branches. Subregions along the centerline of larger branches were ignored because no leakages were

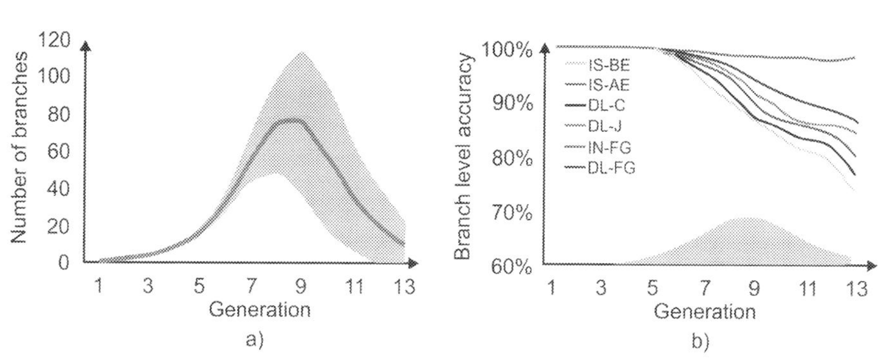

FIG. 7 Quantitative accuracy of airway tree segmentation at standard radiation CT scans. (A) Mean and variance of airway counts in reference segmentation masks at different generations. (B) Percentage of true branches captured by different methods.

observed at larger branches using the DL and FG-based segmentation. As a result, obtained accuracy is as shown in Fig. 7.

2.8 Computational fluid dynamics (CFD)

Different discretization techniques, namely finite volume and finite element methods can be used in various simulations to solve the governing equations for the flow. In order to model the turbulence in the airways, we are using large eddy simulation (LES) models and Reynolds-averaged-Navier-Stokes (RANS) models. The governing equations are given by the filtered Navier-Stokes equations (Filipovic et al., 2014, 2016; Filipovic, Nikolic, et al., 2013; Filipovic, Teng, et al., 2013),

$$\frac{\partial \overline{u}}{\partial t} + (\overline{u} \cdot \nabla)\overline{u} = -\frac{1}{\rho}\nabla \overline{p} + (v + v_r)\nabla^2 \overline{u} \tag{1}$$

$$\nabla \cdot \overline{u} = 0 \tag{2}$$

where $u=(u; v; w)$ is the velocity vector, p is the pressure, ρ and ν are the density and kinematic viscosity of the fluid, respectively, and ν_T is the turbulent viscosity. The overbar denotes resolved quantities.

The Reynolds number at the inlet, based on the inflow bulk velocity and inlet tube diameter, is $\mathrm{Re} = U\mathrm{in}^* D\mathrm{in}/\nu = 3745$, which lies in the turbulent regime.

2.9 Particle tracking

The particle equations of motion will be given by

$$m_p \frac{du_p}{dt} = \frac{3\rho_f m_p}{4\rho_p d_p} C_D |u_f - u_p|(u_f - u_p) + m_p g \frac{\rho_p - \rho_f}{\rho_p} + F_B \tag{3}$$

where the first and second terms on the right-hand side are the drag and gravitational forces respectively, and FB represents the Brownian force included in LES. In the RANS simulations, where only mean velocities are computed, a turbulent dispersion model is required in order to take into account the effect of the velocity fluctuations on the particles. RANS adopted a continuous random walk technique, where the fluctuating velocities are obtained from the generalized Langevin equation (Sommerfeld et al., 1993).

2.10 Aerosol transport and deposition

Aerosol deposition in the upper airways occurs primarily via impaction, due to the high velocities and rapid changes in the flow direction. The inertia of the particles causes them to deviate from the fluid streamlines and collide with the airway walls. The larger the particles, the higher the probability of deposition by impaction. Turbulent dispersion also plays a role in this region, in particular, for small particles the trajectories of which are considerably influenced by the fluctuations in the flow.

Particle transport can be modeled using an Eulerian or a Lagrangian approach. The Eulerian, or two-fluid, approach treats the dispersed phase as a continuum, solving the conservation equations of particle mass and momentum. On the other hand, the Lagrangian approach treats the dispersed phase as a set of individual point-particles in a continuous carrier phase. The particles are tracked through the flow field by solving the equations of motion for each particle with the relevant forces acting on it. Description of turbulent dispersion and collision of particles with the airway walls is more natural with this approach. For this reason, the Lagrangian approach has featured prominently in studies of aerosol deposition in the airways (Debhi, 2011; Jayaraju et al., 2007; Kleinstreuer & Zhang, 2003; Koullapis et al., 2016; Li et al., 2007; Matida et al., 2004; Nicolaou & Zaki, 2016; Radhakrishnan & Kassinos, 2009). In all these studies, the particles are assumed to be spherical, nonrotating, and noninteracting.

The aerosol is considered a dilute suspension and modeled using a one-way coupling approach, where the effect of the particles on the flow and interparticle interactions are neglected. In reality, however, inhaled particles are often nonspherical (e.g., dry powder inhaler formulations), and may possibly collide with each other and aggregate.

The equations of motion that describe the change in position and velocity along the particle trajectory are given by

$$\frac{dx_p}{dt} = u_p \tag{4}$$

$$\frac{du_p}{dt} = \sum F \tag{5}$$

where ΣF represents all the forces acting on the particles. The balance of forces acting on particles as they travel in a fluid was derived from first principles by Maxey and Riley (1983) in the low Reynolds number limit and is given by

$$m_p \frac{du_p}{dt} = \frac{18\mu_f}{\rho_p d_p^2} m_p (u_f - u_p) + m_f \left(\frac{Du_f}{Dt}\right) + \frac{1}{2} m_f \left(\frac{Du_f}{Dt} - \frac{du_p}{dt}\right) + 6a^2 (\pi\rho\mu)^2 \int_{t_{p0}}^{t_p} \frac{\frac{Du_f}{Dt'} - \frac{du_p}{dt'}}{(t - t')} dt' + g(m_p - m_f) \tag{6}$$

In shear flow, particles may also experience Saffman lift force. This force is the most prominent for large particles and small particle-to-fluid density ratios, ρ_p/ρ_f (Kallio & Reeks, 1989; Young & Leeming, 1997). For aerosol particles in turbulent channel flow, McLaughlin (1989) found that the Saffman lift force had virtually no effect on particle trajectories, except within the viscous sublayer where it played a significant role both in the inertial deposition of particles and in the accumulation of trapped particles.

Submicrometer particles are also subjected to Brownian motion, caused by random collisions with the gas molecules (Ounis et al., 1991). The Brownian force can be modeled as a Gaussian white-noise random process (Li & Ahmadi, 1992). Li and Ahmadi (1992) studied the effects of Brownian diffusion on particle dispersion and deposition in turbulent channel flow. Very near the wall, where turbulent fluctuations die down, Brownian motion was shown to be the dominant mechanism for diffusion of particles smaller than 0.1 μm. For particles larger than 0.5 μm, the effect of Brownian diffusion was negligibly small. The majority of studies focused on micron-sized particles in the upper airways have taken into account the drag and gravitational force, and discounted all other forces acting on the particles (Matida et al., 2004; Jayaraju et al., 2007; Debhi, 2011; Lambert et al., 2011; Nicolaou et al., 2015; Nicolaou & Zaki, 2016). A couple of studies have considered only the aerodynamic drag, based on order of magnitude arguments (Kleinstreuer & Zhang, 2003; Li et al., 2007; Zhang et al., 2002). Simulations with and without the gravitational force were performed by Ma and Lutchen (2009) at an inhalation flow rate of 15 L/min. When gravity was ignored, there was a maximum 10% reduction in a total deposition which showed that, while inertial impaction is the dominant deposition mechanism for micron-sized particles in the upper airways, gravitational sedimentation is also important and should be taken into account. The effect of gravity is more significant for larger particles and diminishes as the flow rate increases. Due to the large particle-to-fluid density ratios for aerosol particles in the air, the Saffman lift, pressure gradient, added mass, and Basset forces are typically considered insignificant in the upper airways (Finlay et al., 1996). Other effects on aerosol transport and deposition in the airways such as humidity, temperature, and electrostatic charge are less understood. Vaporization, as well as particle growth due to hygroscopicity, can both occur in the airways. Zhang et al. (2004) examined fuel droplet deposition in an upper airway model, with and without evaporation. Their results demonstrated that thermal effects were significant in the oral airways at low inhalation rates ($Q = 15$ L/min). Evaporation increased with higher ambient temperatures and lower inspiratory flow rates, resulting in lower deposition fractions. Longest and Xi (2008) evaluated the effect of condensation particle growth on the transport and deposition of cigarette smoke particles in the upper respiratory tract, under various relative humidity and temperature conditions. For the inhalation of warm saturated air 3°C above body temperature, 200 and 400 nm particles were observed to increase in size to above 3 μm near the trachea inlet, leading to an increase in deposition compared to the nonhygroscopic case. Recently, Koullapis et al. (2016) investigated the effect of electrostatic charge on aerosol deposition in a realistic geometry of the upper airways. Electrostatic charge was shown to increase deposition of smaller particles by as much as sevenfold, with most of the increase located in the mouth-throat region. The impact of inhalation flow rate on the deposition of charged particles was negligible for sizes smaller than 1 μm, whereas inertial impaction prevailed over electrostatic deposition for particles above 2.5 μm, with deposition increasing as the flow rate increased. Overall, the authors reported a significant interplay between particle size, electrostatic charge, and flow rate.

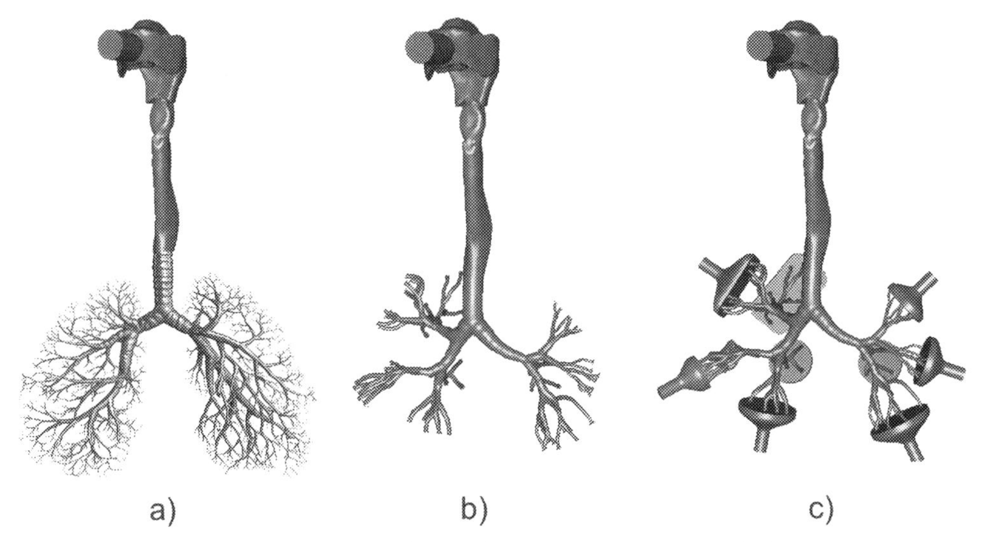

FIG. 8 Geometry of the respiratory airways: (A) original realistic airway geometry; (B) inner geometry adopted for the benchmark case; (C) physical segmented model for deposition measurements.

3 Results

3.1 Airway geometry

The realistic airway geometry adopted in the benchmark case is shown in Fig. 8. The tracheobronchial tree was acquired from a human lung of an adult male, excised at autopsy, and fixed with a liquid rubber solution at nearly end-inspiratory volume. The lung tissue was removed and the rubber cast of the bronchial tree was scanned using high-resolution computed tomography (HRCT) (Schmidt et al., 2004).

The extrathoracic airways were obtained from the Lovelace Respiratory Research Institute (LRRI) upper airway model. The oral cavity was molded from an in vivo dental impression of a Caucasian male at approximately 50% of the full opening, and the remaining of the model was acquired from a cadaver (Cheng et al., 1997). The LRRI geometry was obtained as a wax cast, which was scanned by an Atos (GOM, Braunschweig, Germany) device at Brno University of Technology, converted to STL format, and concatenated with the bronchial tree model at the trachea. The complete realistic geometry comprising the oral cavity, larynx, and tracheobronchial airways down to the 12th generation of branching is shown in Fig. 8A. Further details on the construction of the airway model can be found in Lizal et al. (2012). A 3-mm-thick envelope was created around the original airway geometry to obtain a negative cast of the airways. Only branches with a diameter above 3 mm were used, and the terminal bronchi segments were connected to 10 outlets. The model was also divided into sections to facilitate the measurement of regional deposition by various methods, such as optical microscopy and gravimetry (see Fig. 8C). The same airway geometry was employed in the numerical simulations.

3.2 Results of the CFD simulation

The realistic airway geometry from a patient with CT images (Fig. 8) was used for CFD analysis and particle tracking distribution. Numerical results are validated with experimental measurements by Lizal et al. (2015). In this study, we used tetrahedral elements. Mesh consists of 347.433 nodes and 1.320.246 elements. 3D mesh distribution is presented in Fig. 9. Velocity result for case 1–15 L/min for upper airway is presented in Fig. 10. Velocity and turbulence kinetic energy results for case 2–30 L/min are shown in Fig. 11. Velocity and turbulence kinetic energy results for case 3–60 L/min are shown in Fig. 12 with our open software PAK (Koullapis et al., 2018).

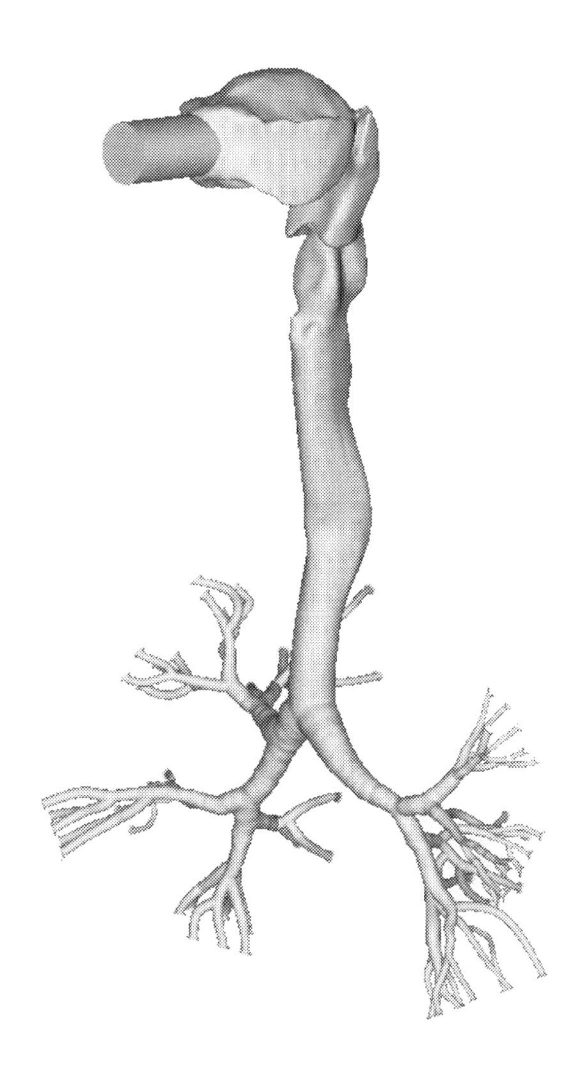

FIG. 9 3D mesh.

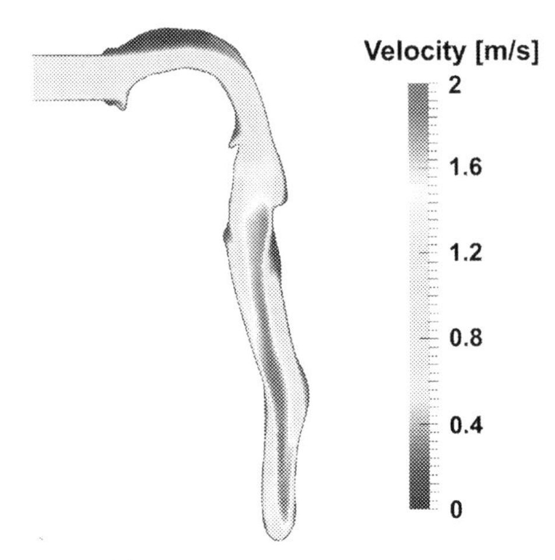

FIG. 10 Velocity results for case 1–15 L/min.

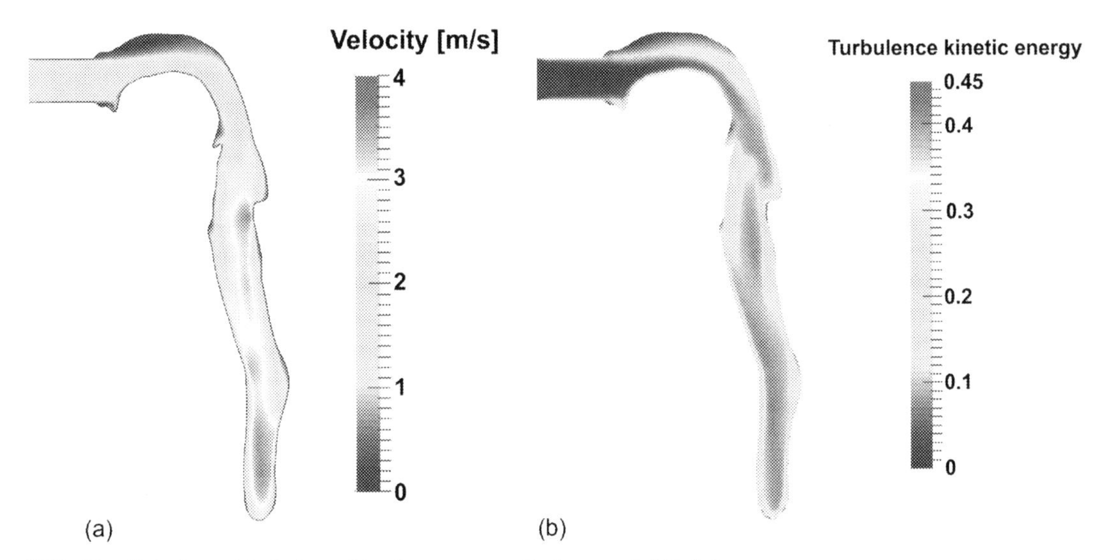

FIG. 11 Velocity and turbulence kinetic energy results for case 2–30 L/min.

3.3 A strategy in order to increase the efficiency of drug delivery

The right and left lung anatomy are similar but asymmetrical. The right lung consists of three lobes: the right upper (RU), the right middle (RM), and the right lower (RL). The left lung consists of two lobes: the left upper (LU) and the left lower (LL). The right lobe is divided by an oblique and horizontal fissure, where the horizontal fissure divides the upper and middle lobes, and the oblique fissure divides the middle and lower lobes. In the left lobe, there is only an oblique fissure that separates the upper and lower lobe (Fig. 13).

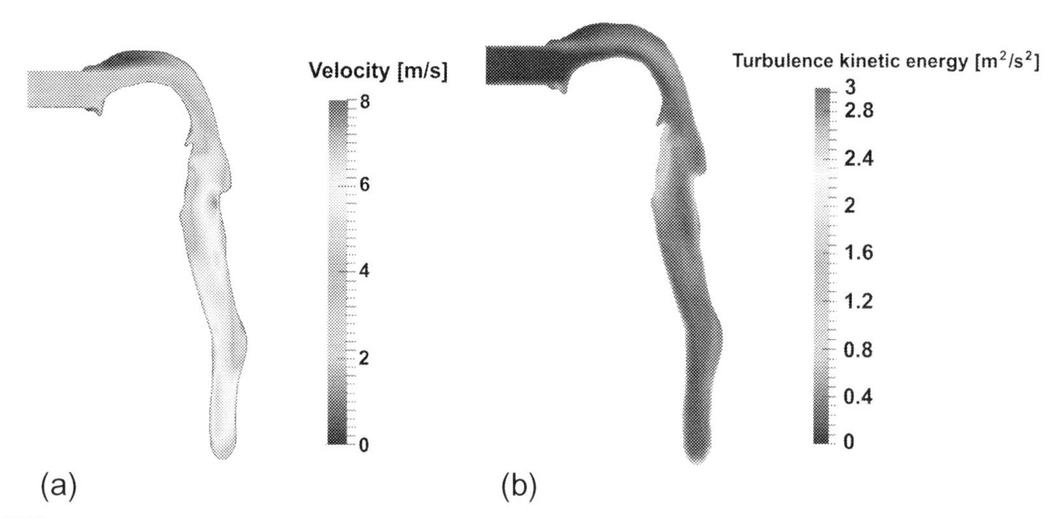

FIG. 12 Velocity and turbulence kinetic energy results for case 3–60 L/min.

We use the particle tracking algorithm which eliminates the virtual gap problem (addressed in Kuang et al., 2008), and will preserve the total number of particles throughout computations (based on our previous work, Cohen Stuart et al., 2011). The algorithm is suitable for complex geometries and is computationally efficient and numerically robust. The particle tracking method for 3D unstructured grids is based on the tetrahedral walk method (Kenwright & Lane, 1995; Lohner & Ambrosiano, 1990; Sadarjoen et al., 1998) and the directed-search path method (Chen & Pereira, 1999; Chordá et al., 2002).

The developed particle tracking method can now be summarized as follows (steps 1–5):

1. Decompose the mesh into tetrahedra. Pyramids and prisms are decomposed in two different ways, while the hexahedra are decomposed in three different ways in order to solve the virtual gap problem.
2. Brute force search over all cells to locate the initial host cell of the particle.
3. Particle displacement.
4. Deploy directed-search method: (a) Check whether particle still resides in the host cell, if not, continue. Else, return to 3. (b) Find exit face and calculate the intersection point. (c) Continue to the connecting tetrahedron and check whether the particle resides in the new tetrahedron. If so, return to 3. Else, go to 4(b). Note: Choose the right decomposition when a particle crosses from one element to another in order to rule out the possibility of a virtual gap.
5. End.

Another advantage of the directed-search method is that it is easy to implement particle-wall interactions. This is because the directed-search method follows the path of the particle exactly and therefore immediately obtains information about the time and location of the particle's impact with the wall of the domain (Fig. 14).

We assumed that we can determine where the flow from the inlet cross section will go and which lobe will be occupied. If we assume that we can design a drug device in which we can

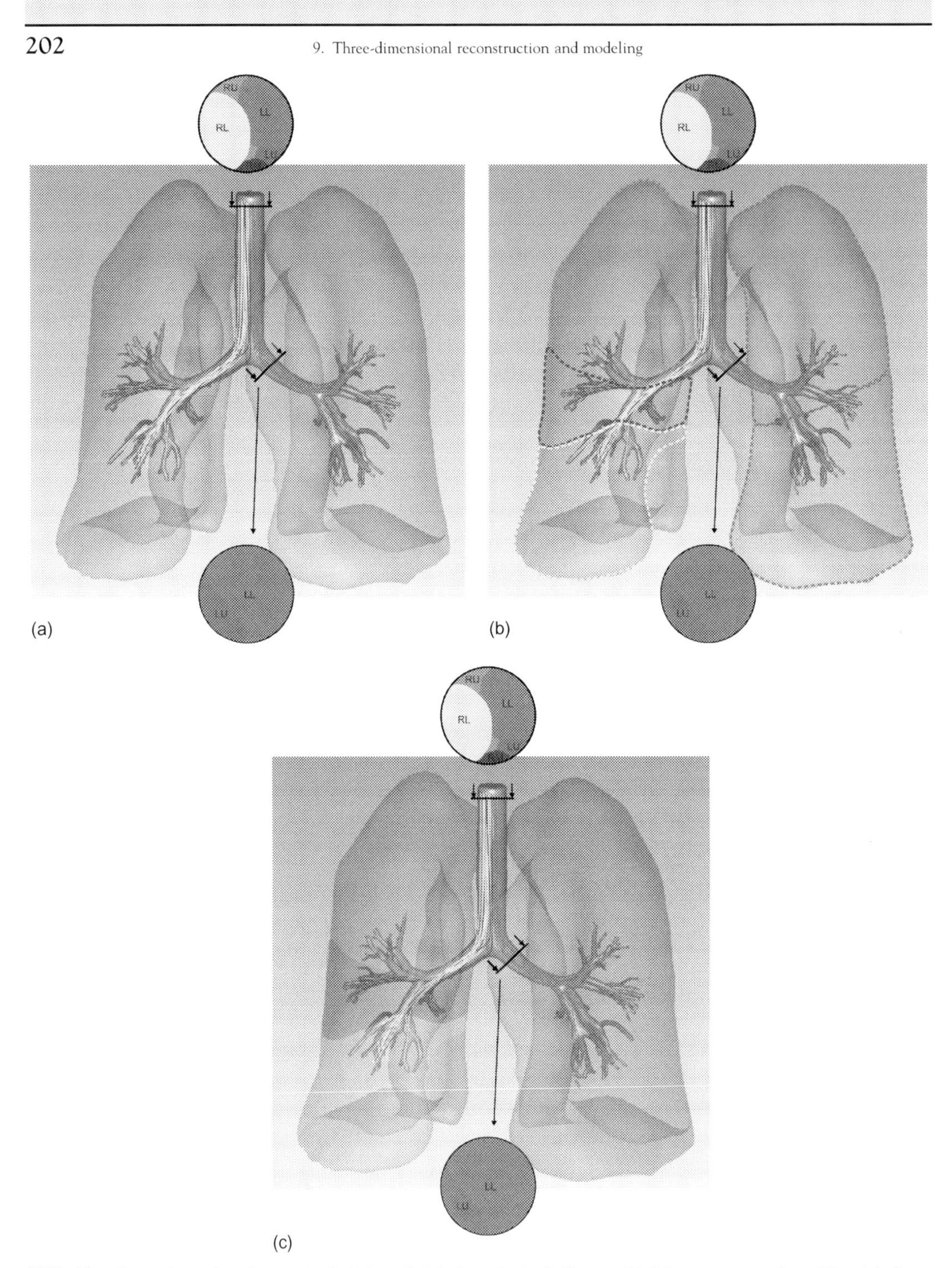

FIG. 13 Streamlines distribution in the left and right lung in the inflow and left lung cross sections. The right lung consists of three lobes: the right upper (RU), the right middle (RM), and the right lower (RL). The left lung consists of two lobes: the left upper (LU) and the left lower (LL).

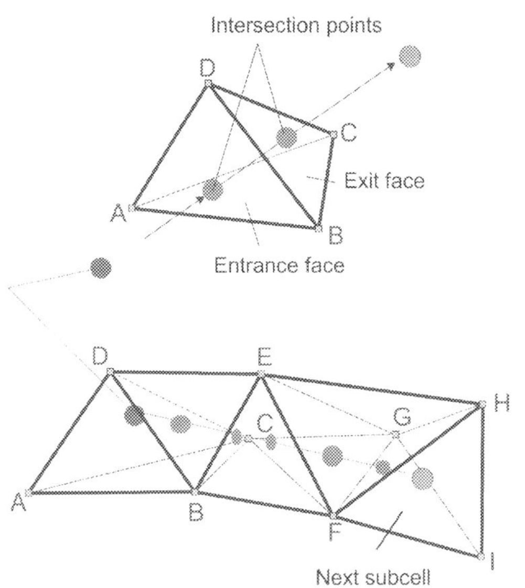

FIG. 14 The path of a particle is traced through the tetrahedron until the hosting tetrahedron is found. The face through which the particle leaves and enters a tetrahedron is called the exit face and entrance face, respectively (additional details provided in Cohen Stuart et al., 2011).

control the lobe in which drug will be distributed, we can optimize that device to precisely distribute drug particle only in the lobe where airway constriction has occurred.

If we assume a very simplified model of lung airway with only five outlet boundary conditions, we can determine from outlet streamlines where the input position of each streamline is located (Fig. 15). Even in the more complex geometry of the airway as it is presented in

FIG. 15 Simplified model of the airway with only five outlets.

Fig. 13, we can determine the location of the streamline starting position in order to go to the diseased lobe with our open software PAK (Koullapis et al., 2018).

3.4 Inhaled drug-device design

Inhaled medication in the form of drug formulation is the most common treatment of pulmonary diseases such as asthma. This field of pharmacy includes a good understanding of the physiology of human airways (Suwandecha, 2014). Some of the widely used inhalers include small-volume nebulizers (SVN), pressurized metered-dose inhalers (pMDI), or dry-powder inhalers (DPI) (Gardenhire, 2013) which enable aerosol particles (aerosolized drugs) to distribute to the lungs. Dry powder inhalers (DPI) were designed to administer individual doses of drugs in the form of powder contained in capsules that should be pierced before the inhalation process (unidose systems), or in blisters that move around in a device or have powder reservoirs (multidose systems) (Fernández & Clarà, 2012).

We focus on the dry powder inhaler Aerolizer (Fig. 16) because it is a readily available commercial device. Working with Aerolizer includes removing protecting cap and the inhaler opens by twisting the mouthpiece. A capsule filled with the drug is placed in a capsule chamber. After closing, the capsule is pierced by pressing buttons with "winglets." During inhalation, air enters the inhaler through tangential air inlets, rotating the capsule and ejecting the drug powder into the circulation chamber. Deagglomerating forces provided by the device flow field disperse the drug agglomerates to produce a fine respirable aerosol cloud. It should be also emphasized that an existing grid area at the bottom of the mouthpiece prevents the capsule from exiting the inhaler (Coates et al., 2006).

Many researchers have examined different dry powder inhalers in order to have a better understanding of their overall performance. Milenkovic et al. (2014a, 2014b) examined commercial Dry Powder Inhaler Turbohaler and considered different flow models combined with particle motion and deposition. Coates et al. (2004), Coates, Chan, et al. (2005a, 2005b), Coates, Fletcher, et al. (2005), and Coates et al. (2006) investigated the effects of the grid structure, mouthpiece length, inlet size, and air flow on Aerolizer DPI performance. They have shown that the structure of the inhaler grid has a significant effect on the amount of powder retained in the device without affecting the inhaler dispersion performance. Since the size of the inlet controls the air velocity that enters the device, the changes in dimensions will influence both the levels of turbulence generated in the device and the intensity of particle collisions with the

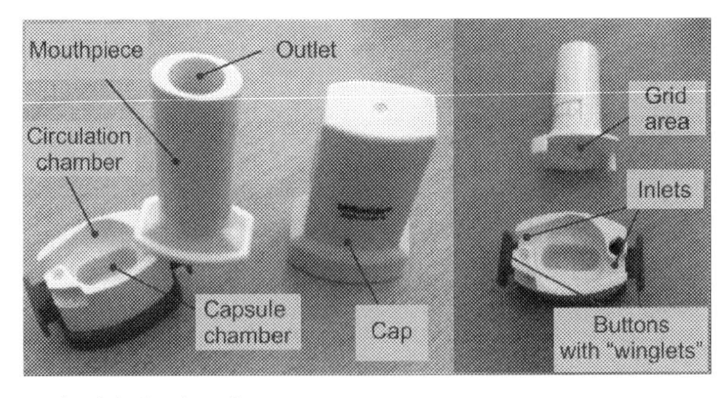

FIG. 16 Parts of dry powder inhaler Aerolizer.

device walls (Coates et al., 2006). Some experimental investigations have further investigated particle deposition as well as the fine particle fraction and particle size distribution of particles that reach the outlet (de Koning et al., 2001; Hoe et al., 2009). Tong et al. (2012) used the combined CFD and DPM method to better understand dispersion mechanisms in commercial DPIs, while Zhou, Tong, Tang, Citterio, et al. (2013) and Zhou, Tong, Tang, Yang, and Chan (2013) analyzed how device design affects aerosolization, and Yang et al. (2014) investigated air flow-induced detachment of particles.

Some of the advantages DPIs offer are that they do not require propellants for their administration, which makes them more respectful of the environment, as well as that many of them have an indicator of the doses remaining. The main drawback, that the research showed, is the loss of powder/drug due to deposition within the device, which leads to their overall inefficiency; less than 30% of the dose reaches the lungs (Islam & Gladki, 2008). The performance is related to the combination of the powder formulation and delivery device utilization (Chew et al., 2000). Donovan et al. found that in Aerolizer, in comparison to the HandiHaler, aerosolization performance was more dependent on the carrier particle size, which appeared to be related to a greater number of carrier particle-inhaler collisions (Donovan et al., 2012). Powder properties such as cohesion, size, and size distribution, influence powder dispersion and the breakage of particle aggregates (Milenkovic et al., 2014a, 2014b). Other factors that influence DPI performance include storage in a dry setting (humidity favors the agglomeration of the powder that can obstruct the inhalation system), the angle of the inhaler when distributing the drug, etc.

We design the device so that drug particles could be administered only through the black area (example in Fig. 16) of the device outlet and during a particular period of time to increase spatiotemporal fluid dynamics relationship between device outlet and target site (lobe). For the administration of drug particles, we apply specific CFD calculations to find in which lobe airway constriction exists and define the black area in this particular DPI device (Fig. 17).

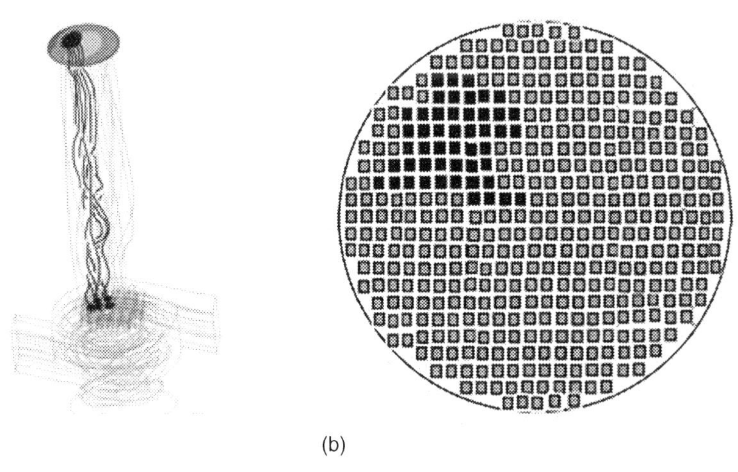

FIG. 17 (A) *Black* area presents assumption where drugs can follow streamlines. (B) Drug particles can go through those black grids, air goes through via other grids (*nonblack* ones).

Since a few studies have been conducted to determine of the weak points in construction of the inhaler, we will validate our idea with a forward CFD to determine the parts of Aerolizer inhaler in which most particles deposit. Additionally, the circulation chamber has not yet been thoroughly investigated. Some modifications in dimensions of the standard, original inhaler will be made in order to optimize and maximize the efficiency of drug delivery to the lungs.

Geometry and dimensions for commercial dry powder inhaler Aerolizer will be adopted by taking detailed measurements from the marketed device using a micrometer. Fluid domain will be taken into account when simulating air flow with particles. Some simplifications on the model will include chamfered and curved edges in areas where those modifications would not have any effect on the process of inhalation. The first step in simulating powder dispersion is to create 3D models of original and modified inhalers using commercial software for computer-aided design (e.g., CATIA V5R21).

3.5 Geometrical definition

To investigate where drug particles will be administered through the only black area (example in Fig. 16) of the device outlet, we modified a standard device Aerolizer (SD), and used four modified devices (MD1-MD4). Geometries of the devices adopted are shown in Fig. 18.

Compared with SD, the height of the modified inhalers was two times greater—MD1, 1.5 times greater—MD2, 1.5 times smaller—MD3, and 2 times smaller—MD4 than the original.

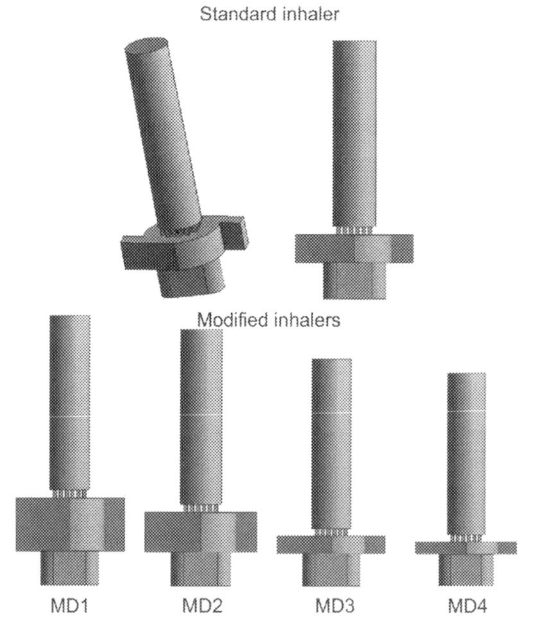

FIG. 18 Standard (SD) and modified devices (MD1-MD4).

3.6 Combined CFD and DPM models

In order to simulate flow fields generated in devices, commercial CFD software ANSYS Fluent (6.3) will be used to simulate the fluid flow. The computational grids will be refined in regions near the wall and in the grid area. Further refinements will be conducted within FLUENT based on actual velocity gradients observed in initial solutions.

A standard k-ω model with low Reynolds number (LRN) will be used to describe the turbulent flow of air inside inhaler based on its ability to accurately predict pressure drop, velocity profiles, and shear stress for transitional and turbulent flow (Ansys Userguide, 2021; Ghalichi et al., 1998) and our open software PAK (Koullapis et al., 2018). All transport equations will be discretized to be at least second-order accuracy in space. A second-order upwind scheme will be used to discretize the pressure equations, and quadratic upwind interpolation for convection kinetics scheme was used to discretize the momentum and turbulence equations (Shur et al., 2012).

In order to implement Fluent's discrete phase model, Rosin-Rammler logarithmic diameter distribution of inert particles will be injected using face normal direction. Particles are assumed to be released instantaneously at $t = 0$ from the bottom surface in the capsule chamber. This is a simplification in comparison to the real situation when a capsule that contains particles is placed at the bottom of the capsule chamber at the beginning of the inhalation process. Powder dispersion is assumed to occur instantaneously after which no more breakage or aggregation occurs. This assumption corresponds to the limiting case of very weak particle cohesion forces but provides valuable information of flow rates, particle sizes, and adhesion forces on the local and total particle deposition in the DPI Aerolizer. After initial powder release and aggregate breakage, particles in motion within the device are considered to be constant in size. Gravity will not be included in simulations as it was shown that its impact is insignificant due to small aerosol diameter sizes and short residence time in the inhaler itself (Milenkovic et al., 2013). Simplified conditions in the model include assumptions that: (1) air in the model is incompressible, (2) flow in the inlets is steady and parallel to the axis, and (3) the friction heat is negligible in this process.

Boundary conditions for this simulation included velocity inputs at two inlets and pressure outlets. The normalized Reynolds stress residuals in the range of 10e − 4 were applied as the convergence criteria to ensure full convergence. Inlet velocity of 11,79,166 m/s is calculated based on the flow rate of 28.3 L/min (Djokic et al., 2014). Values for particle mass at the beginning of simulation and density of particles are 20 mg and 1.68 g/cm^3, respectively.

4 Conclusions

In this chapter we have presented modeling of the regional deposition at the same level of resolution as the in vivo data, using CFD study. Three-dimensional reconstruction of the upper and central parts of airways was done using 3D computer tomography (CT).

Machine learning algorithms were used to generate a 3D surface defining personalized boundary conditions that can be employed as input for computational fluid dynamics simulations.

Results for the CFD study, deposition of the particles, and optimization of drug delivery efficacy have been reported. Also, optimal design of the drug delivery device through the only particular area of the device outlet and particular timing are presented.

References

Adams, R., & Bischof, L. (1994). Seeded region growing. *IEEE Transactions on Pattern Analysis and Machine Intelligence, 16*(6), 641–647.

Ansys Userguide. (2021). *Turbulent models*. from http://www.afs.enea.it/project/neptunius/docs/fluent/html/th/node330.htm. (Accessed 10 February 2021).

Ball, C., Uddin, M., & Pollard, A. (2008). Mean flow structures inside the human upper airway. *Flow, Turbulence and Combustion, 81*, 155–188.

Bauer, C., Bischof, H., & Beichel, R. (2009). Segmentation of airways based on gradient vector flow. In *International workshop on pulmonary image analysis, medical image computing and computer assisted intervention* (pp. 191–201). Citeseer.

Bauer, C., Pock, T., Bischof, H., & Beichel, R. (2009). Airway tree reconstruction based on tube detection. In *Proc. of second international workshop on pulmonary image analysis* (pp. 203–213).

Chen, X. Q., & Pereira, J. C. F. (1999). A new particle-locating method accounting for source distribution and particle-field interpolation for hybrid modeling of strongly coupled two-phase flows in arbitrary coordinates. *Numerical Heat Transfer Part B, 35*, 41–63.

Cheng, K., Cheng, Y., Yeh, H. C., & Swift, D. L. (1997). Measurements of airway dimensions and calculation of mass transfer characteristics of the human oral passage. *Journal of Biomechanical Engineering-Transactions of the ASME, 119*, 476–482.

Chew, N. Y. K., Bagster, D., & Chan, H. K. (2000). Effect of particle size, air flow and inhaler device on the aerosolisation of disodium cromoglycate powders. *International Journal of Pharmaceutics, 206*, 75–83.

Choi, J., Tawhai, M. H., Hoffman, E. A., & Lin, C. L. (2009). On intra- and inter subject variabilities of airflow in the human lungs. *Physics of Fluids, 21*, 101901.

Chordá, R., Blasco, J. A., & Fueyo, N. (2002). An efficient particle-locating algorithm for application in arbitrary 2D and 3D grids. *International Journal of Multiphase Flow, 28*, 1565–1580.

Coates, M. S., Chan, H. K., Fletcher, D. F., & Raper, J. A. (2005a). Influence of air flow on the performance of a dry powder inhaler using computational and experimental analyses. *Pharmaceutical Research, 22*(9), 923–932.

Coates, M. S., Chan, H.-K., Fletcher, D. F., & Raper, J. A. (2005b). Influence of air flow on the performance of a dry powder inhaler using computational and experimental analyses. *Pharmaceutical Research, 22*, 1445–1453.

Coates, M. S., Chan, H. K., Fletcher, D. F., & Raper, J. A. (2006). Effect of design on the performance of a dry powder inhaler using computational fluid dynamics. Part 2: Air inlet size. *Journal of Pharmaceutical Sciences, 95*(6), 1382–1392.

Coates, M. S., Fletcher, D. F., Chan, H. K., & Raper, J. A. (2004). Effect of design on the performance of a dry powder inhaler using computational fluid dynamics. Part 1: Grid structure and mouthpiece length. *Journal of Pharmaceutical Sciences, 93*(11), 2863–2876.

Coates, M. S., Fletcher, D. F., Chan, H.-K., & Raper, J. A. (2005). The role of capsule on the performance of a dry powder inhaler using computational and experimental analyses. *Pharmaceutical Research, 22*, 923–932.

Cohen Stuart, D. C., Kleijn, C. R., & Kenjeres, S. (2011). An efficient and robust method for Lagrangian magnetic particle tracking in fluid flow simulations on unstructured grids. *Computers and Fluids, 40*(1), 188–194. https://doi.org/10.1016/j.compfluid.2010.09.001.

Cui, X., & Gutheil, E. (2011). Large eddy simulation of the unsteady flow-field in an idealized human mouth-throat configuration. *Journal of Biomechanics, 44*, 2768–2774.

de Koning, J. P., Visser, M. R., Oelen, G. A., de Boer, A. H., van der Mark, T. W., Coenegracht, P. M. J., et al. (2001). Effect of peak inspiratory flow and flow increase rate on in vitro drug deposition from four dry powder inhaler devices. In *Dry powder inhalation, technical and physiological aspects, prescribing and use* (pp. 83–94). Rijksuniversiteit Groningen (Thesis). (chapter 6).

Debhi, A. (2011). Prediction of extrathoracic aerosol deposition using RANS-random walk and LES approaches. *Aerosol Science and Technology, 45*, 555–569.

DeHaan, W., & Finlay, W. (2001). In vitro monodisperse aerosol deposition in a mouth and throat with six different inhalation devices. *Journal of Aerosol Medicine, 14*, 361–367.

Djokic, M., Kachrimanis, K., Solomun, L., Djuris, J., Vasiljevic, D., & Ibric, S. (2014). A study of jet-milling and spray-drying process for the physicochemical and aerodynamic dispersion properties of amiloride HCl. *Powder Technology, 262*, 170–176.

Donovan, M. J., Kim, S. H., Raman, V., & Smyth, H. D. (2012). Dry powder inhaler device influence on carrier particle performance. *Journal of Pharmaceutical Sciences*, *101*(3), 1097–1107.

Fernández, T. A., & Clarà, P. C. (2012). Deposition of inhaled particles in the lungs. *Archivos de Bronconeumología (English Edition)*, *48*(7), 240–246.

Filipovic, N., Ghimire, K., Saveljic, I., Milosevic, Z., & Ruegg, C. (2016). Computational modeling of shear forces and experimental validation of endothelial cell responses in an orbital well shaker system. *Computer Methods in Biomechanics and Biomedical Engineering*, *19*(6), 581–590.

Filipovic, M., Nikolic, D., Saveljic, I., Milosevic, Z., Exarchos, T., Pelosi, G., & Parodi, O. (2013). Computer simulation of three dimensional plaque formation and progression in the coronary artery. *Computers and Fluids*, *88*, 826–833.

Filipovic, N., Teng, Z., Radovic, M., Saveljic, I., Fotiadis, D., & Parodi, O. (2013). Computer simulation of three dimensional plaque formation and progression in the carotid artery. *Medical & Biological Engineering & Computing*, *51*(6), 607–616.

Filipovic, N., Zivic, M., Obradovic, M., Djukic, T., Markovic, Z., & Rosic, M. (2014). Numerical and experimental LDL transport through arterial wall. *Microfluidics and Nanofluidics*, *16*(3), 455–464.

Finlay, W., Stapleton, K., & Yokota, J. (1996). On the use of computational fluid dynamics for simulating flow and particle deposition in the human respiratory tract. *Journal of Aerosol Science*, *9*, 329–341.

Gardenhire, D. S. (2013). *A guide to aerosol delivery devices for respiratory therapists* (3rd ed.). American Association for Respiratory Care.

Ghalichi, F., Deng, X., Champlain, A. D., Douville, Y., King, M., & Guidoin, R. (1998). Low Reynolds number turbulence modeling of blood flow in arterial stenoses. *Biorheology*, *35*, 281–294.

Grgic, B., Finlay, W., Burnell, P., & Heenan, A. (2004). In vitro intersubject and intrasubject deposition measurements in realistic mouth-throat geometries. *Journal of Aerosol Science*, *35*, 1025–1040.

Heenan, A., Matida, E., Pollard, A., & Finlay, W. (2003). Experimental measurements and computational modeling of the flow filed in an idealized human oropharynx. *Experiments in Fluids*, *35*, 70–84.

Hoe, S., Traini, D., Chan, H. K., & Young, P. M. (2009). Measuring charge and mass distributions in dry powder inhalers using the electrical Next Generation Impactor (eNGI). *European Journal of Pharmaceutical Sciences*, *38*, 88–94.

Islam, N., & Gladki, E. (2008). Dry powder inhalers (DPIs)–A review of device reliability and innovation. *International Journal of Pharmaceutics*, *360*(1–2), 1–11.

Jayaraju, S., Brouns, C., Lacor, C., Belkassem, B., & Verbanck, S. (2008). Large eddy and detached eddy simulations of fluid flow and particle deposition in a human mouth–throat. *Journal of Aerosol Science*, *39*, 862–875.

Jayaraju, S. T., Brouns, M., Verbanck, S., & Lacor, C. (2007). Fluid flow and particle deposition analysis in a realistic extrathoracic airway model using unstructured grids. *Journal of Aerosol Science*, *38*, 494–508.

Johnstone, A., Uddin, M., Pollard, A., Heenan, A., & Finlay, W. H. (2004). The flow inside an idealised form of the human extra-thoracic airway. *Experiments in Fluids*, *37*, 673–689.

Kallio, G. A., & Reeks, M. W. (1989). A numerical simulation of particle deposition in turbulent boundary layers. *International Journal of Multiphase Flow*, *15*, 433–446.

Kenwright, D. N., & Lane, D. A. (1995). Optimization of the time-dependent particle tracing using tetrahedral decomposition. In *Proceedings of the 6th conference on visualization* (pp. 321–328). IEEE Computer Society.

Kleinstreuer, C., & Zhang, Z. (2003). Laminar-to-turbulent fluid-particle flows in a human airway model. *International Journal of Multiphase Flow*, *29*, 271–289.

Koullapis, P. G., Kassinos, S. C., Bivolarova, M. P., & Melikov, A. K. (2016). Particle deposition in a realistic geometry of the human conducting airways: Effects of inlet velocity profile, inhalation flowrate and electrostatic charge. *Journal of Biomechanics*, *49*, 2201–2212.

Koullapis, P., Kassinos, S. C., Muela, J., Perez-Segarra, C., Rigola, J., Lehmkuhl, O., Cui, Y., Sommerfelde, M., Elcnerf, J., Jichaf, M., Saveljic, I., Filipovic, N., Lizal, F., & Nicolaou, L. (2018). Regional aerosol deposition in the human airways: The SimInhale benchmark case and a critical assessment of in silico methods. *European Journal of Pharmaceutical Sciences*, *113*, 77–94.

Kuang, S. B., Yu, A. B., & Zou, Z. S. (2008). A new point-locating algorithm under three-dimensional hybrid meshes. *International Journal of Multiphase Flow*, *34*(11), 1023–1030.

Lambert, A. R., O'shaughnessy, P. T., Tawhai, M. H., Hoffman, E. A., & Lin, C. L. (2011). Regional deposition of particles in an image-based airway model: Large-eddy simulation and left-right lung ventilation asymmetry. *Aerosol Science and Technology*, *45*, 11–25.

Li, A., & Ahmadi, G. (1992). Dispersion and deposition of spherical particles from point sources in a turbulent channel flow. *Aerosol Science and Technology*, *16*, 209–226.

Li, Z., Kleinstreuer, C., & Zhang, Z. (2007). Particle deposition in the human tracheobronchial airways due to transient inspiratory flow patterns. *Journal of Aerosol Science, 38*, 625–644.

Lin, C. L., Tawhai, M. H., McLennan, G., & Hoffman, E. A. (2007). Characteristics of the turbulent laryngeal jet and its effect on airflow in the human intra-thoracic airways. *Respiratory Physiology & Neurobiology, 157*, 295–309.

Lizal, F., Belka, M., Adam, J., Jedelsky, J., & Jicha, M. (2015). A method for in vitro regional aerosol deposition measurement in a model of the human tracheobronchial tree by the positron emission tomography. *Proceedings of the Institution of Mechanical Engineers Part H-Journal of Engineering in Medicine, 229*, 750–757.

Lizal, F., Elcner, J., Hopke, P., Jedelsky, J., & Jicha, M. (2012). Development of a realistic human airway model. *Proceedings of the Institution of Mechanical Engineers Part H-Journal of Engineering in Medicine, 226*, 197–207.

Lo, P., Van Ginneken, B., Reinhardt, J. M., Yavarna, T., De Jong, P. A., Irving, B., et al. (2012). Extraction of airways from CT (EXACT'09). *IEEE Transactions on Medical Imaging, 31*(11), 2093–2107. 22855226.

Lohner, R., & Ambrosiano, J. (1990). A vectorized particle tracer for unstructured grids. *Journal of Computational Physics, 91*, 22–31.

Longest, P. W., & Xi, J. (2008). Condensational growth may contribute to the enhanced deposition of cigarette smoke particles in the upper respiratory tract. *Aerosol Science and Technology, 42*, 579–602.

Ma, B., & Lutchen, K. (2009). Cfd simulation of aerosol deposition in an anatomically based human large–medium airway model. *Annals of Biomedical Engineering, 37*, 271–285.

Matida, E., Finlay, W., Lange, C., & Grgic, B. (2004). Improved numerical simulation of aerosol deposition in an idealized mouth-throat. *Journal of Aerosol Science, 35*, 1–19.

Maxey, M. R., & Riley, J. J. (1983). Equation of motion for a small rigid sphere in a nonuniform flow. *Physics of Fluids, 26*, 883–889.

McLaughlin, J. (1989). Aerosol particle deposition in numerically simulated channel flow. *Physics of Fluids A: Fluid Dynamics, 1*, 1211–1224.

Milenkovic, J., Alexopoulos, A. H., & Kiparissides, C. (2013). Flow and particle deposition in the Turbuhaler: A CFD simulation. *International Journal of Pharmaceutics, 448*, 205–213.

Milenkovic, J., Alexopoulos, A. H., & Kiparissides, C. (2014a). Airflow and particle deposition in a dry powder inhaler: An integrated CFD approach. In *Simulation and modeling methodologies, technologies and applications* (pp. 127–140). Springer International Publishing.

Milenkovic, J., Alexopoulos, A. H., & Kiparissides, C. (2014b). Dynamic flow and particle deposition in the Turbuhaler DPI. A CFD simulation. *International Journal of Pharmaceutics, 461*(1–2), 129–136.

Nadeem, S. A., Hoffman, E. A., Comellas, A. P., & Saha, P. K. (2020, March). Anatomical labeling of human airway branches using a novel two-step machine learning and hierarchical features. In *Vol. 11313. Medical imaging 2020: Image processing* (p. 1131312). International Society for Optics and Photonics.

Nicolaou, L., Jung, S. Y., & Zaki, T. A. (2015). A robust direct-forcing immersed boundary method with enhanced stability for moving body problems in curvilinear coordinates. *Computers & Fluids, 119*, 101–114.

Nicolaou, L., & Zaki, T. A. (2013). Direct numerical simulations of flow in realistic mouth-throat geometries. *Journal of Aerosol Science, 57*, 71–87.

Nicolaou, L., & Zaki, T. A. (2016). Characterization of aerosol stokes number in 90° bends and idealized extrathoracic airways. *Journal of Aerosol Science, 102*, 105–127.

Ounis, H., Ahmadi, G., & McLaughlin, J. (1991). Brownian diffusion of submicrometer particles in the viscous sublayer. *Journal of Colloid and Interface Science, 143*, 266–277.

Radhakrishnan, H., & Kassinos, S. (2009). CFD modeling of turbulent flow and particle deposition in human lungs. In *Engineering in medicine and biology society, EMBC 2009. Annual international conference of the IEEE* (pp. 2867–2870).

Sadarjoen, I. A., Boer, A. J. D., Post, F. H., & Mynett, A. E. (1998). Particle tracing in r-transformed grids using a tetrahedral 6-decomposition. In *Proceedings of the Eurographics workshop, Blauberen, Germany* (pp. 71–80).

Schmidt, A., Zidowitz, S., Kriete, A., Denhard, T., Krassb, S., & Peitgen, H. O. (2004). A digital reference model of the human bronchial tree. *Computerized Medical Imaging and Graphics, 28*, 203–211.

Shur, J., Lee, S., Adams, W., Lionberger, R., Tibbatts, J., & Price, R. (2012). Effect of device design on the in vitro performance and comparability for capsule-based dry powder inhalers. *The AAPS Journal, 14*(4), 667–676.

Smistad, E., Bozorgi, M., & Lindseth, F. (2015). FAST: Framework for heterogeneous medical image computing and visualization. *International Journal of Computer Assisted Radiology and Surgery, 10*(11), 1811–1822. 25684594.

Sommerfeld, M., Kohnen, G., & Rüger, M. (1993). Some open questions and inconsistencies of lagrangian particle dispersion models. In *Ninth symposium on turbulent shear flows, Kyoto, Japan.*

Stapleton, K. W., Guentsch, E., Hoskinon, M. K., & Finlay, W. H. (2000). On the suitability of k-ε turbulence modeling for aerosol deposition in the mouth and throat: A comparison with experiment. *Journal of Aerosol Science, 31*, 739–749.

Suwandecha, T. (2014). *Effect of device design on the performance of a dry powder inhaler using computational fluid dynamics* (Ph.D. thesis). Thailand: Prince of Songkla University.

Tong, Z. B., Zheng, B., Yang, R. Y., Yu, A. B., & Chan, H. K. (2012). CFD-DEM investigation of the dispersion mechanisms in commercial dry powder inhalers. *Powder Technology, 240*, 19–24.

Yang, J., Wu, C. Y., & Adams, M. (2014). A three-dimensional DEM–CFD analysis of air-flow-induced detachment of api particles from carrier particles in dry powder inhalers. *Acta Pharmaceutica Sinica B, 4*, 52–59.

Young, J., & Leeming, A. (1997). A theory of particle deposition in turbulent pipe flow. *Journal of Fluid Mechanics, 340*, 129–159.

Zhang, Z., Kleinstreuer, C., & Kim, C. S. (2002). Micro-particle transport and deposition in a human airway model. *Journal of Aerosol Science, 33*, 1635–1652.

Zhang, Z., Kleinstreuer, C., Kim, C. S., & Cheng, Y. S. (2004). Vaporizing microdroplet inhalation, transport, and deposition in a human upper airway model. *Aerosol Science and Technology, 38*, 36–49.

Zhou, Q. T., Tong, Z., Tang, P., Citterio, M., Yang, R., & Chan, H. K. (2013). Effect of device design on the aerosolization of a carrier-based dry powder inhaler—A case study on Aerolizer® Foradile®. *The AAPS Journal, 15*(2), 511–522.

Zhou, Q., Tong, Z., Tang, P., Yang, R., & Chan, H. K. (2013). CFD analysis of the aerosolization of carrier-based dry powder inhaler formulations. In A. Yu, K. Dong, R. Yang, & S. Luding (Eds.) *AIP Conference Proceedings, 1542*(1), 1146–1149.

CHAPTER

10

Tissue engineering—Electrospinning approach

Marko N Živanović[a] *and Nenad Filipovic*[b]

[a]Department of Science, University of Kragujevac, Institute of Information Technologies Kragujevac, Kragujevac, Serbia [b]Bioengineering Research and Development Center (BioIRC), Kragujevac, Serbia

1 Introduction

Nano- and microscale fibers have been in nature for millions of years. For example, spiders, silkworms, and other insects produce fibers from which they create networks and other protective and functional structures. On the other hand, the network of protein components that make up the structure of the cytoskeleton within cells also makes an outstanding example of natural nanofibers with a certain structure. The microfilaments that cells express outside their membrane form the structure of the extracellular matrix (ECM), which is practically a niche in which cells are located, exist, divide, and perform a certain function.

We have always been inspired by fibers, finding application in various fields of activity. Nano- and microfibers are used in textile and construction industries, pharmacy, applied biotechnology, and bioengineering. Perhaps one of the first communications of Western and Eastern civilizations, which lasts to this day, dates back to the 13th century AC called *Silk Road*. Moreover, today's geopolitical challenges and initiatives are associated with the timing of the start of trade in micro-derived silk and other goods called the New Silk Road Initiative. The focus of this chapter is on the use of electrospun-derived fibers in tissue engineering. The first commercially available artificial microfibers were obtained from nylon in 1938 (Matthies & Seydl, 1986). Following this discovery, many natural and synthetic polymers were used for similar purposes. Electrospinning is a method based on the phenomenon of elongation of a solution of a charged polymer in a strong electric field into micro- and nanoscale thin structures that are immediately transformed into fibers by evaporation. In 1887, Charles V. Boys first observed the phenomenon of stretching a viscoelastic solution in the presence of a strong electric field (Boys, 1887). To date, this phenomenon is widely used to produce ECM-like fiber

https://doi.org/10.1016/B978-0-12-823956-8.00002-X

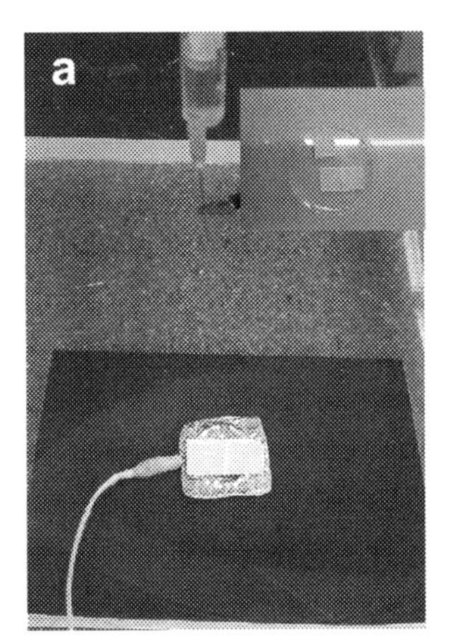

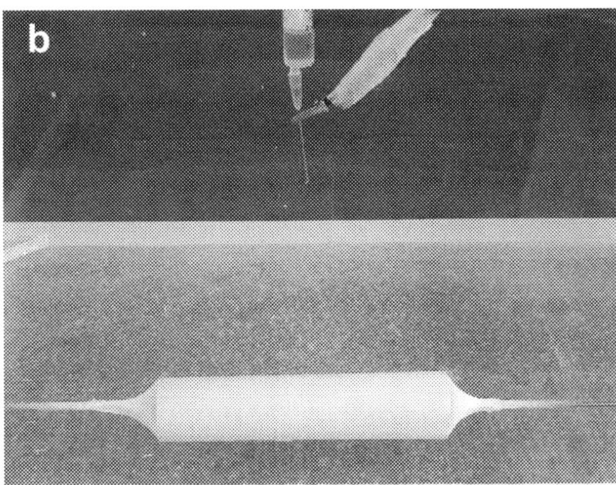

FIG. 1 Electrospinning setup device with flat (A) and rotating (B) negative electrode. *From Živanović, M. N. (2020). Use of electrospinning to enhance the versatility of drug delivery. In W. F. Lai (Eds.),* Systemic delivery technologies in anti-aging medicine: Methods and applications. Healthy ageing and longevity: Vol. 13. *Cham: Springer. https://doi.org/10.1007/978-3-030-54490-4_14, with permission.*

structures in scaffolds production and tissue engineering (Xue et al., 2019). Electrospinning is a complex multiphase and multiphysics phenomenon that includes several processes of mass and heat transfer and diffusion, evaporation, electrohydrodynamics, etc. (Živanović, 2020). The simplest typical electrospinning setup consists of a polymer source (syringe pump or other controlled flow systems), a very high voltage power supply (usually in the range of 5–40kV), and a metal collector on which a fiber network is built making a certain macroscopic structure. The polymer source and the collector are connected to a high potential supply to form a two-electrode system. The polymer generally must be positively charged and attracted during electrospinning by a negatively charged collector. The influence of high field and other controlled parameters (temperature and humidity in the electrospinning system chamber) affect the rapid evaporation of solvents and the formation of dry fibers (Fig. 1). The process of electrospinning involves knitting an endless fiber mainly on a random basis in 2D. As the process lasts, a 3D structure is slowly formed which results in a material intertwined with fibers that can be defined by the distribution of fibers of a certain density and diameter, as well as pores between them that can also be characterized by the same parameters.

The materials created in this way find various applications in industry, pharmacy, and biotechnology. For example, specific filter units can be created that are used to filter different fluids with very high efficiency (Barhate & Ramakrishna, 2007).

The topic of electrospinning is gaining more and more attention in the scientific community worldwide. According to data obtained from the Google Scholar platform, during the 1990s, the search term "Electrospinning" resulted in a relatively small number of scientific publications (below 1000 per year). However, in the years after 2000, the increase in interest and publication grew with logarithmic progression (Fig. 2). In the same timeframe, the term "Tissue engineering" gives a far higher number of results (about 100,000 per year in the period from 2012). However, if the search is filtered with the terms "Electrospinning + Tissue engineering," a significantly smaller number of publications is found, which shows that

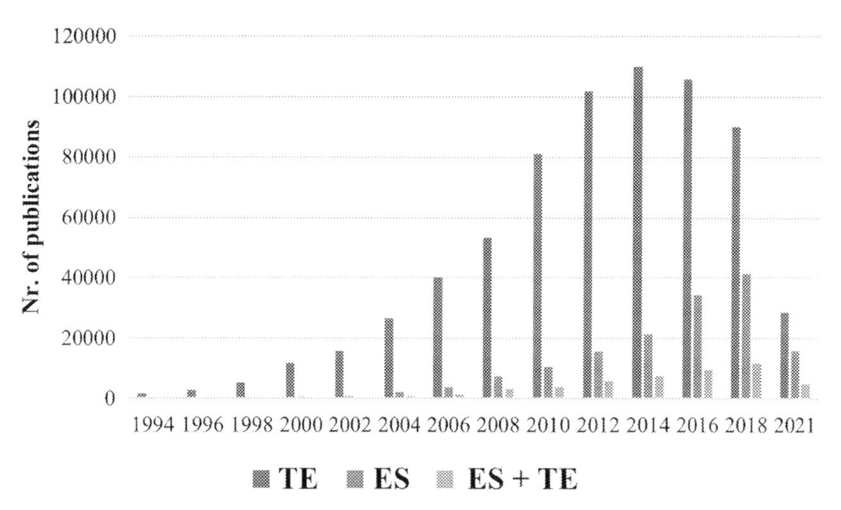

FIG. 2 Number of publications on electrospinning and tissue engineering 1994–2021. *TE*, tissue engineering; *ES*, electrospinning. *Data retrieved from Google Scholar platform.*

electrospinning in tissue engineering began to be applied on a more serious scale only after 2005 (with >1000 publications per year). Every following year, the scientific interest grows at a significant rate. This shows that this topic has just gained relevance and that there are still many segments that have yet to be clarified. In that sense, this review analysis of the situation in this narrow area is an attempt to contribute to the topic.

If this simple analysis on the Google Scholar platform is valid enough, we can assume that from 2016 until today, the total share of electrospinning technology in tissue engineering is very large, which is perhaps a more significant parameter than the number of published papers. This share is presented in Fig. 3.

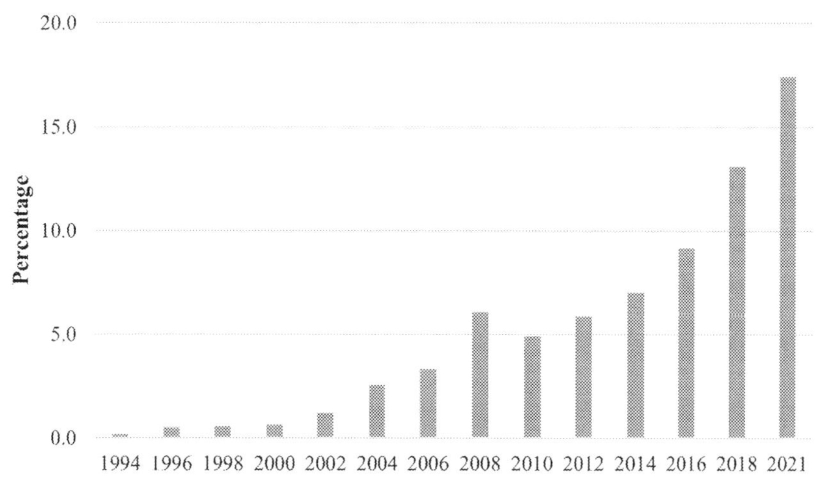

FIG. 3 Share Electrospinning in Tissue Engineering 1994–2021. *Data retrieved from Google Scholar platform.*

Regeneration and repair of human tissues and organs is a scientific discipline that is gaining importance. Comparative development of cell and molecular biology, bioengineering, immunology, materials science, artificial intelligence, and medicine results in solving coupled multiproblem tasks that will in the future lead us to the desired autologous transplantation of damaged body parts. However, to date, there is a major problem with the rejection of allografts, biocompatibility, and the availability of inserted materials (Clark et al., 2007; Crupi et al., 2015). Perhaps a turning point in science is the publication by Langer and Vacanti (1993) and after which the importance of this topic has greatly increased (Langer & Vacanti, 1993). One of the most significant challenges in this is the manipulation of cells in artificially prepared scaffolds that usually mimic the ECM environment of the native tissues. The combination of engineering and regenerative medicine today has resulted in a new scientific term that most closely describes this field: regenerative engineering (Laurencin & Nair, 2015). The focus of regenerative engineering is not only on repairing damaged tissues but also on creating complex physiological systems, such as entire organs. The use of advanced biomaterials certainly contributes to a better outcome. The materials used must possess certain specifics: they are nontoxic, biodegradable (if necessary), easy to manipulate, possess a certain biomechanical strength and resistance, do not cause inflammatory processes in the body after implantation.

2 Electrospinning in tissue engineering

2.1 Materials

In addition to mechanical fabrication of materials, 3D bioprinting, and others, electrospinning is a method that is very often used for these purposes. Laurencin and his group presented in 2002 the results on electrospinning production of poly(D/L-lactide-co-glycolide) fibers (PLGA) (Li et al., 2002). They showed that human bone-marrow-derived mesenchymal stem cells and mouse fibroblasts could be successfully grown on the produced scaffolds in a period of 10 days. This paper was included in the top 25 scientific papers in the Journal of Biomedical Materials Research over a period of 50 years. To date, PLGA is widely used in the field of electrospinning with application in tissue engineering. Google Scholar has so far detected a total of 17,000 scientific papers on this topic. Jose and coworkers presented results on aligned PLGA/hydroxyapatite electrospun-derived scaffolds for bone tissue engineering (Jose et al., 2009). Another approach was achieved in the field of, e.g., tissue engineering of the skin (Ru et al., 2015), nerve tissue (Wang et al., 2011), and vascular tissue (Jia et al., 2012).

Poly(ε-caprolactone) (PCL) is a widely used material for electrospinning of structures that are predominantly used in wound healing (Bui et al., 2014). Other purposes for PCL use could be found in the engineering of bone (Guo et al., 2015), skin (Prasad et al., 2015), neural tissue (Entekhabi et al., 2016), and others. Poly(L-lactic acid) (PLLA) material is widely used nowadays in medicine in general. For the purpose of tissue engineering, PLLA is usually spined for the regeneration of many tissues, such as neural (Yang et al., 2005), bone (Shim et al., 2010), and others. In the text below, one can find many other materials and their combinations that are used in electrospinning processes.

2.2 Characterization of electrospun-derived scaffolds for tissue engineering

In addition to the chemical composition of the fibers that enter the structure of the scaffold, there are many other parameters that must be optimal, and which are crucial for functionality. The basic method for characterization is microscopy. Optical microscopy has its limits and is limited by the lower resolution of nanofibers. That is why scanning electron microscopy (SEM) is most used to obtain the first information. Namely, SEM photographs are suitable for further processing in the *ImageJ* and *DiameterJ* programs to obtain data on the distribution of the diameters of the obtained fibers and on the porosity of the obtained material. As the structure of the ECM has already been processed a lot by SEM (Gillies & Lieber, 2011), it is not difficult to assess the condition of the newly obtained material after electrospinning. In addition to microscopy, scaffolds need to be evaluated by other methods. Determination of water absorption capability is a very important parameter because scaffolds intended for tissue engineering use, involve implantation in the aquatic environment. Water is the natural environment of cells and tissues and is very important due to many factors. One of the most important is the communication of cells and the transport of nutrients. In this sense, the porosity of the material is very important so that water molecules can settle inside the cavities. On the other hand, some materials strongly absorb water due to their nature, such as nanofibers derived from polysaccharides, such as hyaluronic acid (Castro et al., 2021). Water absorption on the other hand gives some stability to the scaffold. Scaffolds after water absorption, in most cases, retain the shape and structure as before absorption. On the other hand, as there is always a balance in chemical and biochemical processes, degradation and dissolution of scaffold is an opposite process of water absorption. Scaffolds intended for tissue engineering are generally created to be biodegradable, i.e., to perform a certain function and to disappear. One of the ways to determine this parameter is the use of PBS (phosphate buffer saline) as an aqueous solution in which spined scaffolds are immersed and whose mass reduction is determined at given time intervals. Compatibility of scaffolds with cells, i.e., their possible toxicity is determined by a standard MTT viability test (Gittens et al., 2011; Petrović et al., 2021). Chemical composition is a very important factor. If any component is toxic to the cells, then such material is not suitable for tissue engineering and implantation. The principle of the MTT method is based on seeding cells on created scaffolds and monitoring their viability over time. On prepared and UV sterilized scaffolds, cells are seeded and propagated under controlled conditions in an incubator at 37°C in a humified 5% CO_2 atmosphere in a cell culture medium. After incubation time (usually 24, 48, and 72h), the cell viability is determined by MTT test based on reduction of 3-(4,5-dimethylthiazol-2-yl)-2,5-diphenyltetrazolium bromide to purple yielded formazan crystals dissolved in dimethyl sulfoxide. The reduction rate is equivalent to the number of viable cells, measured on an ELISA reader. If absorbances measured at 550nm do not decrease or even increase, the scaffold could be considered in vitro compatible. In addition to the MTT test, seeded cells and primitive tissues are mostly photographed by the SEM method (Fig. 4).

Additionally, the process of characterization of the scaffold involves testing of mechanical properties, such as determining stress/strain curves (Fig. 5) and determining Young's module by the formula $E=\sigma/\varepsilon$. Determination of stress relaxation is suitable for proving the viscoelastic nature of the scaffolds.

FIG. 4 SEM images of Adipose-derived mesenchymal stem cells. Samples were fixed 48h in 2.5% glutaraldehyde and rinsed in 3% acetic acid and ethanol in the range of 25%–70%.

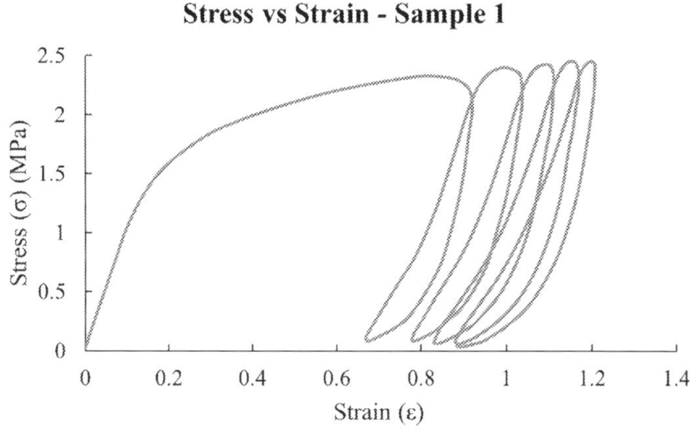

FIG. 5 Hysteresis loop response of sample: 22% solution of polycaprolactone/polyethylene glycol (1:1) in solvent dimethylformamide/chloroform (25:75) electrospinned under 12kV for 10min at flow rate 0.6mL/h.

2.3 Tissue regeneration

2.3.1 Skin regeneration

Although the use of electrospinning methods in tissue engineering is highly exploited, there are many challenges ahead in terms of optimizing these methods and obtaining functional parts of organs ready for transplantation. However, each method used today has its advantages and disadvantages. For example, electrospinning produces mostly uniformly arranged fibers with strong interconnection and porosity of 70%–95%. The fiber thickness ranges from a few dozens to several thousand nanometers, which gives the structure very similar to the ECM. The method is quite simple, but the disadvantage is the use of very high voltage and solvents which can often be toxic. Self-assembly of peptides and other biomacromolecules also gives the ECM a similar structure, but it is hard to reproduce in the same fashion every time, i.e., process automation is quite demanding. The use of 3D printing allows the production of exactly the desired geometry and structure, but in addition to the use of often toxic materials, porosity is a problem that leads to the brittleness of the material.

In addition, electrospinning methods can still be considered as the most optimal in order to create new tissues and organ replacement. Movahedi et al. (2020) has successfully used a combination of polyurethane/starch/hyaluronic acid materials to produce fibers that can be used in wound healing and skin tissue regeneration. Similarly, Koosha et al. presented the use (Koosha et al., 2019) of chitosan and polyvinyl alcohol for enhanced fibroblast cell attachment for skin tissue regeneration. Although there are many publications on the use of electrospinning in the production of skin tissues (Dias et al., 2016; Rahmati et al., 2020; Vázquez & Martínez, 2019), to our knowledge there is still no certified and approved skin production process for clinical purposes by the use of electrospinning-derived scaffolds. This is not surprising, given the scientific novelty of this topic.

2.3.2 *Bone regeneration*

In addition to attempts to produce skin, electrospun-derived scaffolds are widely used in bone tissue engineering. In this area, it is especially important to pay attention to the optimization of the production of scaffolds, which, in addition to biocompatibility and biodegradability, will have a precisely defined porosity and enhanced mechanical strength. The use of electrospinning enables work with various materials and their combinations. The following materials are most used for this purpose: poly(D/L-lactide)/gelatin/glass-ceramics (Bochicchio et al., 2020), silk (Gholipourmalekabadi et al., 2015), starch (Martins et al., 2009), polyurethane (Jaganathan et al., 2018), chitosan/polyethylene oxide (Singh et al., 2020), etc. Shih et al. (2006) used collagen type I to produce scaffolds. They showed that the use of bone marrow-derived mesenchymal stem cells is possible and that their propagation makes it possible to create bone tissue. In bone regeneration, the electrospinning method, in addition to the use of biomacromolecules and chemical synthetic polymers, also allows combination with inorganic substances, which is very important due to the mentioned mechanical strength. Thus Fujihara et al. (2005) used a combination of PCL synthetic polymer and $CaCO_3$ composite. The average fiber thickness was slightly higher (about 1000 nm), but the addition of $CaCO_3$ significantly increased the strength of the obtained material. The use of hydroxyapatite in all its forms is not new in the process of bone regeneration (Zhou & Lee, 2011), however, electrospinning of hydroxyapatite in combination with polyvinyl alcohol is possible and has been shown to regulate the fibrous network of significant mechanical strength (Dai & Shivkumar, 2007). Moreover, many studies reveal the use electrospun-derived nanofibers as reinforcement to strengthen hydrogels that can be used in bone engineering. Sadat-Shojai et al. (2016) developed a hybrid scaffold from electrospun fibers and hydrogels. In addition to mechanical strength and biocompatibility, they recorded cells infiltration into scaffold formation and occurrence of mineralization.

2.3.3 *Cartilage regeneration*

Nowadays, cartilage tissue engineering based on the use of electrospun-derived nanomaterials is also used to a significant extent. Cartilage tissue with its antifriction function consists of calcified parts and cells forming one complex tissue (Kuo et al., 2006). Many people around the world suffer from cartilage defects, and in that sense, every step toward the repair of this tissue is very important. The cartilage scaffolds mimic ECM structure suitable for the seeding and cultivation of chondrocytes. Different materials are used for these purposes. The PCL/gelatin combination was used to obtain slightly larger pores suitable for chondrocyte infiltration (Semitela et al., 2020). The poly(L-lactic acid)/silk combination was also successfully used for the same purpose (Li et al., 2016) where chondrocytes retained their native phenotype after seeding and cultivation. Reboredo et al. (2016) showed that Human mesenchymal stem cells can regenerate into native cartilage structures on electrospun collagen and derived scaffolds. Similarly, as with other organs and tissues, the hybrid cryo-printing and electrospinning technique produced scaffolds suitable for mesenchymal stem cell growth and cartilage formation (Munir et al., 2019). Like bone tissue, cartilage scaffolds must also possess significant mechanical resistance. Therefore, the use of poly(lactic-co-glycolic acid) (PLGA) resulted in the production of mechanically stable, nontoxic scaffolds where cell proliferation and ECM formation were superior to membrane-like produced scaffolds, thus non-electrospinning scaffolds (Shin et al., 2006). As the cartilage is a quite simple tissue, some progress in terms of implants has been made. Xue et al. (2013) were able to use the PCL/gelatin combination for spinning production of ear-shaped cartilage and implanted in vivo, where the structure was maintained for 6 weeks. Engineered cartilage showed good elasticity and impressive mechanical strength.

2.3.4 *Vascular regeneration*

Electrospun vascular nanofibrous scaffolds also represent a significant application in preclinical practice with high ambitions for immediate clinical use. Reconstruction of blood vessels is of utmost importance for clinical applications. The blood vessel has three multifunctional layers: tunica intima, tunica media, and tunica adventitia. In the tunica intima layer lies endothelial cells that can be grown on specially created porous biodegradable scaffolds. On the other hand, the tunica media is composed of smooth muscle cells, which can be similarly cultured on scaffolds, while the outer layer of the tunica adventitia is composed of ECM and fibroblasts. The rotation technique in electrospinning can result in the production of tube-like scaffolds suitable for cell preparation and maintenance (Živanović, 2020). The technique of creating a blood vessel itself is possible in two ways. One is the cultivation of cells layer by layer, where each layer of cells lies on its own scaffold, which is very rapidly degraded so that the cells forming the tunica intima and tunica media can come into contact with each other very quickly. Another possible way is to grow both types of cells on the same scaffold in two zones next to each other, whereby simply twisting the scaffold provides smooth muscle cells in the layer above the endothelial cells. A lot of effort has been made in this direction. Wang et al. (2012) spined a polyurethane/poly(ethylene glycol) combination producing scaffolds with a lower possibility of thrombi formation. HUVEC cells were successfully cultured for a period of 14 days. Huang et al. (2018) produced triple-layer hybrid vascular grafts that were used for in vivo implantation. The produced scaffolds proved to enhance cell growth and infiltration. Similarly, to many authors, Zhao et al. (2021) produced PCL/fibrin vascular grafts implanted in vivo.

2.3.5 Cardiac tissue regeneration

Cardiac tissue engineering is also one of the very important fields for the application of electrospinning. Cardiac scaffolds and grafts also need to possess significant elasticity and mechanical resistance. For example, PCL is a material that is widely used in this area. Balguid et al. (2009) used PCL to produce cardiac scaffolds and showed significant cellular infiltration in cardiovascular tissue in vivo. Many authors had a similar approach and showed that in vivo application of similar grafts is possible (Best et al., 2019; Wu et al., 2018).

3 Conclusion remarks

The application of electrospinning in tissue engineering is very promising. Simple technical setup and proper selection of polymer composition, in many cases, will result in the production of nanofibers that can be used to mimic the ECM. ECM-like structures is crucial for cell growth in vitro and in vivo. Primitive tissues and/or parts of organs produced in the laboratory can be used in clinical practice. In vivo studies prove this daily. However, given that this branch of science is relatively new, full clinical application is yet to be expected. This text briefly describes only a small part of the application of electrospinning in tissue engineering. As for the material, it can be concluded that there are many classes of materials that can be spined: synthetic polymers, proteins and peptides, polysaccharides, combinations of the above with inorganic compounds, there are even studies on spinning of DNA itself. A special part of the application of such scaffolds can be the production of drug-bearing materials. Nowadays, it is necessary to interconnect several branches of science into a single goal. Thus, the use of advanced mathematical models (Sakellarios et al., 2020) can predict, e.g., releasing the drug from some material (Filipović & Živanović, 2020). In other words, the introduction of other branches of science in this special subfield of tissue engineering can lead to significant innovations that will result in a targeted goal—the replacement of damaged parts of the human body.

References

Balguid, A., Mol, A., van Marion, M. H., Bank, R. A., Bouten, C. V., & Baaijens, F. P. (2009). Tailoring fiber diameter in electrospun poly(epsilon-caprolactone) scaffolds for optimal cellular infiltration in cardiovascular tissue engineering. *Tissue Engineering. Part A, 15*(2), 437–444. https://doi.org/10.1089/ten.tea.2007.0294.

Barhate, R. S., & Ramakrishna, S. (2007). Nanofibrous filtering media: Filtration problems and solutions from tiny materials. *Journal of Membrane Science, 296*(1–2), 1–8. https://doi.org/10.1016/j.memsci.2007.03.038.

Best, C. A., Szafron, J. M., Rocco, K. A., Zbinden, J., Dean, E. W., Maxfield, M. W., Kurobe, H., Tara, S., Bagi, P. S., Udelsman, B. V., Khosravi, R., Yi, T., Shinoka, T., Humphrey, J. D., & Breuer, C. K. (2019). Differential outcomes of venous and arterial tissue engineered vascular grafts highlight the importance of coupling long-term implantation studies with computational modeling. *Acta Biomaterialia, 94*, 183–194. https://doi.org/10.1016/j.actbio.2019.05.063.

Bochicchio, B., Barbaro, K., De Bonis, A., Rau, J. V., & Pepe, A. (2020). Electrospun poly(d,l-lactide)/gelatin/glass-ceramics tricomponent nanofibrous scaffold for bone tissue engineering. *Journal of Biomedical Materials Research. Part A, 108*(5), 1064–1076. https://doi.org/10.1002/jbm.a.36882.

Boys, C. V. (1887). On the production, properties, and some suggested uses of the finest threads. *Proceedings of the Physical Society of London*, *9*, 8–19.

Bui, H. T., Chung, O. H., Dela Cruz, J., & Park, J. S. (2014). Fabrication and characterization of electrospun curcumin-loaded polycaprolactone-polyethylene glycol nanofibers for enhanced wound healing. *Macromolecular Research*, *22*, 1288–1296. https://doi.org/10.1007/s13233-014-2179-6.

Castro, K. C., Campos, M. G. N., & Mei, L. H. I. (2021). Hyaluronic acid electrospinning: Challenges, applications in wound dressings and new perspectives. *International Journal of Biological Macromolecules*, *173*, 251–266. https://doi.org/10.1016/j.ijbiomac.2021.01.100.

Clark, R. A. F., Ghosh, K., & Tonnesen, M. G. (2007). Tissue engineering for cutaneous wounds. *The Journal of Investigative Dermatology*, *127*(5), 1018–1029. https://doi.org/10.1038/sj.jid.5700715.

Crupi, A., Costa, A., Tarnok, A., Melzer, S., & Teodori, L. (2015). Inflammation in tissue engineering: The Janus between engraftment and rejection. *European Journal of Immunology*, *45*(12), 3222–3236. https://doi.org/10.1002/eji.201545818.

Dai, X., & Shivkumar, S. (2007). Electrospinning of hydroxyapatite fibrous mats. *Materials Letters*, *6*(13), 2735–2738. https://doi.org/10.1016/j.matlet.2006.07.195.

Dias, J. R., Granja, P. L., & Bártolo, P. J. (2016). Advances in electrospun skin substitutes. *Progress in Materials Science*, *84*, 314–334. https://doi.org/10.1016/j.pmatsci.2016.09.006.

Entekhabi, E., Nazarpak, M. H., Moztarzadeh, F., & Sadeghi, A. (2016). Design and manufacture of neural tissue engineering scaffolds using hyaluronic acid and polycaprolactone nanofibers with controlled porosity. *Materials Science & Engineering. C, Materials for Biological Applications*, *69*, 380–387. https://doi.org/10.1016/j.msec.2016.06.078.

Filipović, N., & Živanović, M. N. (2020). Use of numerical simulation in carrier characterization and optimization. In W. F. Lai (Ed.), *Vol. 13. Systemic delivery technologies in anti-aging medicine: Methods and applications. Healthy ageing and longevity*. Cham: Springer. https://doi.org/10.1007/978-3-030-54490-4_18.

Fujihara, K., Kotaki, M., & Ramakrishna, S. (2005). Guided bone regeneration membrane made of polycaprolactone/calcium carbonate composite nano-fibers. *Biomaterials*, *26*(19), 4139–4147. https://doi.org/10.1016/j.biomaterials.2004.09.014.

Gholipourmalekabadi, M., Mozafari, M., Bandehpour, M., Salehi, M., Sameni, M., Caicedo, H. H., Mehdipour, A., Hamidabadi, H. G., Samadikuchaksaraei, A., & Ghanbarian, H. (2015). Optimization of nanofibrous silk fibroin scaffold as a delivery system for bone marrow adherent cells: In vitro and in vivo studies. *Biotechnology and Applied Biochemistry*, *62*(6), 785–794. https://doi.org/10.1002/bab.1324.

Gillies, A. R., & Lieber, R. L. (2011). Structure and function of the skeletal muscle extracellular matrix. *Muscle & Nerve*, *44*(3), 318–331. https://doi.org/10.1002/mus.22094.

Gittens, R. A., McLachlan, T., Olivares-Navarrete, R., Cai, Y., Berner, S., Tannenbaum, R., Schwartz, Z., Sandhage, K. H., & Boyan, B. D. (2011). The effects of combined micron−/submicron-scale surface roughness and nanoscale features on cell proliferation and differentiation. *Biomaterials*, *32*(13), 3395–3403. https://doi.org/10.1016/j.biomaterials.2011.01.029.

Guo, Z., Xu, J., Ding, S., Li, H., Zhou, C., & Li, L. (2015). In vitro evaluation of random and aligned polycaprolactone/gelatin fibers via electrospinning for bone tissue engineering. *Journal of Biomaterials Science. Polymer Edition*, *26*(15), 989–1001. https://doi.org/10.1080/09205063.2015.1065598.

Huang, R., Gao, X., Wang, J., Chen, H., Tong, C., Tan, Y., & Tan, Z. (2018). Triple-layer vascular grafts fabricated by combined E-jet 3D printing and electrospinning. *Annals of Biomedical Engineering*, *46*(9), 1254–1266. https://doi.org/10.1007/s10439-018-2065-z.

Jaganathan, S. K., Mani, M. P., Palaniappan, S. K., & Rathanasamy, R. (2018). Fabrication and characterisation of nanofibrous polyurethane scaffold incorporated with corn and neem oil using single stage electrospinning technique for bone tissue engineering applications. *Journal of Polymer Research*, *25*, 146. https://doi.org/10.1007/s10965-018-1543-1.

Jia, X., Zhao, C., Li, P., Zhang, H., Huang, Y., Li, H., Fan, J., Feng, W., Yuan, X., & Fan, Y. (2012). Sustained release of VEGF by coaxial electrospun dextran/PLGA fibrous membranes in vascular tissue engineering. *Journal of Biomaterials Science. Polymer Edition*, *22*(13), 1811–1827. https://doi.org/10.1163/092050610X528534.

Jose, M. V., Thomas, V., Johnson, K. T., Dean, D. R., & Nyairo, E. (2009). Aligned PLGA/HA nanofibrous nanocomposite scaffolds for bone tissue engineering. *Acta Biomaterialia*, *5*(1), 305–315. https://doi.org/10.1016/j.actbio.2008.07.019.

Koosha, M., Raoufi, M., & Moravvej, H. (2019). One-pot reactive electrospinning of chitosan/PVA hydrogel nanofibers reinforced by halloysite nanotubes with enhanced fibroblast cell attachment for skin tissue regeneration. *Colloids and Surfaces. B, Biointerfaces*, *179*, 270–279. https://doi.org/10.1016/j.colsurfb.2019.03.054.

Kuo, C. K., Li, W. J., Mauck, R. L., & Tuan, R. S. (2006). Cartilage tissue engineering: Its potential and uses. *Current Opinion in Rheumatology, 18*(1), 64–73. https://doi.org/10.1097/01.bor.0000198005.88568.df.

Langer, R., & Vacanti, J. P. (1993). Tissue engineering. *Science, 260*(5110), 920–926. https://doi.org/10.1126/science.8493529.

Laurencin, C. T., & Nair, L. S. (2015). Regenerative engineering: Approaches to limb regeneration and other grand challenges. *Regenerative Engineering and Translational Medicine, 1,* 1–3. https://doi.org/10.1007/s40883-015-0006-z.

Li, W.-J., Laurencin, C. T., Caterson, E. J., Tuan, R. S., & Ko, F. K. (2002). Electrospun nanofibrous structure: A novel scaffold for tissue engineering. *Journal of Biomedical Materials Research, 60*(4), 613–621. https://doi.org/10.1002/jbm.10167.

Li, Z., Liu, P., Yang, T., Sun, Y., You, Q., Li, J., Wang, Z., & Han, B. (2016). Composite poly(l-lactic-acid)/silk fibroin scaffold prepared by electrospinning promotes chondrogenesis for cartilage tissue engineering. *Journal of Biomaterials Applications, 30*(10), 1552–1565. https://doi.org/10.1177/0885328216638587.

Martins, A., Chung, S., Pedro, A. J., Sousa, R. A., Marques, A. P., Reis, R. L., & Neves, N. M. (2009). Hierarchical starch-based fibrous scaffold for bone tissue engineering applications. *Journal of Tissue Engineering and Regenerative Medicine, 3*(1), 37–42. https://doi.org/10.1002/term.132.

Matthies, P., & Seydl, W. F. (1986). History and development of nylon 6. In R. B. Seymour, & G. S. Kirshenbaum (Eds.), *High performance polymers: Their origin and development*. Dordrecht: Springer.

Movahedi, M., Asefnejad, A., Rafienia, M., & Khorasani, M. T. (2020). Potential of novel electrospun core-shell structured polyurethane/starch (hyaluronic acid) nanofibers for skin tissue engineering: In vitro and in vivo evaluation. *International Journal of Biological Macromolecules, 146,* 627–637. https://doi.org/10.1016/j.ijbiomac.2019.11.233.

Munir, N., McDonald, A., & Callanan, A. (2019). A combinatorial approach: Cryo-printing and electrospinning hybrid scaffolds for cartilage tissue engineering. *Bioprinting, 16*. https://doi.org/10.1016/j.bprint.2019.e00056, e00056.

Petrović, A. Z., Ćoćić, D. C., Bockfeld, D., Živanović, M., Milivojević, N., Virijević, K., Janković, N., Scheurer, A., Vraneš, M., & Bogojeski, J. V. (2021). Biological activity of bis(pyrazolylpyridine) and terpiridine Os(II) complexes in the presence of biocompatible ionic liquids. *Inorganic Chemistry Frontiers, 8,* 2749–2770. https://doi.org/10.1039/D0QI01540G.

Prasad, T., Shabeena, E. A., Vinod, D., Kumary, T. V., & Kumar, P. R. A. (2015). Characterization and in vitro evaluation of electrospun chitosan/polycaprolactone blend fibrous mat for skin tissue engineering. *Journal of Materials Science. Materials in Medicine, 26*(1), 5352. https://doi.org/10.1007/s10856-014-5352-8.

Rahmati, M., Blaker, J. J., Lyngstadaas, S. P., Mano, J. F., & Haugen, H. J. (2020). Designing multigradient biomaterials for skin regeneration. *Materials Today Advances, 5*. https://doi.org/10.1016/j.mtadv.2019.100051, 100051.

Reboredo, J. W., Weigel, T., Steinert, A., Rackwitz, L., Rudert, M., & Walles, H. (2016). Investigation of migration and differentiation of human mesenchymal stem cells on five-layered collagenous electrospun scaffold mimicking native cartilage structure. *Advanced Healthcare Materials, 5*(17), 2191–2198. https://doi.org/10.1002/adhm.201600134.

Ru, C., Wang, F., Pang, M., Sun, L., Chen, R., & Sun, Y. (2015). Suspended, shrinkage-free, electrospun PLGA nanofibrous scaffold for skin tissue engineering. *ACS Applied Materials & Interfaces, 7*(20), 10872–10877. https://doi.org/10.1021/acsami.5b01953.

Sadat-Shojai, M., Khorasani, M. T., & Jamshidi, A. (2016). A new strategy for fabrication of bone scaffolds using electrospun nano-HAp/PHB fibers and protein hydrogels. *Chemical Engineering Journal, 289,* 38–47. https://doi.org/10.1016/j.cej.2015.12.079.

Sakellarios, A., Correia, J., Kyriakidis, S., Georga, E., Tachos, N., Siogkas, P., Sans, F., Stofella, P., Massimiliano, V., Clemente, A., Rocchiccioli, S., Pelosi, G., Filipovic, N., & Fotiadis, D. I. (2020). A cloud-based platform for the noninvasive management of coronary artery disease. *Enterprise Information Systems, 14*(8), 1102–1123. https://doi.org/10.1080/17517575.2020.1746975.

Semitela, Â., Girão, A. F., Fernandes, C., Ramalho, G., Bdikin, I., Completo, A., & Marques, P. A. (2020). Electrospinning of bioactive polycaprolactone-gelatin nanofibres with increased pore size for cartilage tissue engineering applications. *Journal of Biomaterials Applications, 35*(4–5), 471–484. https://doi.org/10.1177/0885328220940194.

Shih, Y. R. V., Chen, C. N., Tsai, S. W., Wang, Y. V., & Lee, O. K. (2006). Growth of mesenchymal stem cells on electrospun type I collagen nanofibers. *Stem Cells, 24*(11), 2391–2397. https://doi.org/10.1634/stemcells.2006-0253.

Shim, I. K., Jung, M. R., Kim, K. H., Seol, Y. J., Park, Y. J., Park, W. H., & Lee, S. J. (2010). Novel three-dimensional scaffolds of poly(L-lactic acid) microfibers using electrospinning and mechanical expansion: Fabrication and bone regeneration. *Journal of Biomedical Materials Research. Part B, Applied Biomaterials*, *95*(1), 150–160. https://doi.org/10.1002/jbm.b.31695.

Shin, H. J., Lee, C. H., Cho, I. H., Kim, Y. J., Lee, Y. J., Kim, I. A., Park, K. D., Yui, N., & Shin, J. W. (2006). Electrospun PLGA nanofiber scaffolds for articular cartilage reconstruction: Mechanical stability, degradation and cellular responses under mechanical stimulation in vitro. *Journal of Biomaterials Science. Polymer Edition*, *17*(1–2), 103–119. https://doi.org/10.1163/156856206774879126.

Singh, Y. P., Dasgupta, S., Nayar, S., & Bhaskar, R. (2020). Optimization of electrospinning process & parameters for producing defect-free chitosan/polyethylene oxide nanofibers for bone tissue engineering. *Journal of Biomaterials Science. Polymer Edition*, *31*(6), 781–803. https://doi.org/10.1080/09205063.2020.1718824.

Vázquez, J. J., & Martínez, E. S. M. (2019). Collagen and elastin scaffold by electrospinning for skin tissue engineering applications. *Journal of Materials Research*, *34*, 2819–2827. https://doi.org/10.1557/jmr.2019.233.

Wang, H., Feng, Y., Fang, Z., Yuan, W., & Khan, M. (2012). Co-electrospun blends of PU and PEG as potential biocompatible scaffolds for small-diameter vascular tissue engineering. *Materials Science and Engineering: C*, *32*(8), 2306–2315. https://doi.org/10.1016/j.msec.2012.07.001.

Wang, G., Hu, X., Lin, W., Dong, C., & Wu, H. (2011). Electrospun PLGA–silk fibroin–collagen nanofibrous scaffolds for nerve tissue engineering. *In Vitro Cellular & Developmental Biology. Animal*, *47*, 234–240. https://doi.org/10.1007/s11626-010-9381-4.

Wu, T., Zhang, J., Wang, Y., Li, D., Sun, B., El-Hamshary, H., Yin, M., & Mo, X. (2018). Fabrication and preliminary study of a biomimetic tri-layer tubular graft based on fibers and fiber yarns for vascular tissue engineering. *Materials Science & Engineering. C, Materials for Biological Applications*, *82*, 121–129. https://doi.org/10.1016/j.msec.2017.08.072.

Xue, J., Feng, B., Zheng, R., Lu, Y., Zhou, G., Liu, W., Cao, Y., Zhang, Y., & Zhang, W. J. (2013). Engineering ear-shaped cartilage using electrospun fibrous membranes of gelatin/polycaprolactone. *Biomaterials*, *34*(11), 2624–2631. https://doi.org/10.1016/j.biomaterials.2012.12.011.

Xue, J., Wu, T., Dai, Y., & Xia, Y. (2019). Electrospinning and electrospun nanofibers: Methods, materials, and applications. *Chemical Reviews*, *119*(8), 5298–5415. https://doi.org/10.1021/acs.chemrev.8b00593.

Yang, F., Murugan, R., Wang, S., & Ramakrishna, S. (2005). Electrospinning of nano/micro scale poly(L-lactic acid) aligned fibers and their potential in neural tissue engineering. *Biomaterials*, *26*(15), 2603–2610. https://doi.org/10.1016/j.biomaterials.2004.06.051.

Zhao, L., Li, X., Yang, L., et al. (2021). Evaluation of remodeling and regeneration of electrospun PCL/fibrin vascular grafts in vivo. *Materials Science & Engineering. C, Materials for Biological Applications*, *118*. https://doi.org/10.1016/j.msec.2020.111441, 111441.

Zhou, H., & Lee, J. (2011). Nanoscale hydroxyapatite particles for bone tissue engineering. *Acta Biomaterialia*, *7*(7), 2769–2781. https://doi.org/10.1016/j.actbio.2011.03.019.

Živanović, M. N. (2020). Use of electrospinning to enhance the versatility of drug delivery. In W. F. Lai (Ed.), *Vol. 13. Systemic delivery technologies in anti-aging medicine: Methods and applications. Healthy ageing and longevity*. Cham: Springer. https://doi.org/10.1007/978-3-030-54490-4_14.

CHAPTER

11

Application of numerical methods for the analysis of respiratory system

Aleksandra Vulović and Nenad Filipovic

Bioengineering Research and Development Center (BioIRC), Kragujevac, Serbia

1 Introduction

The respiratory system represents a system of tissues and organs that help us breathe. The surface of the respiratory tract is the largest interface between humans and their environments and as such is constantly exposed to a spectrum of contaminants and particulates dispersed in the air.

The primary function of the respiratory system is to

- transport oxygen from the external environment to cells,
- remove carbon dioxide produced by cell metabolism from the body,
- protect airways from harmful substances and irritants that we breathe in, and
- maintain acid-base balance.

The sections of the respiratory system are also responsible for nonvital functions such as speech production and sensing odors. Noninvasive approach to the analysis of the respiratory system has been employed for the analysis of airways, which are responsible for delivering air to our lungs. The airways are divided into two sections:

- upper airways (from the nose to the vocal cords) and
- lower airways (from trachea and the bronchial structures to the alveolus).

The upper respiratory tract (URT) consists of the anterior nares, nasal cavity, sinuses, nasopharynx, Eustachian tube, middle ear cavity, oral cavity, oropharynx, and larynx (Kumpitsch et al., 2019). Air transports to and from the lower airways through the mouth or nose. The breathing through the nose or mouth depends on the activity that is performed (Bennett et al., 2003). When at rest, the air is usually passing through the nose, while during activities such as exercise, up to 70% of air is inhaled through the mouth (Foster & Costa,

https://doi.org/10.1016/B978-0-12-823956-8.00005-5

2005). Nasal breathing is a better option as it allows for filtering of the particulate matter as well as the lung defence. It was shown that total deposition of 0.5–3 μm particles is higher for nasal breathing (Heyder et al., 1975). Experimental research has shown that fine and coarse particle deposition increases with increased flow rate and ellipticity of nostril dimensions and decreased minimal cross-sectional area of the nasal passage (Kesavanathan et al., 1998).

The tracheobronchial tree or the lower respiratory tract begins at the larynx (Generation 0) and ends in the alveoli. The trachea or larynx divides into left and right main-stem bronchi which are known as Generation 1. Starting from Generation 1, airways penetrate the lung tissue known as lung parenchyma. Due to the angle of bifurcation and the size, foreign bodies more often enter the right main-stem bronchus. Generation 1 bronchi branch into Generation 2 bronchi known as lobar bronchi which branch to the Generation 3 bronchi (segmental bronchi). Up until the seventh generation, the airway branching can be considered symmetric in terms of branching angle, size, number of branches, and the number of subsequent generations. Starting from the seventh generation, the branching is considered asymmetric, that is, the number of generations can significantly vary between individuals. Each new airway generation has smaller and numerous airways which penetrate deeper into the lung tissue.

2 Air transport and particle deposition

Transport of air in the lungs occurs by diffusion and convection. Depending on a region in the lungs one of these two mechanisms dominates. In the bronchi, convection is the dominant mechanism, while in the bronchioles diffusion is the dominant mechanism. As we move to the transitional and respiratory zone, transport of the air happens predominately by diffusion (Foster & Costa, 2005).

Deposition of particles in the respiratory tract depends on (Agnew, 1984):

- physical characteristics (size, shape, and density) and
- chemical (hygroscopicity and charge) characteristics of the inhaled particles

Particle size is one of the most important parameters for particle deposition in the lung. Particles between 0.3 and 0.7 μm diameter have minimal deposition in the lung while particles with a diameter above and below this range have higher deposition rates.

The deposition also depends on (Foster & Costa, 2005):

- breathing pattern (volume and rate),
- route of breathing (mouth vs. nose), and
- the anatomy of the airways.

Breathing patterns that include tidal volumes, breathing rates, and the route of inhalation are different among people. Breathing patterns depend on age, race, gender, activity, and respiratory health (Brown et al., 2002). The route of inhalation is related to the performed activities and it is noticed that different activities have different effects on tidal volumes and breathing rates. In the 1980s a large analysis was performed in a laboratory in order to determine empirical equations for particle deposition in correlation to breathing patterns and the route of breathing (Stahlhofen et al., 1989). During this study particles from 0.005 to 15 μm were analyzed. Their findings indicated that deposition increases with the increase of the tidal volume during the constant breathing period. It was also noticed that increased tidal volume leads to an increased number of particles that reach distal airways.

With aging, changes in airway structure and breathing patterns happen, even for healthy individuals. The effect of aging has been analyzed and the results indicate that there is no difference in tidal volume between young and old individuals with normal lung function (Tobin et al., 1983). Also, age had no effect on the whole lung deposition fraction of fine particles (2 μm) when analyzed using spontaneous breathing conditions in adults (aged 18–80) with normal lung function. Results for fixed breathing patterns indicated a mild decrease in the whole lung deposition of fine particles with aging (Bennett et al., 1996). Difference between genders was noticed when analyzing the particle deposition. Whole lung deposition in adult females is increased compared to adult males across a large range of inhaled particle sizes for a fixed breathing pattern (the same tidal volume and frequency) (Foster & Costa, 2005). For comparable particle sizes (2.5–7.5 μm size range) and fixed inspiratory flow rates, females had smaller alveolar and higher extrathoracic and tracheobronchial deposition compared to males (Pritchard et al., 1986). A greater deposition fraction in females was also noticed under normal resting breathing patterns.

Deposition of the particles occurs primarily by the following mechanisms (Foster & Costa, 2005):

- diffusion,
- impaction, and
- sedimentation.

Sedimentation and diffusion mainly occur in the smaller distal airways where the distance to an airway surface is short, while the impaction occurs in the large proximal airways where linear velocities are high. Impaction and sedimentation depend on a particle's aerodynamic diameter (d_{ae}). Impaction occurs when a particle, due to its inertia, is unable to follow a change in the flow direction, while sedimentation occurs by the gravitational settling of particles to an airway wall. Diffusive deposition happens when a particle reaches an airway surface by random Brownian movement. The diffusive and sedimentary deposition is important for particles in the range of 0.1–1 μm, while impaction and sedimentation predominate above and diffusion predominates below this range (Foster & Costa, 2005). Data on the regional deposition of inhaled particles are available for the nose, mouth, tracheobronchial airways as well as alveolar region. Out of these four regions, only deposition in the nose and mouth could be directly measured or calculated based on different breathing maneuvers, while for the other two regions it cannot be measured in vivo. Data on the tracheobronchial and alveolar deposition has been acquired using radiolabeled particles (Foster & Costa, 2005).

3 Respiratory system models

The complexity of the respiratory system models has changed a lot during the past decades. Recent technological development has allowed for the increase in the number and complexity of in silico respiratory models. The first models that were used were significantly

FIG. 1 Simplified airway structure.

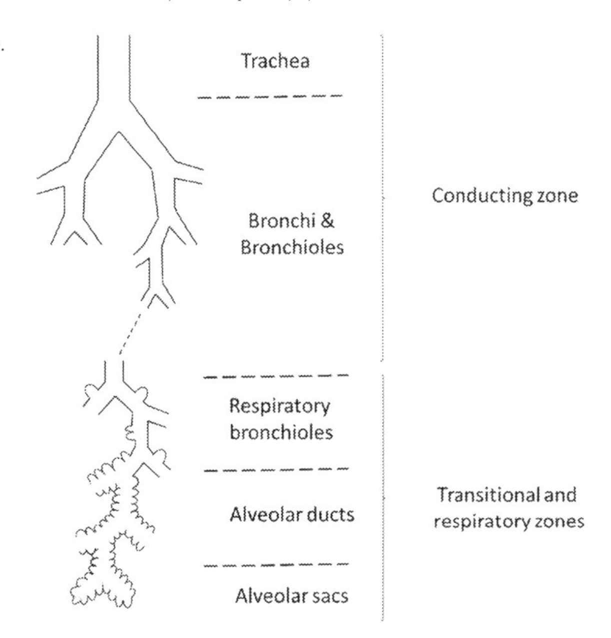

simpler than the models in use today. Most of the early models included simplistic geometry such as the models by Weibel (1963) or Horsfield et al. (1971). The model presented by Weibel starts with the trachea which branches into 23 symmetric and dichotomously branching generations (z) of the airways (Fig. 1).

The trachea is considered to be generation 0. The total number of airways in a generation can be calculated as 2^z, where z is the generation we are calculating the number of airways for.

These airway generations are divided into (Table 1):

- Conducting zone—Generations 0–16 and
- Transitional and respiratory zones—Generations 17–23.

Conducting zone is responsible for air transport to and from the distal airways. It consists of bronchi and bronchioles. Cartilage maintains the size and shape of bronchi and bronchioles start to diminish after the 12th generation and disappears at the end of conducting zone (Foster & Costa, 2005).

A couple of decades after models presented by Weibel and Horsfield, a more complex model was suggested. It was based on the Weibel model, but the parent and daughter tubes were considered to lie in different planes (Balásházy & Hofmann, 1995). Newer airway models have been able to integrate geometry based on images (Ma & Lutchen, 2006); physiological airway curvature and half-mouth opening (Xi & Longest, 2007) realistic respiration maneuvers (Miyawaki et al., 2016), lung mechanics (Oakes et al., 2014) as well as all pathways from the first bifurcation to the subacinus unit (Koullapis et al., 2020; Koullapis, Hofemeier, et al., 2018). One of the most commonly used models is the model which includes 1453 bronchi extending to the 17th generation (Schmidt et al., 2004). Generations of human airways

TABLE 1 Airway generations and their approximate dimensions (Chovancová & Elcner, 2014).

	Generation		Number	Diameter (cm)	Length (cm)	Total cross sectional area (cm^2)
Conducting zone	Trachea	0	1	1.8	12	2.54
	Bronchi	1	2	1.22	4.8	2.33
		2	4	0.83	1.9	2.13
		3	8	0.56	0.8	2.0
	Bronchioles	4	16	0.45	1.3	2.48
	Terminal bronchioles	5	32	0.35	1.07	3.11
		16	6×10^4	0.06	0.17	180.0
Transitional and respiratory zones	Respiratory bronchioles	17	↓	↓	↓	↓
		18				
		19	5×10^5	0.05	0.10	10^3
	Alveolar ducts	20	↓	↓	↓	↓
		21				
		22				
	Alveolar sacks	23	8×10^6	0.04	0.05	10^4

containing both the upper airways and the TB tree down to the 6th–7th, or to the 17th have recently been created (Atzeni et al., 2021). The SimInhale benchmark case model has been developed as a reference and guidance for future computational fluid-particle dynamics (CFPD) models related to human airway modeling (Koullapis, Kassinos, et al., 2018). The aim of this benchmark case model was to provide information related to a lack of a full description of the flow field from the oral cavity down to the upper conducting airways of the bronchial tree; effect of turbulent variations on particle deposition; and employment of RANS and large eddy simulations (LES) models.

The main issue with imaging data is the quality of the image at or above the third generation of airways. Due to the heartbeat that occurs during the acquisition of the images, a significant reduction in the image quality is noticeable. Analysis of the respiratory system is extremely complex due to the variety of airflow regimes that need to be taken into consideration. These regimes range from intermittent to turbulent in the extrathoracic airways to almost constant in the acinar depths (Koullapis et al., 2019). MRI images of the lung did not have sufficient signal and contrast from the lung parenchyma and/or the airway wall which was a problem for the model development (Altes et al., 2007).

The model reconstruction consists of the following steps:

- Preprocessing
- 3D airway segmentation
- Airway skeletonization
- Airway cross-section estimation

The first step in this process is to enhance the airway wall using filtering methods. Two approaches have been noticed—an approach based on the morphological operators (in order to highlight black structure surrounded by white structure) and based on the Hessian-based operators in order to highlight dark tubular structure (Baldacci et al., 2019). The second step in this process is related to the extraction of the 3D airway tree. The main problem during this phase is to find bronchi that are significantly getting smaller with each generation (Table 1). A commonly used approach for segmentation is to use a region growing algorithm. This algorithm is initialized in the trachea and will combine voxels that belong to the airway lumen (appear as a dark structure on a CT/MRI scan data) until they reach voxels that correspond to the airway wall (appear as a bright structure on a CT/MRI scan data) (Baldacci et al., 2019). The next step in this process is related to the computation of the curve-skeleton of the binary volume (Cornea et al., 2007). The extraction of the curve-skeleton of a bronchial tree is commonly performed using thinning approach (Couprie et al., 2007) where the aim is to remove the volume until only the central line remains. This approach can lead to false positive branches or increased noise of the curve which then needs to be smoothed (Grélard et al., 2017). The last step is focused on the reconstruction of a bronchial tree orthogonal cross section in order to estimate its diameter. The cross section is usually estimated from the skeleton. Additional approaches have been developed and employed such as trying to recognize the local digital straight segments passing through a point of the curve (Postolski et al., 2012), or by analyzing the shape of the Voronoï cells defined by the skeleton points using a covariance measure (Grélard et al., 2015).

Manual segmentation and extraction of lung airways from CT scans take a long time and can often be influenced by the person performing the segmentation. Automatic segmentation of the airways allows for faster extraction and can be also used to obtain information such as airway lumen size and wall thickness. The key issue with automatic airway segmentation is the need to prevent false positive detections, as many algorithms can segment sections of the lung parenchyma (Nadeem et al., 2020). This is known as the "leakage." A variety of methods have been developed and employed in order to avoid false positive detections (Bian et al., 2018; van Ginneken, 2017). CT-based algorithms for the segmentation of airway trees have been well documented in the literature (Pu et al., 2012; Van Rikxoort & Van Ginneken, 2013). A number of fully automated methods have been developed (i.e., using shape, boundary, region-growing, etc.) (Nadeem et al., 2020). The regional growth strategies have been commonly employed for the segmentation of anatomical tubular structures, such as lung airways, due to the high complexity of their shapes and branching patterns (Aysola et al., 2008). Various approaches have been considered for the airway segmentation: increasing and thresholding regions (Graham et al., 2010); morphological and geometric model-based approaches (Fetita et al., 2009); methods incorporating the previous two approaches (Meng et al., 2017).

Artificial intelligence (AI) methods have been commonly used in the field of image processing as they have shown unique capabilities in extracting the optimal information embedded in data (Deng & Yu, 2014). These methods have found significant application in the area of medical imaging, such as applications related to pulmonary imaging. AI methods have been used to identify leakage using three orthogonal 2-D slices as input (Charbonnier et al., 2017); detect airway structures by combining an airway centerline with 3-D U-Net (Jin et al., 2017); integrate voxel communication into a deep-learning airway segmentation (Qin et al., 2019). Additionally, AI methods have been used for the analysis of drugs (Ingrande et al., 2020; Lim et al., 2019; Madzarevic et al., 2019).

4 Numerical analysis of human respiratory system

Computational fluid dynamics (CFD) is a numerical method used to analyze and solve problems that involve fluid flows. This methodology has been used to analyze and describe fluid flow patterns in the human respiratory system. It was shown that the CFD-based stochastic human route modeling can be used to estimate the delivery of pharmaceutical aerosols through tracheobronchial airways (Tian, Longest, Su, Walenga, & Hindle, 2011). The application of CFD for the human respiratory system allows for the analysis of various parameters. This methodology has also been used for the analysis of fluid flow and particle tracking in the inhalers (Ponzini et al., 2021; Ruzycki et al., 2013; Sommerfeld et al., 2019; Tong et al., 2013; Vulović et al., 2018) as a first step before tracking particles in the respiratory system.

TABLE 2 Research of respiratory system using CFD.

Disease	Details regarding simulation	Conclusion	Reference
Asthma	Four models (two healthy subjects and two severe asthmatics) were analyzed. Subject-specific flow boundary conditions obtained by image registration. A large eddy simulation (LES) model for transitional and turbulent flows. Particle transport simulations for 2.5, 5, and 10 μm particles.	Severe asthmatics had a smaller air-volume change in the lower lobes and a larger air-volume change in the upper lobes. With increasing particle size, particles are distributed more toward lower lobar regions due to inertial effects.	Choi et al. (2018)
	14 mild to moderately severe asthmatics analyzed. Air flow was considered to be laminar. RANS Solver was used	Changes in the CFD-based calculated airway resistance significantly correlate with the observed changes in spirometric values	De Backer et al. (2008)
	Deposition of particulate matters in G3-G6 and G9-G12 triple bifurcation models of airways among 4-year-old asthmatic children was analyzed. The severity of asthma was indicated by the airway constriction in terms of diameter reduction. Air flow was considered to be laminar. Lagrangian method was used for particle tracking	Airway narrowing due to asthma can increase particle deposition and make asthmatic children more susceptible to the effect of particulate air pollution. The constricted airways increase the particle deposition in the upper airways by inertial impaction which can lead to the hospitalization of asthmatic children.	Zhang et al. (2020)
COPD	10 COPD patients (six male/four female) were analyzed. Flow simulations solved the Reynolds-averaged Navier-Stokes (RANS) equations with steady flow 30 L/min.	Results have shown that changes in airway geometry after inhalation of inhalation therapy in severe COPD can be described with an image-based approach compared to traditionally used spirometry.	De Backer et al. (2012)
	Three-dimensional numerical model was used to study the gas/solid flow behaviors in a COPD pulmonary airway. The gas phase was modeled with laminar computational fluid dynamics (CFD) model and the particle phase was modeled with discrete-phase model (DPM)	Results show that air flow rates of the obstruction and its downstream generations reduce due to the stagnation and recirculation zones development with unsteady inhalation. Secondary flow may contribute to the particle deposition in the inflamed airway for real inhalation.	Chen et al. (2012)

A number of respiratory diseases have been analyzed using CFD (Table 2): asthma, chronic obstructive pulmonary disease (COPD), lung cancer, pulmonary fibrosis, etc.

Fleming et al. (2011) used 3D radionuclide imaging with CT scanning of the airways to help report the location of aerosol accumulation in the respiratory system. During the same time, measurements of deposition of micrometer-sized particles (0.5–5.3mm) in replicas of older children (4–14years of age) have been published (Golshahi et al., 2011). The aim of this research was to establish associations, such as geometric proportions of airways and breathing patterns. These models were ideal and did not contain any anomalies which is the main limitation of these two papers. However, a number of researchers (Kleinstreuer et al., 2008; Kleinstreuer & Zhang, 2010; Tian, Longest, Su, & Hindle, 2011) employed patient-specific lung models which led to more realistic results. Rahimi-Gorji et al. (2015) have used human airways geometry developed from the CT scans in order to analyze the transport and deposition of inhaled microparticles. (Oakes et al., 2018) explored variations in the airflow trends for three distinct age groups: babies, children, and adults. Kolanjiyil and Kleinstreuer (2017) have employed a computational technique for predicting particle deposition in the entire respiratory tract model which included patient-specific upper airways from nose/mouth to generation 3, airway generations 4 through 21 depicted as customizable triple bifurcation units, and spherical alveoli added with increasing density.

Numerical methods have been used to analyze COVID-19 virus. Wedel et al. (2021) have analyzed aerosol transport and deposition in a human respiratory tract using digital replicas of human airways. These replicas included the oral cavity, larynx, and tracheobronchial airways down to the 12th generation of branching. They have used Lagrangian particle tracking in Reynolds-Averaged Navier-Stokes resolved turbulent flow and were able to conclude that a higher aerosol deposition in the upper airways of children, leads to a reduction of virus load in the lower airways. Pan et al. (2021) analyzed critically ill patients with COVID-19. Their results indicated that the change of airway resistance correlated well with the volume changes of the bronchial.

5 Conclusions

Numerical methods have allowed us to analyze complex problems such as fluid flow patterns in the human respiratory system and particle deposition in the upper and lower respiratory tract. These noninvasive methods provide an opportunity to analyze a healthy and diseased lung in order to get better insight into how different lung diseases affect airway geometry and fluid flow but also to understand the effect that disease has on the total deposition efficiency of drugs that are used for treatments. By analyzing drugs deposition in the upper and lower respiratory tract, we are able to suggest possible improvements such as recommended particle dimensions. On the other side, the same methodologies could be used for the analysis of air pollution and the deposition of different pollutants in our upper and lower respiratory tract. As we are witnessing increased air pollution, utilization of such methodologies will significantly increase in the coming years.

Acknowledgments

This research was funded by the Serbian Ministry of Education, Science, and Technological Development [451-03-9/2021-14/200107 (Faculty of Engineering, University of Kragujevac)].

References

Agnew, J. E. (1984). Physical properties and mechanisms of deposition of aerosols. In S. W. Clarke, & D. Pavia (Eds.), *Aerosols and the lung: Clinical and experimental aspects* (pp. 49–70). Butterworth-Heinemann.

Altes, T. A., Eichinger, M., & Puderbach, M. (2007). Magnetic resonance imaging of the lung in cystic fibrosis. *Proceedings of the American Thoracic Society*, *4*(4), 321–327.

Atzeni, C., Lesma, G., Dubini, G., Masi, M., Rossi, F., & Bianchi, E. (2021). Computational fluid dynamic models as tools to predict aerosol distribution in tracheobronchial airways. *Scientific Reports*, *11*(1), 1–13.

Aysola, R. S., Hoffman, E. A., Gierada, D., Wenzel, S., Cook-Granroth, J., Tarsi, J., Zheng, J., Schechtman, K. B., Ramkumar, T. P., Cochran, R., Xueping, E., Christie, C., Newell, J., Fain, S., Altes, T. A., & Castro, M. (2008). Airway remodeling measured by multidetector CT is increased in severe asthma and correlates with pathology. *Chest*, *134*(6), 1183–1191.

Balásházy, I., & Hofmann, W. (1995). Deposition of aerosols in asymmetric airway bifurcations. *Journal of Aerosol Science*, *26*(2), 273–292.

Baldacci, F., Laurent, F., Berger, P., & Dournes, G. (2019). 3D human airway segmentation from high hesolution MR imaging. In *Eleventh international conference on machine vision (ICMV 2018)*. *11041* (p. 110410Y). International Society for Optics.

Bennett, W. D., Zeman, K. L., & Jarabek, A. M. (2003). Nasal contribution to breathing with exercise: Effect of race and gender. *Journal of Applied Physiology*, *95*(2), 497–503.

Bennett, W. D., Zeman, K. L., & Kim, C. (1996). Variability of fine particle deposition in healthy adults: Effect of age and gender. *American Journal of Respiratory and Critical Care Medicine*, *153*(5), 1641–1647.

Bian, Z., Charbonnier, J. P., Liu, J., Zhao, D., Lynch, D. A., & van Ginneken, B. (2018). Small airway segmentation in thoracic computed tomography scans: A machine learning approach. *Physics in Medicine & Biology*, *63*(15), 155024.

Brown, J. S., Zeman, K. L., & Bennett, W. D. (2002). Ultrafine particle deposition and clearance in the healthy and obstructed lung. *American Journal of Respiratory and Critical Care Medicine*, *166*(9), 1240–1247.

Charbonnier, J. P., Van Rikxoort, E. M., Setio, A. A., Schaefer-Prokop, C. M., van Ginneken, B., & Ciompi, F. (2017). Improving airway segmentation in computed tomography using leak detection with convolutional networks. *Medical Image Analysis*, *36*, 52–60.

Chen, X., Zhong, W., Sun, B., Jin, B., & Zhou, X. (2012). Study on gas/solid flow in an obstructed pulmonary airway with transient flow based on CFD–DPM approach. *Powder Technology*, *217*, 252–260.

Choi, S., Miyawaki, S., & Lin, C. L. (2018). A feasible computational fluid dynamics study for relationships of structural and functional alterations with particle depositions in severe asthmatic lungs. *Computational and Mathematical Methods in Medicine*, *2018*, 6564854.

Chovancová, M., & Elcner, J. (2014). The pressure gradient in the human respiratory tract. In *Vol. 67*. *EPJ web of conferences* (p. 02047). EDP Sciences.

Cornea, N. D., Silver, D., & Min, P. (2007). Curve-skeleton properties, applications, and algorithms. *IEEE Transactions on Visualization and Computer Graphics*, *13*(3), 530.

Couprie, M., Coeurjolly, D., & Zrour, R. (2007). Discrete bisector function and Euclidean skeleton in 2D and 3D. *Image and Vision Computing*, *25*(10), 1543–1556.

De Backer, L. A., Vos, W., De Backer, J., Van Holsbeke, C., Vinchurkar, S., & De Backer, W. (2012). The acute effect of budesonide/formoterol in COPD: A multi-slice computed tomography and lung function study. *European Respiratory Journal*, *40*(2), 298–305.

De Backer, J. W., Vos, W. G., Devolder, A., Verhulst, S. L., Germonpré, P., Wuyts, F. L., Parizel, P. M., & De Backer, W. (2008). Computational fluid dynamics can detect changes in airway resistance in asthmatics after acute bronchodilation. *Journal of Biomechanics*, *41*(1), 106–113.

Deng, L., & Yu, D. (2014). Deep learning: Methods and applications. *Foundations and Trends in Signal Processing*, *7*(3–4), 197–387.

Fetita, C., Ortner, M., Brillet, P. Y., Prêteux, F., & Grenier, P. (2009). A morphological-aggregative approach for 3D segmentation of pulmonary airways from generic MSCT acquisitions. In *Second international workshop on pulmonary image analysis* (pp. 215–226).

Fleming, J., Conway, J., Majoral, C., Tossici-Bolt, L., Katz, I., Caillibotte, G., Perchet, D., Pichelin, M., Muellinger, B., Martonen, T., Kroneberg, P., & Apiou-Sbirlea, G. (2011). The use of combined single photon emission computed tomography and X-ray computed tomography to assess the fate of inhaled aerosol. *Journal of Aerosol Medicine and Pulmonary Drug Delivery*, *24*(1), 49–60.

Foster, M. W., & Costa, D. L. (2005). *Air pollutants and the respiratory tract*. Boca Raton: CRC Press.

Golshahi, L., Noga, M. L., Thompson, R. B., & Finlay, W. H. (2011). In vitro deposition measurement of inhaled micrometer-sized particles in extrathoracic airways of children and adolescents during nose breathing. *Journal of Aerosol Science*, *42*(7), 474–488.

Graham, M. W., Gibbs, J. D., Cornish, D. C., & Higgins, W. E. (2010). Robust 3-D airway tree segmentation for image-guided peripheral bronchoscopy. *IEEE Transactions on Medical Imaging*, *29*(4), 982–997.

Grélard, F., Baldacci, F., Vialard, A., & Domenger, J. P. (2017). New methods for the geometrical analysis of tubular organs. *Medical Image Analysis*, *42*, 89–101.

Grélard, F., Baldacci, F., Vialard, A., & Lachaud, J. O. (2015). Precise cross-section estimation on tubular organs. In *International conference on computer analysis of images and patterns* (pp. 277–288). Cham: Springer.

Heyder, J., Armbruster, L., Gebhart, J., Grein, E., & Stahlhofen, W. (1975). Total deposition of aerosol particles in the human respiratory tract for nose and mouth breathing. *Journal of Aerosol Science*, *6*(5), 311–328.

Horsfield, K., Dart, G., Olson, D. E., Filley, G. F., & Cumming, G. (1971). Models of the human bronchial tree. *Journal of Applied Physiology*, *31*(2), 207–217.

Ingrande, J., Gabriel, R. A., McAuley, J., Krasinska, K., Chien, A., & Lemmens, H. J. (2020). The performance of an artificial neural network model in predicting the early distribution kinetics of propofol in morbidly obese and lean subjects. *Anesthesia & Analgesia*, *131*(5), 1500–1509.

Jin, D., Xu, Z., Harrison, A. P., George, K., & Mollura, D. J. (2017). 3D convolutional neural networks with graph refinement for airway segmentation using incomplete data labels. In *International workshop on machine learning in medical imaging* (pp. 141–149). Cham: Springer.

Kesavanathan, J., Bascom, R., & Swift, D. L. (1998). The effect of nasal passage characteristics on particle deposition. *Journal of Aerosol Medicine*, *11*(1), 27–39.

Kleinstreuer, C., & Zhang, Z. (2010). Airflow and particle transport in the human respiratory system. *Annual Review of Fluid Mechanics*, *42*, 301–334.

Kleinstreuer, C., Zhang, Z., & Li, Z. (2008). Modeling airflow and particle transport/deposition in pulmonary airways. *Respiratory Physiology & Neurobiology*, *163*(1–3), 128–138.

Kolanjiyil, A. V., & Kleinstreuer, C. (2017). Computational analysis of aerosol-dynamics in a human whole-lung airway model. *Journal of Aerosol Science*, *114*, 301–316.

Koullapis, P. G., Hofemeier, P., Sznitman, J., & Kassinos, S. C. (2018). An efficient computational fluid-particle dynamics method to predict deposition in a simplified approximation of the deep lung. *European Journal of Pharmaceutical Sciences*, *113*, 132–144.

Koullapis, P., Kassinos, S. C., Muela, J., Perez-Segarra, C., Rigola, J., Lehmkuhl, O., Cui, Y., Sommerfeld, M., Elcner, J., Jicha, M., Saveljic, I., Filipovic, N., Lizal, F., & Nicolaou, L. (2018). Regional aerosol deposition in the human airways: The SimInhale benchmark case and a critical assessment of in silico methods. *European Journal of Pharmaceutical Sciences*, *113*, 77–94.

Koullapis, P., Ollson, B., Kassinos, S. C., & Sznitman, J. (2019). Multiscale in silico lung modeling strategies for aerosol inhalation therapy and drug delivery. *Current Opinion in Biomedical Engineering*, *11*, 130–136.

Koullapis, P. G., Stylianou, F. S., Sznitman, J., Olsson, B., & Kassinos, S. C. (2020). Towards whole-lung simulations of aerosol deposition: A model of the deep lung. *Journal of Aerosol Science*, *144*, 105541.

Kumpitsch, C., Koskinen, K., Schöpf, V., & Moissl-Eichinger, C. (2019). The microbiome of the upper respiratory tract in health and disease. *BMC Biology*, *17*(1), 1–20.

Lim, J., Ryu, S., Park, K., Choe, Y. J., Ham, J., & Kim, W. Y. (2019). Predicting drug–target interaction using a novel graph neural network with 3D structure-embedded graph representation. *Journal of Chemical Information and Modeling*, *59*(9), 3981–3988.

Ma, B., & Lutchen, K. R. (2006). An anatomically based hybrid computational model of the human lung and its application to low frequency oscillatory mechanics. *Annals of Biomedical Engineering*, *34*(11), 1691–1704.

Madzarevic, M., Medarevic, D., Vulovic, A., Sustersic, T., Djuris, J., Filipovic, N., & Ibric, S. (2019). Optimization and prediction of ibuprofen release from 3D DLP printlets using artificial neural networks. *Pharmaceutics*, *11*(10), 544.

Meng, Q., Kitasaka, T., Nimura, Y., Oda, M., Ueno, J., & Mori, K. (2017). Automatic segmentation of airway tree based on local intensity filter and machine learning technique in 3D chest CT volume. *International Journal of Computer Assisted Radiology and Surgery*, *12*(2), 245–261.

Miyawaki, S., Choi, S., Hoffman, E. A., & Lin, C. L. (2016). A 4DCT imaging-based breathing lung model with relative hysteresis. *Journal of Computational Physics*, *326*, 76–90.

Nadeem, S. A., Hoffman, E. A., Sieren, J. C., Comellas, A. P., Bhatt, S. P., Barjaktarevic, I. Z., Abtin, F., & Saha, P. K. (2020). A CT-based automated algorithm for airway segmentation using freeze-and-grow propagation and deep learning. *IEEE Transactions on Medical Imaging*, *40*(1), 405–418.

Oakes, J. M., Marsden, A. L., Grandmont, C., Shadden, S. C., Darquenne, C., & Vignon-Clementel, I. E. (2014). Airflow and particle deposition simulations in health and emphysema: From in vivo to in silico animal experiments. *Annals of Biomedical Engineering*, *42*(4), 899–914.

Oakes, J. M., Roth, S. C., & Shadden, S. C. (2018). Airflow simulations in infant, child, and adult pulmonary conducting airways. *Annals of Biomedical Engineering*, *46*(3), 498–512.

Pan, S. Y., Ding, M., Huang, J., Cai, Y., & Huang, Y. Z. (2021). Airway resistance variation correlates with prognosis of critically ill COVID-19 patients: A computational fluid dynamics study. *Computer Methods and Programs in Biomedicine, 106257*.

Ponzini, R., Da Vià, R., Bnà, S., Cottini, C., & Benassi, A. (2021). Coupled CFD-DEM model for dry powder inhalers simulation: Validation and sensitivity analysis for the main model parameters. *Powder Technology*, *385*, 199–226.

Postolski, M., Janaszewski, M., Kenmochi, Y., & Lachaud, J. O. (2012). Tangent estimation along 3D digital curves. In *21st International conference on pattern recognition (ICPR2012)* (pp. 2079–2082). IEEE.

Pritchard, J. N., Black, A., & Jefferies, S. J. (1986). Sex differences in the regional deposition of inhaled particles in the 2.5–7.5 mm size range. *Journal of Aerosol Science*, *17*, 385–389.

Pu, J., Gu, S., Liu, S., Zhu, S., Wilson, D., Siegfried, J. M., & Gur, D. (2012). CT based computerized identification and analysis of human airways: A review. *Medical Physics*, *39*(5), 2603–2616.

Qin, Y., Chen, M., Zheng, H., Gu, Y., Shen, M., Yang, J., Huang, X., Zhu, Y. M., & Yang, G. Z. (2019). Airwaynet: A voxel-connectivity aware approach for accurate airway segmentation using convolutional neural networks. In *International conference on medical image computing and computer-assisted intervention* (pp. 212–220). Cham: Springer.

Rahimi-Gorji, M., Pourmehran, O., Gorji-Bandpy, M., & Gorji, T. B. (2015). CFD simulation of airflow behavior and particle transport and deposition in different breathing conditions through the realistic model of human airways. *Journal of Molecular Liquids*, *209*, 121–133.

Ruzycki, C. A., Javaheri, E., & Finlay, W. H. (2013). The use of computational fluid dynamics in inhaler design. *Expert Opinion on Drug Delivery*, *10*(3), 307–323.

Schmidt, A., Zidowitz, S., Kriete, A., Denhard, T., Krass, S., & Peitgen, H. O. (2004). A digital reference model of the human bronchial tree. *Computerized Medical Imaging and Graphics*, *28*(4), 203–211.

Sommerfeld, M., Cui, Y., & Schmalfuß, S. (2019). Potential and constraints for the application of CFD combined with Lagrangian particle tracking to dry powder inhalers. *European Journal of Pharmaceutical Sciences*, *128*, 299–324.

Stahlhofen, W., Rudolf, G., & James, A. C. (1989). Intercomparison of experimental regional aerosol deposition data. *Journal of Aerosol Medicine*, *2*(3), 285–308.

Tian, G., Longest, P. W., Su, G., & Hindle, M. (2011). Characterization of respiratory drug delivery with enhanced condensational growth using an individual path model of the entire tracheobronchial airways. *Annals of Biomedical Engineering*, *39*(3), 1136–1153.

Tian, G., Longest, P. W., Su, G., Walenga, R. L., & Hindle, M. (2011). Development of a stochastic individual path (SIP) model for predicting the tracheobronchial deposition of pharmaceutical aerosols: Effects of transient inhalation and sampling the airways. *Journal of Aerosol Science*, *42*(11), 781–799.

Tobin, M. J., Chadha, T. S., Jenouri, G., Birch, S. J., Gazeroglu, H. B., & Sackner, M. A. (1983). Breathing patterns: 1. Normal subjects. *Chest*, *84*(2), 202–205.

Tong, Z. B., Zheng, B., Yang, R. Y., Yu, A. B., & Chan, H. K. (2013). CFD-DEM investigation of the dispersion mechanisms in commercial dry powder inhalers. *Powder Technology*, *240*, 19–24.

van Ginneken, B. (2017). Fifty years of computer analysis in chest imaging: Rule-based, machine learning, deep learning. *Radiological Physics and Technology*, *10*(1), 23–32.

Van Rikxoort, E. M., & Van Ginneken, B. (2013). Automated segmentation of pulmonary structures in thoracic computed tomography scans: A review. *Physics in Medicine & Biology*, *58*(17), R187.

Vulović, A., Šušteršič, T., Cvijić, S., Ibrić, S., & Filipović, N. (2018). Coupled in silico platform: Computational fluid dynamics (CFD) and physiologically-based pharmacokinetic (PBPK) modelling. *European Journal of Pharmaceutical Sciences*, *113*, 171–184.

Wedel, J., Steinmann, P., Štrakl, M., Hriberšek, M., & Ravnik, J. (2021). Can CFD establish a connection to a milder COVID-19 disease in younger people? Aerosol deposition in lungs of different age groups based on Lagrangian particle tracking in turbulent flow. *Computational Mechanics*, *67*(5), 1497–1513.

Weibel, E. R. (1963). *Morphometry of the human lung*. Berlin Heidelberg: Springer-Verlag.

Xi, J., & Longest, P. W. (2007). Transport and deposition of micro-aerosols in realistic and simplified models of the oral airway. *Annals of Biomedical Engineering*, *35*(4), 560–581.

Zhang, W., Xiang, Y., Lu, C., Ou, C., & Deng, Q. (2020). Numerical modeling of particle deposition in the conducting airways of asthmatic children. *Medical Engineering & Physics*, *76*, 40–46.

CHAPTER

12

Artificial intelligence approach toward analysis of COVID-19 development—Personalized and epidemiological model

Tijana Šušteršič[a,b] *and Anđela Blagojević*[a,b]

[a]University of Kragujevac, Faculty of Engineering, Kragujevac, Serbia [b]Bioengineering Research and Development Center (BioIRC), Kragujevac, Serbia

1 Introduction

The emergence of novel and reoccurring illnesses has contributed to a growing interest in infectious diseases. Mathematical models have been shown to be tools useful in the research and control of infectious illnesses. They are essential experimental approaches for developing and testing ideas, generating mathematical hypotheses, answering particular questions, characterizing sensitivity to changes in parameter values, and estimating crucial data parameters. Model creation defines expectations, variables, and parameters; models also include quantitative results such as thresholds, particular replication numbers, contact numbers, and replacement numbers (Hethcote, 2000). Predictive mathematical models for epidemics are critical for understanding outbreak dynamics and developing efficient response tactics (Giordano et al., 2020).

The World Health Organization (WHO) declared the illness (COVID-19) caused by SARS-CoV-2 a pandemic and implemented steps to halt the spread of SARS-CoV-2 globally. In patients with a severe clinical condition, the disease progresses to acute respiratory distress syndrome (ARDS) requiring mechanical ventilation for 8–20 days on average, while the findings on MSCT lungs are most pronounced after 10 days, and sometimes in serologically negative patients, the characteristic findings on the lungs are sufficient to make a diagnosis (Böhmer et al., 2020; Pan et al., 2020). SARS and MERS had substantially higher fatality rates, with 8096

https://doi.org/10.1016/B978-0-12-823956-8.00013-4

cases and 774 fatalities from SARS in 29 countries accounting for 9.6%, and 2494 MERS patients with 858 deaths in 27 countries accounting for 34.4% (Wu & McGoogan, 2020). Despite the fact that fatality rates in SARS and MERS were much higher, SARS-CoV-2 has a substantially higher number of deaths in proportion to the larger number of patients (Kucharski et al., 2020; Wilder-Smith et al., 2020). It is currently unknown how SARS-CoV-2 causes such a wide range of clinical manifestations, ranging from asymptomatic patients to acute respiratory distress syndromes, multiple organ failure, and death (Assandri et al., 2020; Guan et al., 2020a; Huang et al., 2020a; Wang et al., 2020). As a result, researchers all around the globe have been attempting to identify and stratify predictors of COVID-19 disease severity in order to correctly advise medical therapy. A basic understanding of disease pathophysiology and techniques for detecting and evaluating COVID-19 infection has been established. Common blood hematology and clinical biochemistry tests are inexpensive, straightforward, and commonly available biomarkers. As a result, they became the primary technique of tracking and forecasting illness impacts (Aronson & Ferner, 2017). Recognizing the diversity and profile of specific biomarkers as a result of different COVID-19 outcomes may aid in the development of a risk-stratified strategy for treating individuals with this illness. Among other things, it is critical to forecast the adverse course of the illness quickly, accurately, and in a timely manner. The ability to recognize patients who are on the verge of death has become a pressing yet important task (Yan et al., 2020a).

Additionally, while there have been considerable developments in statistical epidemiology forecasts, the prediction of disease events remain fundamentally error prone. Indeed, calls for national disease prediction centers have emerged from the critical need to train policy makers at all levels about how to incorporate predictive analytics into decision-making processes (Poletto et al., 2020). One main approach in predictive modeling is the creation of personalized models with the aim of tracking blood biomarkers and the development of disease in specific patients infected with COVID-19. The second approach represents the epidemiological modeling that includes the tracking of COVID-19 spreading in the population (i.e., at national level).

1.1 Review of literature on personalized model

As mentioned above, a personalized model studies the development and prediction of patient condition on the level of a specific individual. Several studies have been conducted to investigate determinants of illness severity in COVID-19 patients. According to some studies, severe or fatal cases of COVID-19 disease are associated with elevated white blood cell count, creatinine, blood urea nitrogen, markers of the liver and kidney function, interleukin-6 (IL-6), C-reactive protein (CRP), lower lymphocyte ($1 \times 109/L$), platelet counts ($100 \times 109/L$), and albumin levels when compared to milder cases of patients who survived the disease (Goyal et al., 2020; Guan et al., 2020b; Ruan et al., 2020). These findings offered an early understanding of the consequences of SARS-CoV-2 infection, but due to geographical limitations, particular experience of the clinical center, and limited cohort sizes, the findings cannot be generalized (Malik et al., 2020). In a meta-analysis of 32 studies including 10,491 COVID-19 patients, Malik et al. discovered that certain biomarkers were associated with poor end results in COVID-19 hospitalized patients. Low lymphocytes, a decreased platelet count, and increased CRP, creatine kinase (CK), procalcitonin (PCT), D-dimer, lactate dehydrogenase (LDH), alanine aminotransferase (ALT), aspartate aminotransferase (AST), and creatinine were among the biomarkers (Malik et al., 2020). Assandri et al. discovered that several laboratory tests were abnormally changed in COVID-19 patients and provided sensitive alternatives for detecting probable COVID-19 cases (Assandri et al., 2020). Specific biomarkers were found to be associated with the clinical outcome (Benelli et al., 2020). The neutrophil/lymphocyte ratio was discovered to be one of the greatest predictive indicators (in patients with a more serious illness) (Liu et al., 2020), CRP, as well as inflammatory cytokines such as IL-6, IL-1, and tumor necrosis factor alpha (TNF) (Sarzi-Puttini et al., 2020). Other investigations have discovered that CRPO7 mg/dL can predict who would have a severe condition (Assandri et al., 2020; Benelli et al., 2020). Following literature research, it is possible to conclude that the findings indicate inconsistency, since biomarkers have proven to be predictors of COVID-19 disease development range from study to study and are sometimes contradictory (Cheng et al., 2020a). As a result, there is a great need to study alternative methodologies, such as machine learning methods (Šušteršič et al., 2019, 2020), to gain a better understanding of prognostic biomarkers.

In the field of COVID-19, machine learning (ML) algorithms have been investigated for a variety of purposes, including epidemiological and clinical issues such as early detection of disease outbreaks, rapid diagnosis, classification and stratification of radiological images, risk factor analysis, and prediction of final clinical outcomes (Bai et al., 2020; Cho, 2020; Gao et al., 2020; Mei et al., 2020; Wu et al., 2020a; Wynants et al., 2020). The majority of articles relating to the use of ML in the COVID-19 study deal with medical image analysis (Lorencin, 2021; Pereira et al., 2020; Shaban et al., 2020). Another element being investigated is the use of blood biomarkers as predictors of clinical outcomes. For instance, Yan et al. (2020b) conducted a study that employed a blood sample database of 404 infected patients in the Wuhan region of China to categorize significant indicators of disease severity in order to assist decision-makers and logistical planners of health systems. To do this, an ML model with more than 90% accuracy was developed that uses the three most prognostic biomarkers: LDH, lymphocytes, and high sensitivity CRP (hs-CRP). This finding is confirmed by existing medical understanding that high LDH levels are linked to tissue degradation in pneumonia and other infectious and inflammatory disorders (Jurisic et al., 2015). The major benefit of this work is that it offers a simple and practical method for simply forecasting and assessing clinical outcomes—death/survival—allowing essential patients to be prioritized and potentially lowering the mortality rate. The primary limitation of this study is that its categorization is binary—survival/death, which may not be the ideal form of classification in cases where health-care systems are overburdened. The same group of researchers, led by Yan, analyzed blood samples from 485 patients in Wuhan, China, in order to identify strong and meaningful risk mortality indicators (Yan et al., 2020a). The objective was to use state-of-the-art interpretable machine learning algorithms to find the best discriminating indicators of patient mortality. Inputs included detailed information, symptoms, blood samples, and laboratory findings on the liver, renal functions, coagulation activity, electrolytes, and inflammatory variables obtained from patients at various stages of illness, as well as linked mortality or survival results. The same biomarkers (features) as in the previous study were chosen experimentally with significant predicted degradation values or illness fatality, which also matched those in the literature (Huang et al., 2020a; Li et al., 2020a, 2020b; Tan et al., 2020). As in the previous research, the categorization was binary, and apart from predicting survival, it did not contribute to lessening the load placed on the health-care system. According to Huang et al., while Yan et al. show that death outcomes in COVID-19 patients are related to three biomarkers using a single-tree XGboost model, the forecast findings are overly optimistic.

Gao et al. also proposed a COVID-19 mortality risk prediction (MRPMC) model that uses hospital admission data to classify individuals based on mortality risk, allowing prediction of physiological deterioration and death up to 20 days in advance. They incorporated four methods of machine learning, including logistic regression, support vector machine, gradient boosted decision tree, and neural network. The MRPMC was tested on internal and external evaluation cohorts, and it achieved an average area under the curve (AUC) of 95% on internal and 97% and 92% on two different external cohorts, indicating that it has the potential to be of high utility in future clinical practice (Gao et al., 2020). Within the Central European Initiative (CEI)-funded project named "Use of Regressive Artificial Intelligence (AI) and Machine Learning (ML) Methods in Modelling of COVID-19 Spread"—COVIDAi, one of the main tasks was the analysis of blood biomarkers and the development of a personalized model (COVIDAi, n.d.). The methodology developed included several aspects: (1) classification of patients into several classes of clinical condition, (2) segmentation of human lungs in X-ray images, and (3) finite element simulation to investigate the spreading of SARS-COV-2 virion in the lungs (Blagojevic et al., 2021). This chapter presents the extension of the work from this project and the main findings.

1.2 Review of literature on an epidemiological model

With advancements in patient clinical care and the implementation of tougher methods of disease prevention, many researchers are studying the impact of such advancements using statistical reasoning (Chinazzi et al., 2020; Jin, Yu, et al., 2020) and stochastic simulations (Hellewell et al., 2020; Quilty et al., 2020). Compared to statistical methods (Huang & Qiao, 2020; Zeng et al., 2020), mathematical models based on dynamic equations (Peng et al., 2020; Read et al., 2020; Tang et al., 2020) attract comparatively less consideration, although they may provide more insight into the dynamics of epidemics (Peng et al., 2020). Infectious illness modeling is most easily accomplished with deterministic compartmental models. These models will be studied since they are based on a system of initial value problems of ordinary differential equations. By adjusting the parameters of the equations, it is possible to better simulate environmental features such as social constraints (Piccolomini & Zama, 2020). Because these models are based on flow patterns between compartments such as susceptible (S), exposed (E), infected (I), recovered (R), and so on, their names are frequently abbreviated as SEIR, SIT, SIRS, and so on. The standard susceptible, expose, infectious, recovered (SEIR) model is the most frequently utilized paradigm for describing the COVID-19 pandemic in both China and other nations. (Peng et al., 2020).

The SIR model for human-to-human transmission is a commonly used model that specifies the migration of people through three mutually incompatible stages of infection: prone, contaminated, and recovered (Giordano et al., 2020). More sophisticated models will precisely describe the dynamic distribution of certain epidemics. However, according to Yi-Cheng et al. the conventional SIR model ignores time-varying characteristics such as transmission and recovery rate (Yi-Cheng et al., 2020). They argue that the time-independent SIR model is too simplistic to accurately and efficiently anticipate the trajectory of the disease. As a result, they proposed a time-dependent SIR model in which both rates are functions of time t. In addition, numerous models for the COVID-19 pandemic have been constructed. Lin and colleagues created the SEIR (susceptible, released, infectious, removed) model, which takes risk assessment and the overall number of cases into consideration (Lin et al., 2020). Anastassopoulou and colleagues proposed a discrete SIR model including deceased persons (Anastassopoulou et al., 2020), Casella created a control-oriented SIR model that emphasizes the costs of delays and assesses the effectiveness of various containment methods (Casella, 2020), and Wu and colleagues used propagation dynamics to measure the clinical magnitude of COVID-19 (Wu et al., 2020b). Stochastic transmission models were also considered (Hellewell et al., 2020; Kucharski et al., 2020). Giordano et al. propose a novel epidemiological mean-field method for the COVID-19 epidemic in Italy, building on the traditional SIR model proposed by Gumel and colleagues for SARS (Gumel et al., 2004). SIDARTHE, the previously mentioned extended SIR model, incorporates eight stages of infection: susceptible, infected, diagnosed, ill, recognized, threatened, healed, and extinct (Giordano et al., 2020). The SIDARTHE model distinguishes between infected persons based on whether they have been diagnosed and the severity of their symptoms. The distinction between diagnosed and undiagnosed persons is significant because those who have been diagnosed positively are isolated and have a lower risk of spreading the virus. This demarcation also helps to understand why the pandemic spread and case fatality rate were misinterpreted (Giordano et al., 2020).

Since epidemiological methods are not producing the expected and desired effects, and the pandemic is continuously spreading and altering its course, it is clear that a new strategy is required to enhance the present procedures to combat the epidemic. In this regard, prognostic models can at least roughly anticipate the tendency in the number of patients who are critical. Different approaches of deep learning have been included in the analysis of COVID-19 disease in order to get more reliable insight into how the epidemic will spread (Andjelic et al., 2021; Baressi Segota et al., 2021; Musulin et al., 2021). Approaches like recurrent neural networks (RNNs) are ideal for modeling temporal sequences (Schmidhuber, 2015). However, the major drawback of RNNs is that they cannot learn long-term dependencies in huge sequences with hundreds or thousands of steps. Long short-term memory networks resolve these constraints (LSTMs) (Chandra, Shaurya, & Rishabh, 2021). For Canada, LSTM was employed for COVID-19 forecasting and obtained an accuracy of 93.4% for short-term predictions and 92.67% for long-term predictions (Chimmula & Zhang, 2020). For COVID-19 forecasting in China, LSTM was also used and in comparison to the dynamic SEIR model, LSTM achieved promising results (Yang, 2020). The results revealed that the LSTM performed well in predicting owing to its ability to handle time-dependent information. Ismail et al. used autoregressive integrated moving average (ARIMA), nonlinear autoregressive neural network (NARNN), and LSTM to model data from Denmark, Belgium, Germany, France, the United Kingdom, Finland, Switzerland, and Turkey. They said that LSTM is the most accurate model when compared to the other two algorithms, and they offered LSTM in order to create 14-day forecasts (Kırbaş, 2020). Based on the present circumstances, this model can generate reasonable predictions and properly anticipate the number of confirmed and recovered cases. This provides decision-makers with accurate information about the predicted scenario and the necessary enforcement actions. Chandra and colleagues examined RNN, LSTM networks, bidirectional LSTM networks, encoder-decoder LSTM networks, and convolutional neural networks (CNN) for multistep ahead prediction on univariate time series. The findings show that the encoder-decoder LSTM network, in addition to the bidirectional LSTM network, gives the greatest performance for the specified time series issues (Chandra, Shaurya, & Rishabh, 2021). Also, mentioned project COVIDAi considered methodology for the development of the epidemiological model, in addition to the personalized model (COVIDAi, n.d.; Blagojevic et al., 2021). The proposed methodology included a SEIRD model to predict the development of an epidemic (Sustersic et al., 2021). Also, the epidemiological model presented in this chapter includes the results built upon the results developed during this project.

1.3 Motivation for the study

Although many papers deal with implementing ML in proposing valid prognostic biomarkers and predictors of survival several days in advance, there is limited research in the field focused on the classification of patients in more subtle severity-of-illness categories (e.g., mild, moderate, severe, critical) during the hospital stay. Such knowledge could assist physicians and hospital administrators in making decisions that aim to improve not only the patient's final unfavorable outcome but also other important secondary treatment endpoints and institutional performances that are harmed by inappropriate measures such as unnecessary prescription of adjunctive drugs and overutilization of sophisticated and invasive diagnostic procedures. As a result, in this work, we offer an ML-based approach for categorizing patients and predicting outcomes in advance (change of severity of the clinical condition).

Furthermore, because epidemiological methods are not yielding the expected and desired outcomes and the pandemic is continually spreading in the lack of a causative medication or vaccine, it is clear that a new strategy is required to enhance the present tactics to combat the epidemic. In this regard, prognostic models that can at least roughly anticipate the tendency in the number of critical patients. By anticipating and averting epidemiological peaks, we may create a "flattened curve" of disease transmission, preventing such a fast spread of illness that could lead to overburdening and collapse of national health systems (Jin, Yang, & Ji, 2020). Because there was a scarcity of official data and numerous unknown factors in COVID-19 epidemic propagation, development, and control, most early published models were prone to overfitting, or the parameters were obtained from literature based on limited and less exact information. This leads to confusing outcomes, especially because many articles are released before they have been peer-reviewed. In this article, we systematically gather epidemic data from reputable sources at the state, regional, and local levels and incorporate it into the created model. The obtained results are compared with the real situation in Belgium.

2 Materials and methods

This section explains the dataset and the proposed methodology for both models, personalized and epidemiological, respectively.

2.1 Personalized model

This section first explains the dataset and the data preprocessing and after that the proposed methodology.

2.1.1 Dataset

Clinical data of the COVID-19 patients were obtained from two hospitals—Clinical Center of Kragujevac, Serbia (45 patients) and Clinical Center of Rijeka, Croatia (60 patients). In total, the results of blood analyses of 105 COVID-19 positive patients were collected, with gender distribution—42% of total patients were women, and 58% were men. The age distribution of the patients in the form mean $\pm$ standard deviation was 52.77 ± 16.63. From the aspect of the most common symptoms, fever was the symptom in 83% of cases, followed by cough in 74.6% of cases and fatigue in 45.7% of cases. A described dataset is shown in Fig. 1.

We divide the clinical data into three subgroups:

1. demographic data (gender and age)
2. symptoms (fever, cough, fatigue, chest pain, muscle pain, headache, dyspnea, loss of taste or smell)
3. blood analysis
   - *Complete blood count* (*CBC*): *erythrocytes* (red blood cells (RBC))—red cell indices: hemoglobin (HGB), mean corpuscular volume (MCV), mean corpuscular hemoglobin (MCH), mean corpuscular hemoglobin concentration (MCHC), red cell distribution width (RDW); *leucocytes* (*white blood cells* (*WBC*))—white cell differentials: neutrophils, lymphocytes, monocytes, eosinophils (EOS) and basophils (BASO), *platelet indices*: platelets (PLT) platelet distribution width (PDW), mean platelet volume (MPV)
   - *Coagulation*: prothrombin time (PT), international normalized ratio (INR), D-dimer

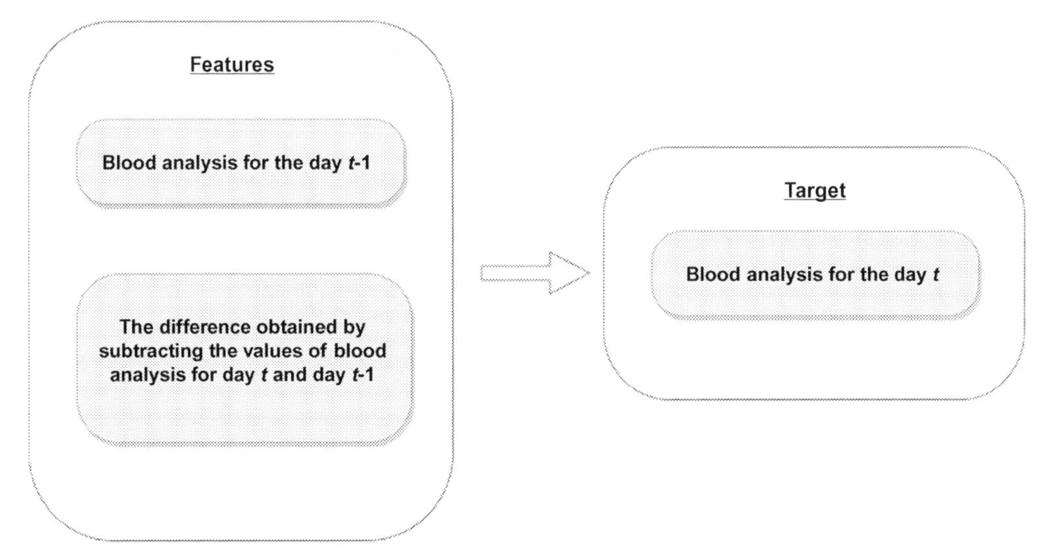

FIG. 1 Schematic representation of principles which was used for database organization.

- *Kidney function*: urea, creatinine (CREA)
- *Hepatic function*: bilirubin—direct and total, alanine transaminase (ALT), aspartate transaminase (AST), gamma-glutamyltransferase (also γ-glutamyltransferase, GGT), albumin
- *Enzymes*: creatine kinase (CK), also known as creatine phosphokinase (CPK) or phosphocreatine kinase; lactate dehydrogenase (LDH)
- *Electrolytes*: sodium (Na), potassium (K)
- *Oxygenation and acid-base balance*: arterial blood gas (ABG) analyses/tests: partial pressure of oxygen (pO_2), arterial partial pressure of oxygen (PaO_2) partial pressure of carbon dioxide (pCO_2), arterial partial pressure of carbon dioxide ($PaCO_2$), SpO_2 (peripheral oxygen saturation s. oxygen saturation as measured by pulse oximetry), pH
- *Inflammation indices*: C-reactive protein (CRP), procalcitonin (PCT)
- *Carbohydrate metabolism (glycemia)*: GLUC—glucose

Blood samples were taken from most patients throughout their hospital stay, but not all patients spent the same number of days in the hospital, so we considered the day of admission to the hospital and days 2, 5, 7, 9, 11, and 14 after admission to the hospital. To deal with the missing data, we used the data imputation method. To fill in the missing values, we used the mean of four different values (values of 2 days before and 2 days after). The goal of this type of data imputation is to include values of analyses of adjacent days that are not directly included in the prediction.

2.1.2 Data preprocessing

The first step in the proposed data preprocessing stage is to perform the clustering method. Since there were a lot of instances without labeled output—severity of the clinical condition and omitting these instances have significantly reduced the size of the dataset that we have performed clustering. This section is particularly important, as without enough output labels, no ML method will be capable to perform classification into different categories. One group clustered around an already known output label was assumed to be associated with the same output label meaning the whole cluster was labeled as the same clinical condition. Further, feature selection in preprocessing stage was performed to extract the most important blood biomarkers as predictors of the severity of the clinical condition.

Clustering

In order to solve the problem of a small number of output labels, unsupervised clustering analysis was used. We have applied several different clustering methods such as k-means, agglomerative, and mean shift (Altman & Krzywinski, 2017). All methods gave similar results, so as the final choice we have adopted K-means algorithm. K-means algorithm requires a predefined number of clusters and for the determination of the optimal number of clusters, we used the Elbow method. This method will start from two clusters and increasing it in each step by one until the optimal number of clusters is determined. For each given number of clusters, it computes the sum of intra-cluster distances. Also, it computes the variance of the intra-cluster distances between two consecutive numbers of clusters. Then, the Elbow method investigates the percentage of variance explained as a function of the number of clusters. One should choose a number of clusters so that adding another cluster in the analysis does not give a much better explanation of the variance. The optimal number of clusters that was calculated by the Elbow method was four, which is in accordance with the manually marked target outputs by clinicians. Clinicians categorized the patients into four categories (mild, moderate, severe, and critical). This confirms the right use of such ML methodology.

Finally, the dataset was divided into four clusters and as we expected, cluster distribution is as follows:

- 31.80% of total data belongs to the cluster of mild clinical conditions,
- 50.90% to moderate,
- 13.88% to severe,
- 3.42% belongs to a cluster of critical clinical conditions.

Feature selection

We used Analysis of Variance (ANOVA) F-statistics feature selection process to evaluate which biomarkers are crucial for the determination of the severity of the clinical condition. The 10 features that had the greatest influence in making the effective model were chosen and considering 10 features instead of 44 original ones is also important in order to reduce the computational complexity.

2.1.3 Supervised ML model development

In this section, we implement regression analysis to predict the values of important features (biomarkers) in the following days, to be able to further perform the classification with the aim to determine the output class—mild, moderate, severe or critical, which corresponds to the severity of the clinical condition.

Regression

We propose a regression-based methodology in order to track and predict the change in blood analysis values of COVID-19 patients. Based on changes in blood analysis for 2 weeks period, we aim to predict the patient's clinical condition. The main limitation was the lack of data for complete blood analysis in time. In the following, the number of patients with complete blood test analyses will be presented by days:

- on the admission day, complete data for all 105 patients was available,
- on the 2nd day, complete data from 104 patients was available,
- on the 5th day, complete data for 67 patients was available,
- on the 7th day, complete data for 68 patients was available,
- on the 9th day, complete data for 59 patients was available,
- on the 11th day, complete data for 51 patients was available, and
- on the 14th day, complete data for only 44 patients was available.

It should be emphasized that not always were the data for one patient available at each time point, but rather spread across some days. For example, for the first patient blood analysis was available on days 0, 5, 11, while for the second patient on days 0, 7, 9, 14, etc. To get accurately predicted blood analysis values, we decided to select only patients with a complete blood analysis for all days. This refers to a total of 34 patients. For the prediction in time, we had to achieve time dependencies between features. This was accomplished by creating an additional feature that represents the difference obtained by subtracting blood analysis for day t and blood analysis for the previously observed day $t-1$.

In order to predict the values of blood biomarkers for the day t, as input of the model, we have put this additional feature and blood analysis for the day $t-1$. This means that for day 14, day 11 as $t-1$ and day 14 as t were taken as part of the analysis. In this way, we have expanded the dataset for the regression problem and created time dependencies between a patient's blood analysis throughout time. Fig. 1 schematically represents the described methodology.

After adding an additional feature, we have created training and testing sets. The training set consists of data from days 2, 5, 7, 9, and 11, meaning that day 14 belongs to the test set. As a proposed supervised methodology for the prediction of blood analysis, a Gradient boosting regressor (GBR) was used. Boosting is a powerful strategy of learning firstly designed for classification problems, but it has been extended to regression as well. The motivation for applying boosting is to combine the output of many other methods, i.e., decision trees into a "powerful" structure (Hastie et al., 2009). The optimal hyperparameter settings were obtained using the grid search method and the final setting for the learning process included the number of tree estimators set to 500, max depth equals 4, min samples split equals 5 and learning rate to 0.05.

Classification

The aim of this research task is to assess how the COVID-19 disease develops in patients throughout time, actually the main aim is the determination of the patient's condition 14 days after hospital admission. This was accomplished by assessing the values of the blood analyses using the already explained methodology, after that the patients were classified into one of the four classes of the severity of the clinical condition (mild, moderate, severe and critical). In the classification dataset, we included all patients in whom we performed a blood test on a particular day. This means that within this dataset, the data for the same patients, but different days, are repeated several times, without any dependence between different days. In this case, it is not important to observe how the patient's condition develops throughout time but to create as many different instances as possible. So, it can be concluded that this modification is justified in order to expand the dataset size. As a result, the dataset was expanded to 497 cases (158 cases belonged to the mild state cluster, 253 to the moderate state, 69 to the severe state, and 17 cases belonged to the critical state cluster).

Within the described classification task, constructing a simplified rule-based decision model was the primary goal and for that purpose, an extreme gradient boosting algorithm (XGBoost) was used. XGBoost is a supervised machine learning algorithm based on the recursive construction of a decision tree, and those trees that most influence the decision of the predictive model can be identified. At each decision step in the XGBoost trees, the significance of each feature is determined by its accumulated use (Chen & Guestrin, 2016; Yan et al., 2020a). This is an advantage of rule-based algorithms because the logic of the model is understandable and it is reflected in the fact that the feature values for decision-making are known, unlike black box models whose strategies are difficult to interpret. XGBoost was trained to obtain optimal hyperparameters using the grid search method and the final setting included a number of tree estimators set to 100, max depth equals 5, learning rate equal to 0.3, gamma equals 1, "subsample" and "colsample bytree" both set to 1.

The model performance was evaluated by assessing different classification metrics: accuracy, recall, precision, and F1 score. Equations of these metrics are defined below:

$$\text{Accuracy} = \frac{TP + TN}{TP + TN + FP + FN} \tag{1}$$

$$\text{Recall} = \frac{TP}{TP + FN} \tag{2}$$

$$\text{Precision} = \frac{TP}{TP + FP} \tag{3}$$

$$F1\,\text{score} = \frac{2 \cdot \text{Precision} \cdot \text{Recall}}{\text{Precision} + \text{Recall}} \tag{4}$$

where TP refers to true positive instances, FP refers to false positive and N refers to negative instances, such as true negative (TN) and false negative (FN).

2.2 Epidemiological model

This section explains the two approaches used in epidemiological modeling, SEIRD model, and ML-based LSTM model.

2.2.1 Compartmental SEIRD epidemiological model

A number of research on the natural prognosis of COVID-19 infection influenced the fundamental SEIRD model framework (Wu & McGoogan, 2020). The compartments utilized in the model are determined by the qualities of the specific illness being represented as well as the model's goal. The SEIRD model, which does not contain an incubation phase, should more precisely depict the development of the epidemic. Infected persons will not experience severe symptoms at once, but will instead go through milder phases of infection initially. In other studies, what we term mild infections are divided into two categories: mild and severe (Yang et al., 2020). The proposed model is presented in Fig. 2.

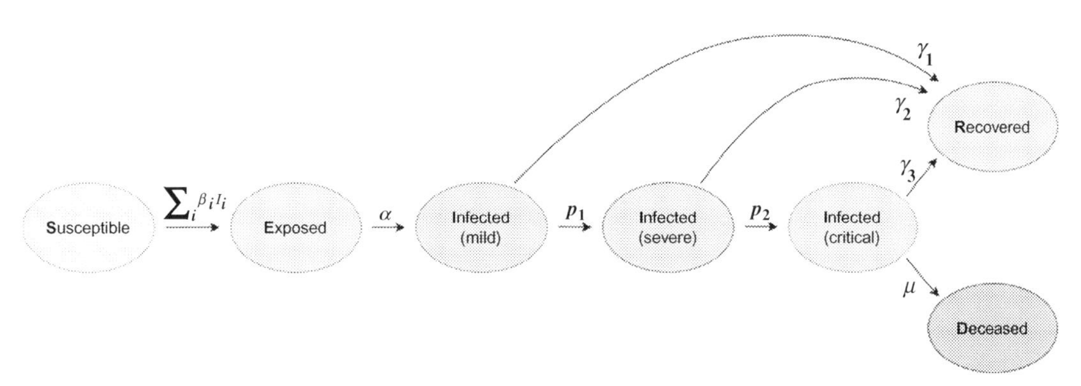

FIG. 2 Compartmental SEIRD epidemiological model, based on the classic susceptible, exposed, infected (mild, severe, critical), recovered, and deceased.

Susceptible (S) individuals are defined as

$$\dot{S} = -(\hat{a}_1 I_1 - \hat{a}_2 I_2 - \hat{a}_3 I_3)S \tag{5}$$

Everything starts in the exposed (E) community, defined as

$$\dot{E} = -(\hat{a}_1 I_1 + \hat{a}_2 I_2 + \hat{a}_3 I_3)S - aE \tag{6}$$

The rate of progression from the exposed to the infected stage I, occurs at a pace a. Infected individuals begin with mild infection (I_{mild}):

$$\dot{I_{\text{mild}}} = aE - (\tilde{a}_2 + p_1) I_{\text{mild}} \tag{7}$$

from which they either recover, at the rate of $\tilde{a}_1$ or advance to severe infection (I_{severe}) at the rate of p_1:

$$\dot{I_{\text{severe}}} = p_1 I_1 - (\tilde{a}_2 + p_2) I_{\text{severe}} \tag{8}$$

Severe infection subjects recover at rate $\tilde{a}_2$ or progresses to critical stage (I_{critical}) at p_2 rate:

$$\dot{I_{\text{critical}}} = p_2 I_2 - (\tilde{a}_3 + \mu) I_{\text{critical}} \tag{9}$$

Recovered persons are assumed to be safe from reinfection for life and are monitored by R class:

$$\dot{R} = \tilde{a}_1 I_{\text{mild}} + \tilde{a}_2 I_{\text{severe}} + \tilde{a}_3 I_{\text{critical}} \tag{10}$$

Individuals with critical infection either recover at a rate of $\tilde{a}_3$ or die at a rate of μ.

$$\dot{D} = \mu I_{\text{critical}} \tag{11}$$

Individuals can transmit the infection at any point, although at different levels. The transmission rate at stage i is defined by $\hat{a}_1$. The assumption is that total population size is constant and given as

$$N = S + E + I_1 + I_2 + I_3 + R + D \tag{12}$$

In order to describe the rates of disease progression from one category to another (susceptible to infected, infected to recovered or deceased, etc.), we use the following notation:

- $\hat{a}_1$ rate at which infected individuals in class I_i contact susceptible and infect them,
- a rate of progression from the exposed to infected class,
- $\tilde{a}_i$ rate at which infected individuals in class I_i recover from disease and become immune,
- p_i rate at which infected individuals in class I_i progress to class I_{i+1} and
- μ death rate for individuals in the most severe stage of the disease.

These coefficients are computed using both literature and real data from reported official websites. All rates are represented per day. Time variables of typical durations are adopted from the literature:

- average incubation period, in days,
- average duration of mild infections, in days,
- average duration of hospitalization (time to recovery) for individuals with severe infection, in days, and
- average duration of ICU admission (until death or recovery), in days.

Mentioned COVID-19 parameters obtained from literature sources are given in Table 1.

TABLE 1 Estimated parameters for COVID-19 clinical progression based on literature sources.

Parameter name	Adopted literature value	Reference
Average incubation period, days	5–6 days 5 days	Milovanovic et al. (2020), Li, Guan, et al. (2020), World Health Organization (2020), European Centre for Disease Prevention and Control (2020), Centers for Disease Control and Prevention (2020), Lauer et al. (2020), Linton et al. (2020), Bi et al. (2020), and Sanche et al. (2020)
Average duration of mild infections, days	7–12 days 5 days	Milovanovic et al. (2020) and Woelfel et al. (2020)
Average duration of hospitalization (time to recovery) for individuals with severe infection, days	5–14 10 days	Milovanovic et al. (2020), Cao et al. (2020), and Liu, Funk, and Flasche (2020)
Average duration of ICU admission (until death or recovery), days	14 days 12–17 days	Milovanovic et al. (2020), Cao et al. (2020), and Liu, Funk, and Flasche (2020)
Reproduction number[a]	2–2.5 2	Milovanovic et al. (2020) and Callaway (2020)
Infective period Mild infection (in days)	5 days	Milovanovic et al. (2020), Woelfel et al. (2020), Hauser et al. (2020), Tindale et al. (2020)
Infective period Severe infection (in days)	7–12 days	Milovanovic et al. (2020), Woelfel et al. (2020), Huang et al. (2020b), Zhou et al. (2020)
Infective period Critical infection (in days)	14 days	Milovanovic et al. (2020), Woelfel et al. (2020), Liu, Funk, and Flasche (2020), and Hauser et al. (2020)
Transmission rate of mild infections	0.4 per day	$R0$/InfPeriodMild
Transmission rate of severe infections	0.2 per day	$R0$/InfPeriodSevere
Transmission rate of critical infections	0.14 per day	$R0$/InfPeriodCritical

[a] *The threshold for many epidemiological models is the specific reproduction number* R0, *which is defined as the average number of secondary infections created when an infectious organism is introduced into a host population where everyone is susceptible (Hethcote, 2000). In many deterministic epidemiological models, infection will begin in a completely susceptible population if and only if* $R0>1$ is present.

When it came to determining which model coefficients were compatible with existing clinical evidence, we computed specific values using real data from the country of Belgium using statistics from Infectious diseases data explorations & visualizations (n.d.). Our focus was Belgium because of the fact that the data were available in a tabular format with clearly reported data (no under or over representation) and no missing values including infection categories (ICU patients, hospitalized patients, etc.). Not many countries provide data in such a format to the public. Available data for Belgium included the number of infected patients, the number of hospitalized patients, the number of patients on ventilators, and the number of deaths. Based on the described methodology, parameters calculated for Belgium were

- Average fraction of (symptomatic) infections that are mild is equal to **95.96%.**
- Average fraction of (symptomatic) infections that are severe is equal to **4.04%.**

- Average fraction of (symptomatic) infections that are critical is equal to **0.86%.**
- Case fatality rate (fraction of infections that eventually result in death) is equal to **8.24%.**

2.2.2 LSTM epidemiological model

Because the path of the epidemic is rapidly shifting and the conventional SEIRD model cannot explain numerous peaks, we used an algorithm that is suited for long-term data fitting and forecasting. Appropriate architecture for this group of issues will be based on recurrent neural networks (RNN). The RNN's uniqueness is represented in the context layer, whose function is to operate as memory in order to combine the current state and inputs for information transmission into subsequent states. Backpropagation through time (BPTT), a type of extension of the backpropagation algorithm, is one approach for training RNNs. BPTT is distinguished by gradient descent, in which the error is backpropagated for a deeper network design with time-defined states. As a result, there is an issue with learning long-term dependencies, which will result in disappearing and inflating gradients. Long short-term memory (LSTM) neural networks were created to overcome the vanishing gradient problem of RNNs. LSTM is a type of circular neural network structure that can learn the specified long-term dependencies and overcome the limitations of RNNs. LSTM comprises memory cells and gates that improve the ability to recall long-term dependence. (Hochreiter & Schmidhuber, 1997).

The encoder-decoder LSTM network (ED-LSTM) was created as a sequence-to-sequence neural network capable of mapping a fixed-length input to a fixed-length output. The benefit of these neural networks is that the two lengths of inputs and outputs described above do not have to be the same. The ED-LSTM has two stages of implementation: encoding in the first phase and decoding in the second. The first phase's goal is to encode an input sequence into a fixed-length vector representation and calculate a series of hidden states, while the goal of the second phase is to decode the encoded sequence and produce an output sequence (Fig. 3 **left**). Each LSTM cell in the encoder-decoder LSTM network consists of the gates shown in Fig. 3 **(right)**, the input gate i_t decides which information can be transferred to the cell, then forget gate f_t decides which information from the previous cell should be neglected. The control gate $\overline{C}_t$ is controlling update of the cell and the output gate o_t control the flow of output activation information.

LSTM calculates hidden layer H_t as

$$i_t = \sigma\left(W^i \times (x_t + h_{t-1}) + b_i\right) \tag{13}$$

$$f_t = \sigma\left(W^f \times (x_t + h_{t-1}) + b_f\right) \tag{14}$$

$$o_t = \sigma\left(W^o \times (x_t + h_{t-1}) + b_o\right) \tag{15}$$

$$\overline{C}_t = \tanh\left(W^c \times (x_t + h_{t-1}) + b_C\right) \tag{16}$$

$$C_t = \sigma\left(f_t * C_{t-1} + i_t * \overline{C}_t\right) \tag{17}$$

$$H_t = \tanh(C_t) * o_t \tag{18}$$

The number of the input features is presented as x_t and H_t is the number of hidden units. Learning started with the zero initial values of C_0 and H_0. Also, during the learning process some parameters were adjusted, such as bias given as b and weight given as W. The internal

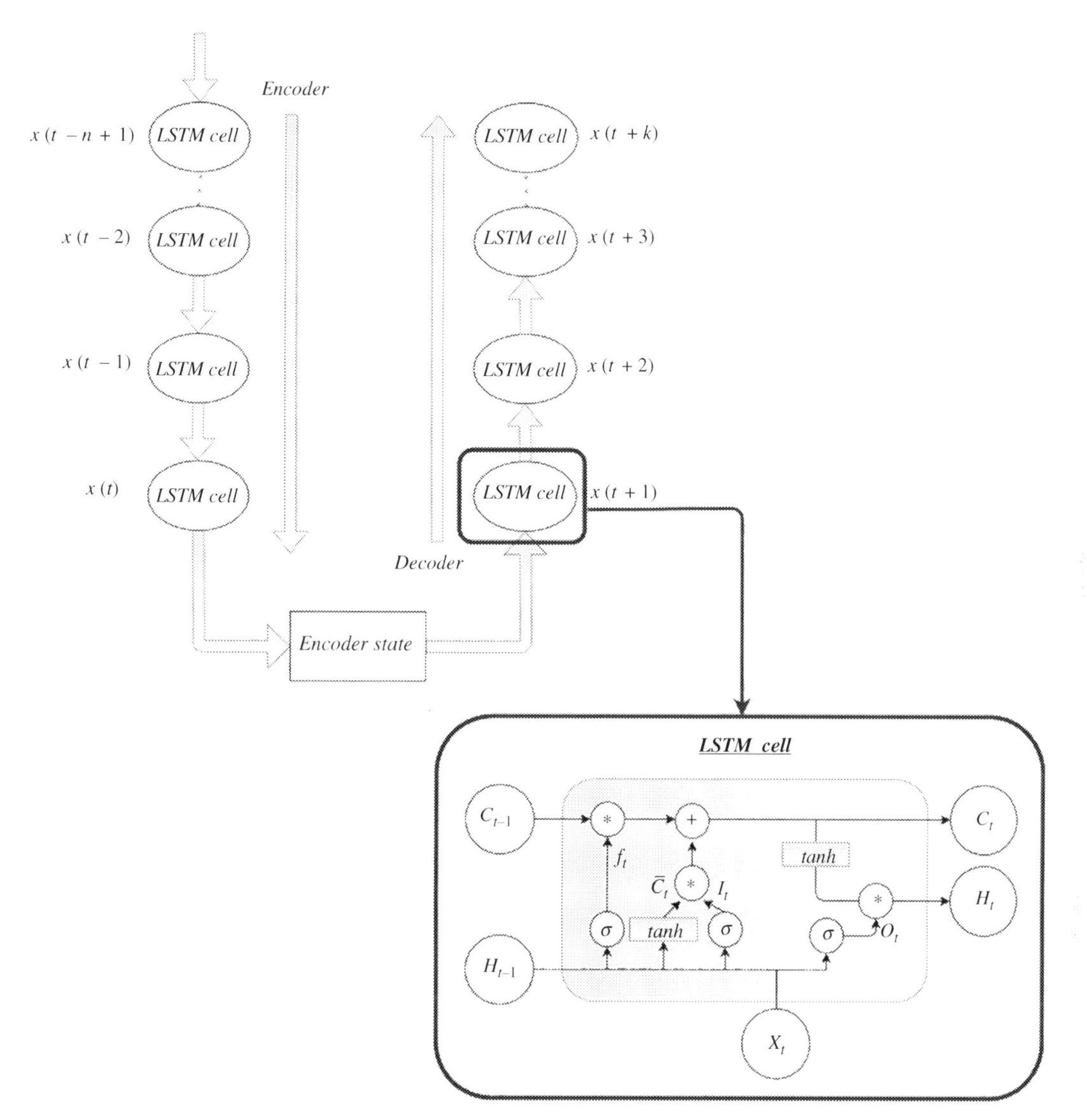

FIG. 3 Proposed encoder-decoder LSTM structure with the architecture of one LSTM cell.

memory of the unit is given as C_t and it should be emphasized that all gates have the same dimension as the size of your hidden state (Chandra, Jain, & Chauhan, 2021).

3 Results and discussion

3.1 Results for personalized model

We selected 10 blood analyses as the best features that the most reliably described the development of COVID-19 disease in patients and contributed the most to the classification task,

TABLE 2 Selected blood analyses of the greatest influence in COVID-19 patients' classification.

Blood analysis	Mean ± standard deviation	Normal values	Units
White blood cell count	8.25 ± 4.72	3.70–10.00	10^9/L
Lymphocytes	22.15 ± 11.06	20.00–46.00	%
MCHC	335.7 ± 8.16	320–360	g/L
RDW	13.69 ± 1.07	11.6–14.5	%
Hgb	130.08 ± 21.05	138–175	g/L
Urea	7.35 ± 5.76	3.0–8.0	mmol/L
Creatinine	92.46 ± 69.62	49–106	μmol/L
Albumins	33.57 ± 6.23	35–52	g/L
LDH	424.86 ± 239.26	220–450	U/L
CRP	69.11 ± 93.55	0.0–5.0	mg/L

according to the previously described methodology. The values of the mentioned blood analyses are presented in the form of mean ± standard deviation in Table 2. Additionally, Table 2 contains the range of normal values of the mentioned blood analyses.

In order to evaluate the importance scores of the features, a correlation of each two features was computed. Fig. 4 shows the importance scores of all 10 selected features. The highest importance score belongs to lactate dehydrogenase (LDH) and this is justified due to the fact that this enzyme is widely distributed in tissues and its elevated serum levels could be caused by systemic hypoxemia (Panteghini, 2020). In the early stages of the disease, high serum activity of LDH may be an indicator of lung injury and high-risk outcomes (Yan et al., 2020a).

In addition to the LDH, which was shown to be by far the most important biomarker in predicting the severity of the clinical condition, other parameters which mostly influence the model are urea, creatinine, C-reactive protein (CRP), white blood cell (WBC), etc. The mentioned blood analyses are shown in order from the most important to the less important in Fig. 4. Urea and creatinine are related to kidney function—within one research paper were considered clinical conditions of 701 COVID-19 patients and concluded that acute kidney injury and death was more frequent in patients which had higher serum creatinine levels in comparison to the patients with normal values (Cheng et al., 2020b). The explanation of the high importance score of CRP is reflected in the fact that elevated CRP is directly correlated with the level of inflammation. It is used as an indicator of inflammatory and infectious conditions and therefore could be helpful in predicting the severity of the clinical conditions. Generally, several studies have found that fatal effects of COVID-19 disease are related to liver and kidney functions, CRP, interleukin-6 (IL-6), lower lymphocytes, and albumin blood levels (Malik et al., 2020).

Values of the 10 selected blood analyses are assessed by the Gradient boost regressor model. For each of 34 patients with complete blood analyses, the values of analyses were assessed on day 14 after the admission day, according to the described methodology in the previous section. The root mean square error (RMSE) between assessed and actual values of blood analyses is shown in Table 3.

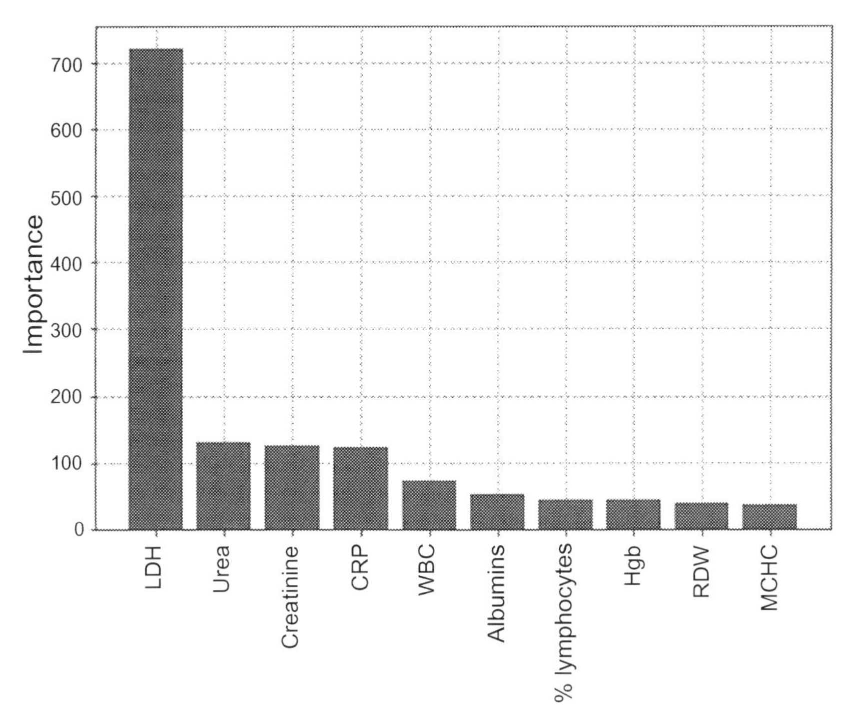

FIG. 4 Importance scores of 10 best features.

TABLE 3 RMSE between assessed and real values of blood biomarkers.

Blood biomarker	RMSE
White blood cell (WBC)	3.03
% Lymphocytes	1.57
MCHC	2.77
RDW	0.35
Hgb	10.12
Urea	2.54
Creatinine	55.23
Albumins	1.63
LDH	35.51
CRP	41.68

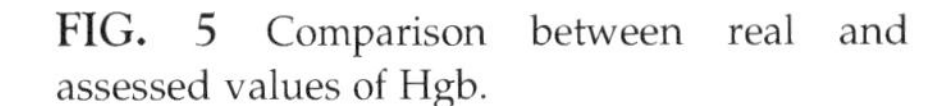

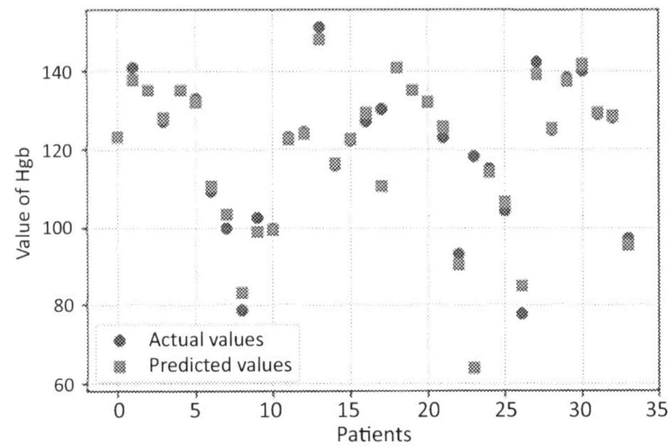

FIG. 5 Comparison between real and assessed values of Hgb.

The smaller values of RMSE are expected for analyses such as WBC, % lymphocytes, MCHC, RDW, urea, and albumins due to the values of these analyses fall under the narrower range as is shown in Table 2. More deviations between real and assessed values were observed in Hgb, creatinine, LDH, and CRP, so the achieved results for each of these four analyses will be discussed individually.

Firstly, a comparison between real and predicted values of Hgb is shown (Fig. 5). Hgb has a slightly higher value of RMSE compared to RMSE of other biomarkers, but a larger deviation from the actual value is expected because the range of actual values (80–150) is wider than the ranges of other biomarkers (e.g., the range of actual urea values is 3.0–8.0). For this reason, an RMSE value that amounts to 10.12 does not affect the model's decision in assessing the clinical condition of COVID-19 patients.

In the following, a comparison between real and assessed values of creatinine was given (Fig. 6). Creatinine has the highest value of RMSE, and the cause can be the existence of some outliers that could be present in the dataset.

The higher value of RMSE for LDH analysis was obtained for the same reasons as in the case of comparison between assessed and actual values of creatinine. The RMSE value of LDH is shown in Fig. 7.

In Fig. 8 comparison between real and assessed values of CRP is shown. Explanation of this RMSE value—41.68 is similar to the RMSE value of Hgb, where CRP has an even wider range of values (0–400) than Hgb.

After assessing the patient's hematology and clinical biochemistry analyses, it is possible to predict the patient's clinical condition in advance and this was achieved by the XGboost classification algorithm. The model successfully classified 34 COVID-19 patients into four classes of clinical condition (mild, severe, moderate, and critical) with an accuracy of 94%. In addition to accuracy, we considered other metrics such as precision, recall, and F1-score. In Table 4, all of these metrics for each class are shown individually.

Further, a confusion matrix with a normalized and regular number of patients is given in Fig. 9.

Our rule-based model consists of hundreds of trees whose rules are understandable and compose the final decision. As one tree is not enough for the final decision, investigation and analysis of several trees is the right path to achieving high accuracy for the classification

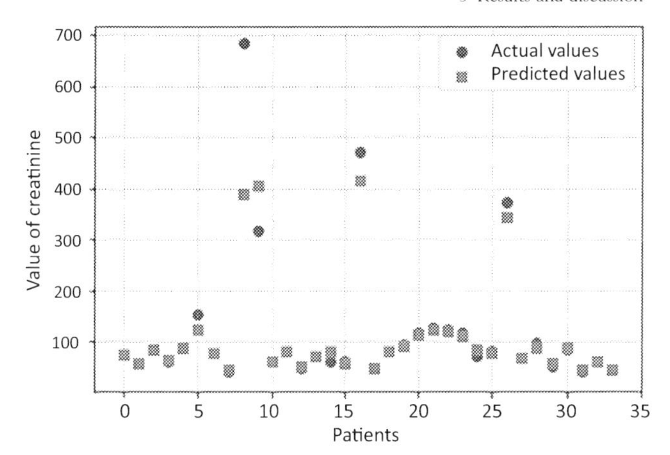

FIG. 6 Comparison between real and assessed values of creatinine.

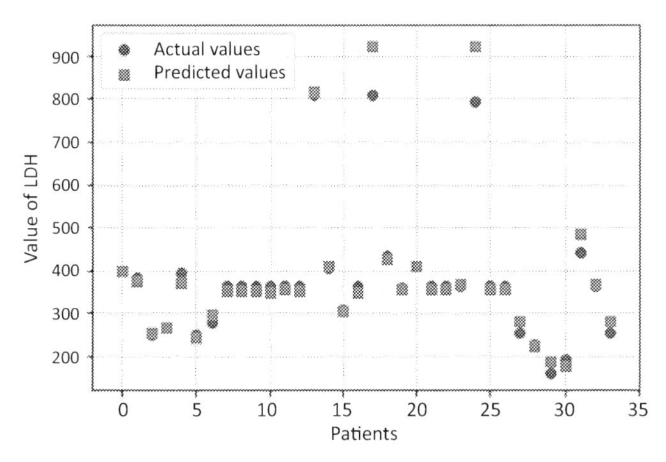

FIG. 7 Comparison between actual and predicted values of LDH.

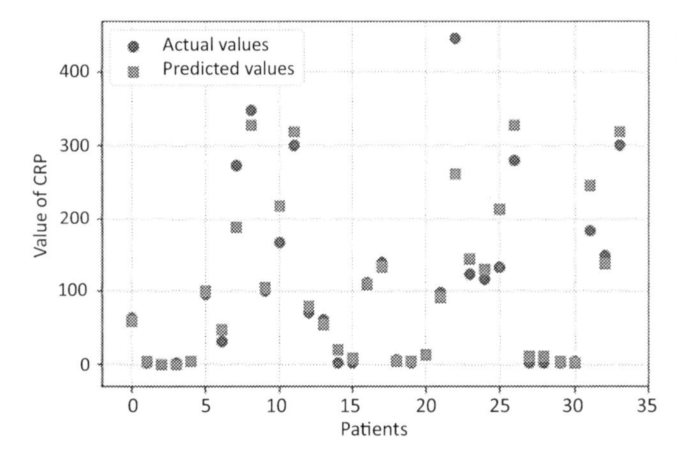

FIG. 8 Comparison between real and assessed values of CRP.

TABLE 4 Overview of classification metrics for each class.

Class	Accuracy	Precision	Recall	F1-score
Mild	1.00	0.80	1.00	0.89
Moderate	0.89	1.00	0.89	0.94
Severe	1.00	1.00	1.00	1.00
Critical	1.00	1.00	1.00	1.00

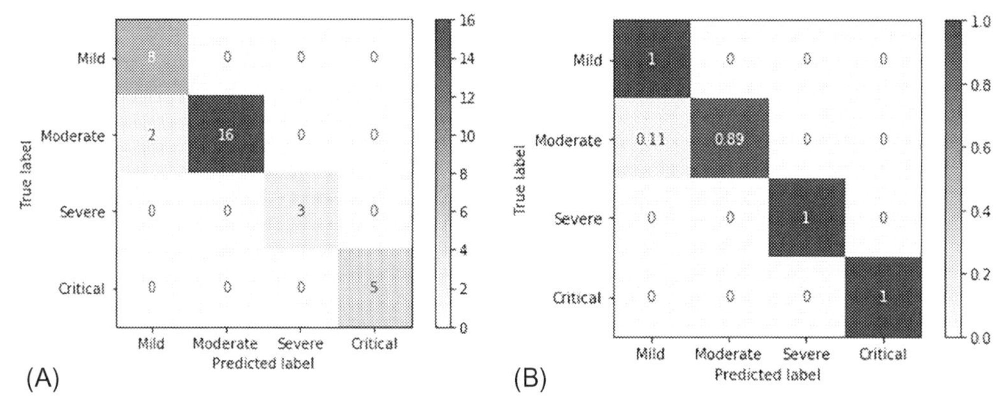

FIG. 9 Confusion matrix with regular (left) and normalized (right) values of patients.

problems. So, due to its simplicity and comprehensibility, this type of model is suitable for application in clinical practice.

3.2 Results for epidemiological model

In this section, we present the results of the applied SEIRD model and LSTM encoder-decoder model to compare the real situation with simulated on regarding COVID-19 outbreak and spread, in Belgium. Data was monitored during the 1-year period from March 15, 2020 to March 15, 2021. Values for the real numbers of people infected, hospitalized, taken in ICU as well as deceased are taken from the official data reports (Infectious diseases data explorations & visualizations, n.d.) for Belgium.

3.2.1 Results for the SEIRD model

The findings related to the SEIRD model imply that the model is successful at forecasting the initial peak of the epidemic. However, owing to the nature of equations and the model's formulation, the model cannot forecast many peaks and should be expanded to include new phenomena. As a result, we only look at this model's behavior during the first 100 days of the claimed pandemic, which corresponds to the initial peak. Figs. 10–13 depict the explained prediction throughout the first 100 days for the Belgian example—mild, severe, critical, and deceased patients, respectively.

As can be observed, following the number of persons who have a serious illness, a critical infection, or have died shows a very strong match. All of the observed curves—mild, severe,

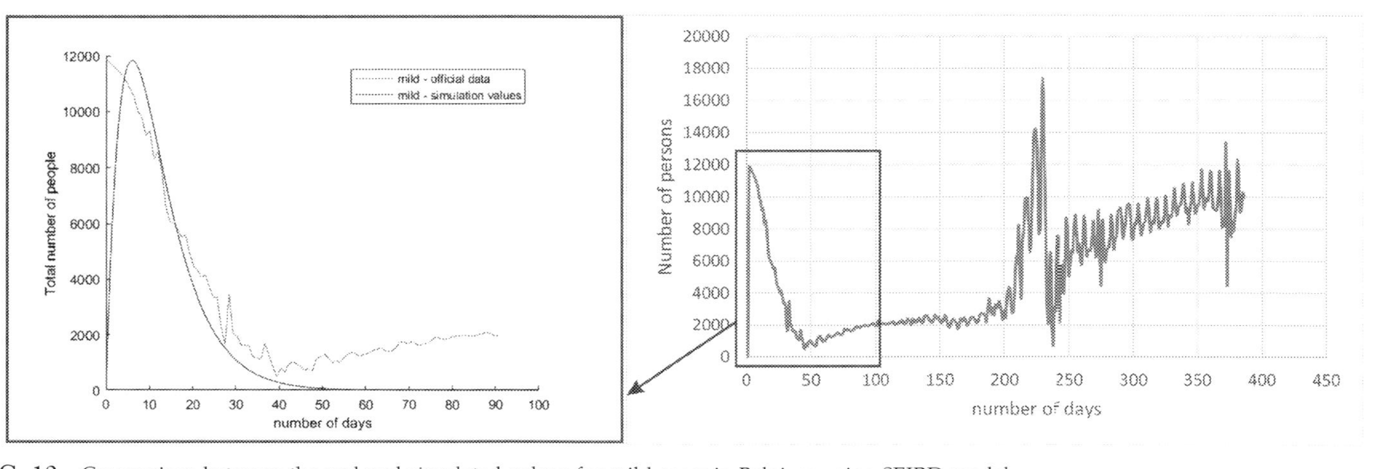

FIG. 10 Comparison between the real and simulated values for mild cases in Belgium using SEIRD model.

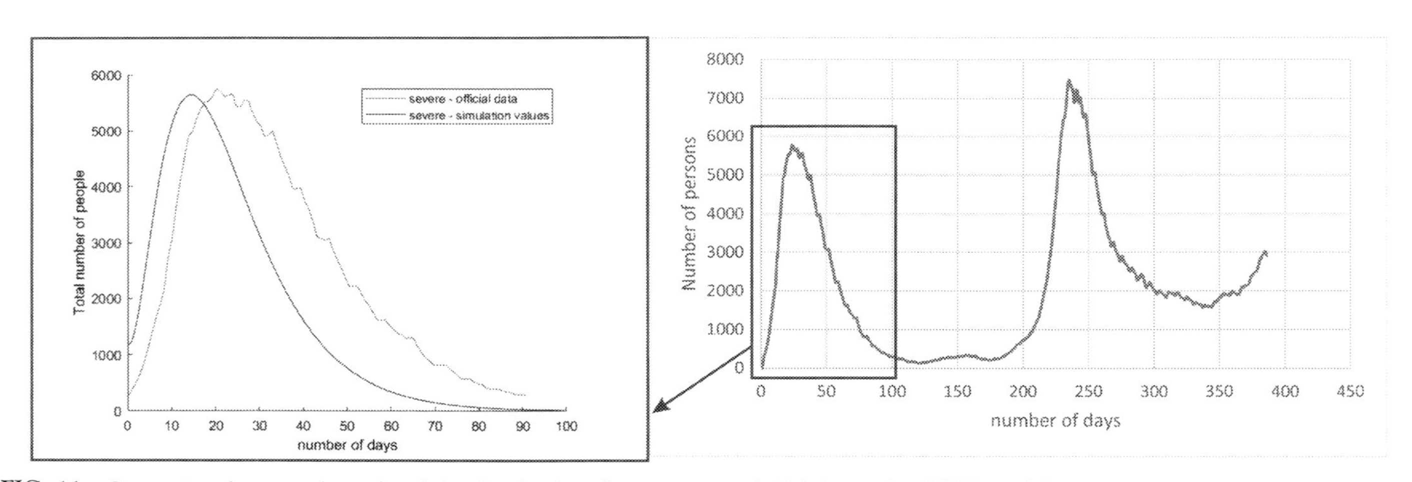

FIG. 11 Comparison between the real and simulated values for severe cases in Belgium using SEIRD model.

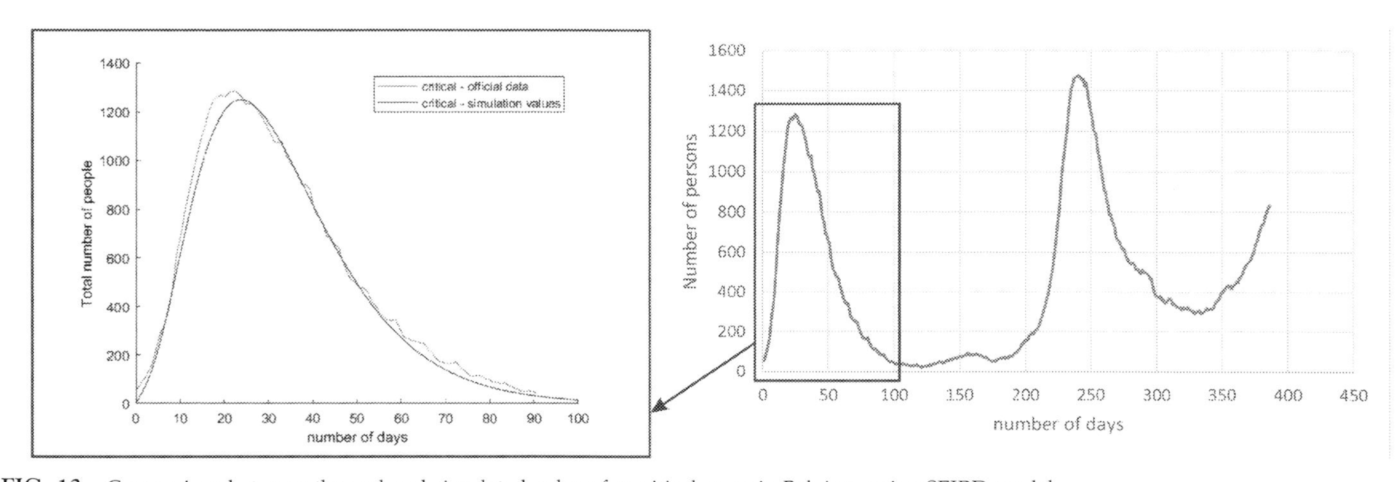

FIG. 12 Comparison between the real and simulated values for critical cases in Belgium using SEIRD model.

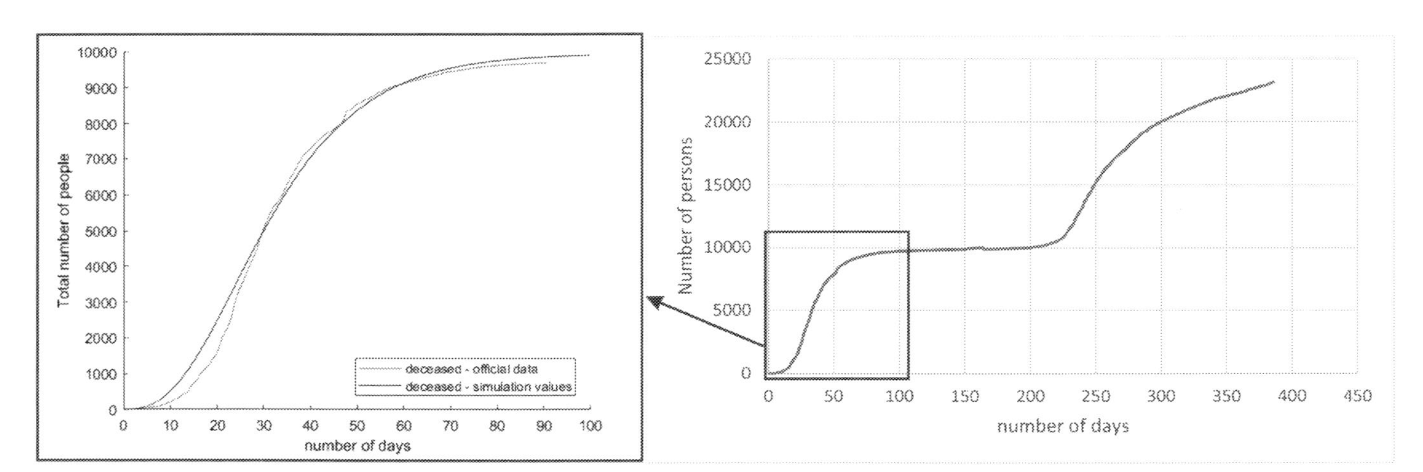

FIG. 13 Comparison between real and simulated values for deceased cases in Belgium using SEIRD model.

critical infected, and deceased—demonstrate a promising match in terms of trend and peak value. There is a variation in peak location between the simulated and actual cases for patients with severe illness. This is most likely due to the starting circumstances given in simulation, which may be further improved, as beginning conditions have a large effect on subsequent prediction. Nonetheless, the trend is sufficient, demonstrating that a technique may be employed to explain epidemiology.

Root mean square error (RMSE) between simulated and real values of investigated curves for Belgium show that RMSE for deceased cases is 20.11, RMSE for mild infected is 363.72, RMSE for severe infected is 17.64 and RMSE for critical infected is 1.52. It can be observed that mild infections had a larger RMSE, which is due to the fact that mild cases were generated from other data (total cases and hospitalized patients), with large day-to-day fluctuations since there is no data on a particular mild category.

All of this demonstrates that there are several elements involved in predicting epidemic propagation, and more factors must be included in order to account for multiple peaks, declining and growing trends, and so on. However, when looking at a shorter period of time, the number of deceased and infected people (mild, severe, and critical) using official statistics and simulated numbers shows a good match, indicating that the models are promising and can be further upgraded to account for different underlying complex phenomena. The primary shortcomings of the SEIRD models can be attributed to a variety of factors:

- Calculated values for parameters based on official statistics are incorrect due to scarce official statistics. It has also been documented in earlier studies that government statistics may be initially underreported, resulting in calculated incorrect parameters, and therefore causing differences between the simulated and real numbers. To anticipate the magnitude of the outbreak in Wuhan, Imai et al. utilized travel volumes from Wuhan and the dates when imported cases first arrived in locations inside China. They claim that there were much more instances in Wuhan than were recorded in official data (Poletto et al., 2020). Paper by Korolev et al. reached the same conclusion (Korolev, 2020). He claims that even if just a portion of all reported instances is detected, it might be beneficial to consider underreporting. If all instances are assumed to be reported and $R0$ estimations are based only on this assumption, the value of $R0$ may be skewed lower. It may result in an overestimation of the number of deaths (Korolev, 2020). This is not the case for Belgium, however, if the model is translated for other countries, the situation may occur.
- The complexity of the COVID-19 epidemic transmission and development has yet to be identified, and the present SEIRD model does not account for medical treatments, hospital bed availability, etc. This indicates that the model predicts what would happen if the original conditions are satisfied and the spread continues to move freely. The true scenario is far more complicated, with many other occurrences included (behavioral reactions to the epidemic, reinfection—no immunity, viral changes, etc.)
- We have computed all of the parameters based on the time of interest, which means that a single value for the rate of moderate, severe, or critical infection may not be able to represent the complexities of the situation presented by COVID-19 in nations. In future studies, it may be preferable to group periods of time when there were no medical interventions (no available hospital beds), when the country was in a lockdown, when a state of emergency was declared, and so on, and calculate all the parameters (i.e., mentioned infection rates) separately for each cluster. This would provide a better

understanding of typical subperiods. Following that, the SEIRD model may be applied to each cluster and the results are superimposed to get more accurate findings.

Although the model produces promising findings, and the curves between simulated trends and values computed using official accessible data are closely matched, there are some discrepancies in peak values and locations, as well as the fact that normal SEIRD cannot characterize more than one pandemic peak. This is largely owing to the fact that modeling disease transmission is difficult and encompasses numerous phenomena, of which just a few major ones are represented in the current model. As a result, the primary drawback of this work is the small number of phenomena simulated (no reinfection, asymptomatic infection, medical intervention, etc.). As a result, we examined deep learning models and present the findings of one such LSTM-based model.

3.2.2 Results for the LSTM encoder-decoder neural network

In addition to analyzing the number of exposed, susceptible, infected, and deceased cases, time series data of daily hospitalized, and patients on ventilators were also considered. For the purpose of LSTM-ED model, we used the number of infected, hospitalized, patients on ventilators, and deceased patients. First, we used the cross-validation process in order to get a more realistic view of the model's error and predictions. This means that we split the dataset on training and testing subsets in several different ratios:

- for the first iteration, we used the 50-day training dataset from March 15 to May 3, 2020, then we predicted each variable of the 50-day test dataset from May 4 to June 22, 2020,
- for the second iteration, we used data from March 15 to June 22, 2020 as a training set and the data from June 23 up to August 12, 2020 as a test set,
- for the third iteration, data up to August 12, 2020 was used as a training set and the data from August 13 up to October 2, 2020 was used as a test set,
- the fourth iteration, data up to October 2, 2020 was used as training set and the test set contains 62 samples instead of 50, and it means that data from October 3 up to December 4, 2020 are included, and
- the last iteration implies the training data from March 15 up to December 4, 2020, and the test set contains 100 samples, from December 5, 2020 to March 15, 2021.

The example of the cross-validation process through mentioned iterations is presented in Fig. 14. In the following example, the number of infected cases in Belgium was predicted.

In order to evaluate LSTM-ED model performances, the same metric—RMSE is used as in the SEIRD model. Average RMSE between simulated and real values for Belgium is given in the following: the RMSE for infected cases is 535.93, the RMSE for hospitalized cases is 20.42, the RMSE for cases on ventilators is 38.97 and RMSE for deceased cases is 8.72. Due to the variability in the real data, we compared the predicted curve with the smoothed version of the real curve. The moving average smoothing technique was applied to real values of the test set in order to reduce the variations between time steps.

During the validation process, we concluded that LSTM-ED is able to predict when another peak of the epidemic will occur based on the position of the first peak. For the case of Belgium, in this 1-year period, only two peaks appeared, therefore, we divided the dataset in such a way that 58% of data is used as a training subset and the remaining 42% of data is used as a testing subset. Then, we decided to show for each variable individually how the model will forecast in the period from October 20 up to March 15, 2021. In Fig. 15 the comparison between

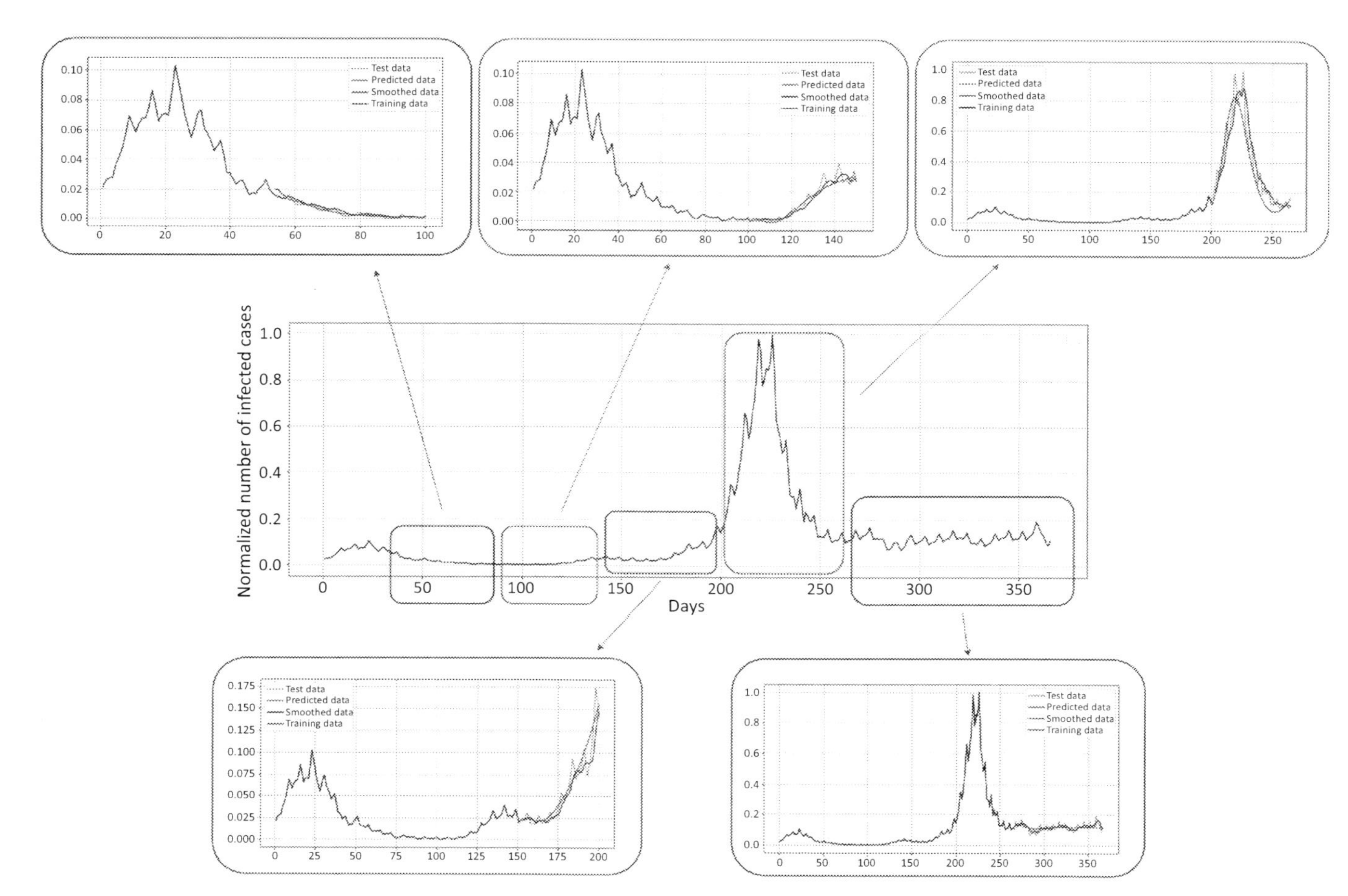

FIG. 14 Cross-validation process for the infected cases in Belgium.

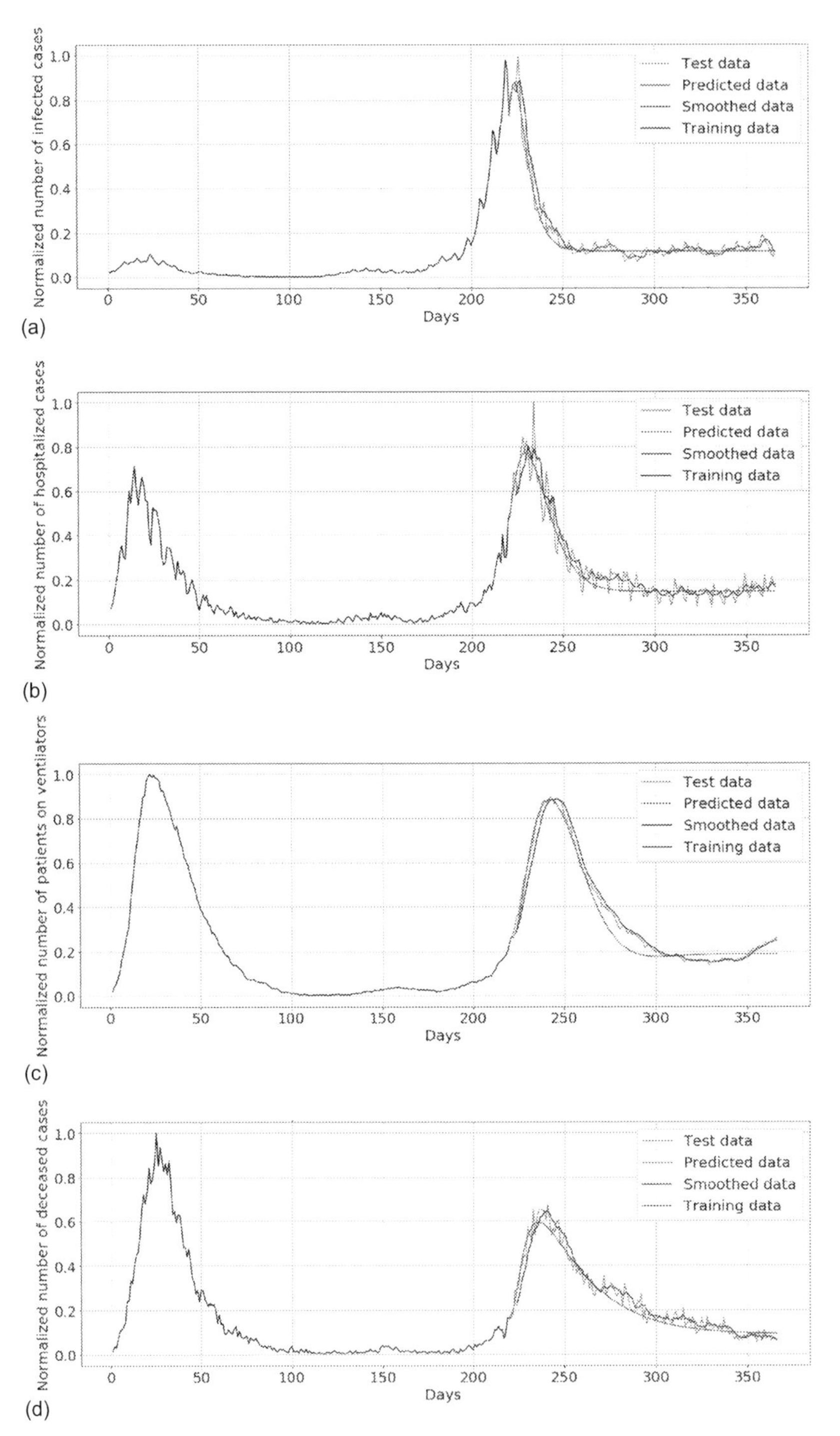

FIG. 15 Comparison between real and forecasted curves for Belgium. (A) number of infected cases, (B) number of hospitalized cases, (C) number of cases on ventilators, and (D) number of deceased cases.

the real and simulated curves is shown. Also, a smoothed curve is presented in order to expose better a trend of the official data curve. The trend of the predicted curve matches well with the smooth curve from real data for all the monitored curves—infected cases, hospitalized cases, cases on ventilators, and deceased cases.

4 Conclusions

This research represents automatic ML methods for predicting the severity of the clinical condition of COVID-19 patients individually and for the modeling of COVID-19 spread and development in the population. From implemented algorithms and hyperparameter optimization for assessing the clinical condition of COVID-19 patients, we can conclude that the proposed methodology using the XGboost classifier is adequate for the classification of patients into four distinct categories of clinical conditions (mild, moderate, severe, critical) with 94% of accuracy. The methodology is more suitable for implementation in real clinical practice since the proposed methodology is rule-based rather than a black box. On the other hand, for the population-based modeling, we proposed SEIRD compartmental epidemiological model with included components—susceptible, exposed, infected (we have divided the infected group into three subgroups—mild, severe, and critical), recovered, deceased and LSTM-ED model with included information about the number of infected cases, hospitalized cases, case on ventilators and deceased cases. In order to calculate the parameters for the model, we have also investigated official statistical data for Belgium. Results show that the SEIRD model is able to accurately predict only the first peak, with larger RMSE for mild cases. In contrast, the LSTM-ED model shows that they are capable to predict the second peak based on the position of the first peaks with the low values of RMSE. Higher values of RMSE are observed in the infected cases in Belgium due to the thousands of infected people per day in those countries. The match between simulated and real values can be affected by several things, such as underreporting of the number of cases, estimating initial conditions, setting parameters, etc. In general, official and simulated values show a good match, which means that the model is showing promising results and can be further upgraded to take into account different underlying complex phenomena. Future research will include more phenomena, especially medical intervention and asymptomatic infection, in order to better describe the COVID-19 spread and development. Also, from the perspective of personalized modeling, further research would be focused on collecting a larger database of patients as well as investigating other ML models that can be coupled with existing ones in order to create a hybrid model which would achieve even larger accuracy. We would also like to include patients from other countries as the geographical position of the country could be taken into account as an additional feature. In such a way we could investigate how geographical area influences the disease development in patients and possible reaction to COVID-19 as a function of geographical location.

Ethics

The work has been approved by the appropriate ethical committees related to the institution(s) in which it was performed and that subjects gave informed consent to the work.

Conflict of interest statement

The authors declare that they have no known competing financial interests or personal relationships that could have appeared to influence the work reported in this paper.

Acknowledgments

This research was funded by Serbian Ministry of Education, Science, and Technological Development [451-03-68/2020-14/200107 (Faculty of Engineering, University of Kragujevac)]. This research was also supported by the CEI project "Use of Regressive Artificial Intelligence (AI) and Machine Learning (ML) Methods in Modeling of COVID-19 spread—COVIDAi" [Grant No. 305.6019-20].

References

Altman, N., & Krzywinski, M. (2017). Points of significance: Clustering. *Nature Methods, 14*, 545–546.

Anastassopoulou, C., Russo, L., Tsakris, A., & Siettos, C. (2020). Data-based analysis, modelling and forecasting of the COVID-19 outbreak. *PLoS One, 15*.

Andjelic, N., Baressi Segota, S., Lorencin, I., Jurilj, Z., Sustersic, T., Blagojevic, A., Protic, A., Cabov, T., Filipovic, N., & Car, Z. (2021). Estimation of covid-19 epidemiology curve of the united states using genetic programming algorithm. *International Journal of Environmental Research and Public Health, 18*(3), 959.

Aronson, J. K., & Ferner, R. E. (2017). Biomarkers—A general review. *Current Protocols in Pharmacology, 76*(1), 9–23.

Assandri, R., Buscarini, E., Canetta, C., Scartabellati, A., Viganò, G., & Montanelli, A. (2020). Laboratory biomarkers predicting COVID-19 severity in the emergency room. *Archives of Medical Research, 51*(6), 598–599.

Bai, H. X., et al. (2020). AI augmentation of radiologist performance in distinguishing COVID-19 from pneumonia of other etiology on chest CT. *Radiology*, 201491.

Baressi Segota, S., Lorencin, I., Andjelic, N., Stifanic, D., Musulin, J., Vlahinic, S., Sustersic, T., Blagojevic, A., & Car, Z. (2021). Automated pipeline for continual data gathering and retraining of the machine learning-based COVID-19 spread models. *EAI Endorsed Transactions on Bioengineering and Bioinformatics, 21*(2), 1–9.

Benelli, G., et al. (2020). SARS-COV-2 comorbidity network and outcome in hospitalized patients in Crema, Italy. *medRxiv*.

Bi, Q., Wu, Y., Mei, S., Ye, C., Zou, X., Zhang, Z., Liu, X., & Wei, L. (2020). Epidemiology and transmission of COVID-19 in Shenzhen China: Analysis of 391 cases and 1,286 of their close contacts. *MedRxiv*.

Blagojevic, A., Sustersic, T., Lorencin, I., Baressi Segota, S., Milovanovic, D., Baskic, D., Baskic, D., Car, Z., & Filipovic, N. (2021). Combined machine learning and finite element simulation approach towards personalized model for prognosis of COVID-19 disease development in patients. *EAI Endorsed Transactions on Bioengineering and Bioinformatics, 21*(2), e6.

Böhmer, M., Buchholz, U., & Corman, V. E. A. (2020). Investigation of a COVID-19 outbreak in Germany resulting from a single travel-associated primary case: a case series. *The Lancet Infectious Diseases, 20*(8), 920–928.

Callaway, E. (2020). Coronavirus vaccines: Five key questions as trials begin. *Nature, 579*, 481.

Cao, B., Wang, Y., Wen, D., et al. (2020). A trial of lopinavir-ritonavir in adults hospitalized with severe Covid-19. *The New England Journal of Medicine, 382*, 1787–1799.

Casella, F. (2020). *Can the COVID-19 epidemic be managed on the basis of daily data?*.

Centers for Disease Control and Prevention. (2020). *Healthcare professionals: Frequently asked questions and answers*. Atlanta: Centers for Disease Control and Prevention. [Online]. Available https://www.cdc.gov/coronavirus/2019-ncov/hcp/faq.html.

Chandra, R., Jain, A., & Chauhan, D. S. (2021). Deep learning via LSTM models for COVID-19 infection forecasting in India. *arXiv preprint*. arXiv:2101.11881.

Chandra, R., Shaurya, G., & Rishabh, G. (2021). Evaluation of deep learning models for multi-step ahead time series prediction. *arXiv Preprint*. arXiv:2103.14250.

Chen, T., & Guestrin, C. (2016). Xgboost: A scalable tree boosting system. In *Proceedings of the 22nd acm sigkdd international conference on knowledge discovery and data mining* (pp. 785–794).

Cheng, A., et al. (2020a). Diagnostic performance of initial blood urea nitrogen combined with D-dimer levels for predicting in-hospital mortality in COVID-19 patients. *International Journal of Antimicrobial Agents, 56*(3), 106110.

Cheng, Y., et al. (2020b). Kidney disease is associated with in-hospital death of patients with COVID-19. *Kidney International, 97*(5), 829–838.

Chimmula, V. K. R., & Zhang, L. (2020). Time series forecasting of COVID-19 transmission in Canada using LSTM network. *Chaos, Solitons & Fractals, 135*, 109864.

Chinazzi, M., et al. (2020). The effect of travel restrictions on the spread of the 2019 novel coronavirus (2019-ncov) outbreak. *Science, 368*, 395–400.

Cho, A. (2020). AI systems aim to sniff out coronavirus outbreaks. *Science, 368*, 810–811.

COVIDAi. *Use of regressive artificial intelligence (AI) and machine learning (ML) methods in modelling of COVID-19 spread.* [Online]. Available http://www.covidai.kg.ac.rs/.

European Centre for Disease Prevention and Control. (2020). *Rapid risk assessment: Novel coronavirus disease 2019 (COVID-19) pandemic: Increased transmission in the EU/EEA and the UK—Seventh update.* Stockholm: European Centre for Disease Preven. 25 March. [Online]. Available https://www.ecdc.europa.eu/sites/default/files/documents/RRA-seventh-update-Outbreak-of-coronavirus-disease-COVID-19.pdf.

Gao, Y., et al. (2020). Machine learning based early warning system enables accurate mortality risk prediction for COVID-19. *Nature Communications, 11*(1), 1–10.

Giordano, G., Blanchini, F., Bruno, R., Colaneri, P., Di Filippo, A., Di Matteo, A., & Colaneri, M. (2020). Modelling the COVID-19 epidemic and implementation of population-wide interventions in Italy. *Nature Medicine*, 1–6.

Goyal, P., et al. (2020). Clinical characteristics of Covid-19 in New York city. *New England Journal of Medicine, 382.*

Guan, W.-J., et al. (2020a). Clinical characteristics of coronavirus disease 2019 in China. *New England Journal of Medicine, 382*(18), 1708–1720.

Guan, W.-J., et al. (2020b). Comorbidity and its impact on 1590 patients with Covid-19 in China: A nationwide analysis. *European Respiratory Journal, 55*(5).

Gumel, A. B., et al. (2004). Modelling strategies for controlling SARS outbreaks. *Proceedings of the Royal Society B: Biological Sciences, 271*(1554).

Hastie, T., Tibshirani, R., & Friedman, J. (2009). *The elements of statistical learning: Data mining, inference, and prediction.* Springer Science & Business Media.

Hauser, A., Counotte, M. J., Margossian, C. C., Konstantinoudis, G., Low, N., Althaus, C. L., & Riou, J. (2020). Estimation of SARS-CoV-2 mortality during the early stages of an epidemic: A modeling study in Hubei, China, and six regions in Europe. *MedRxiv.*

Hellewell, J., et al. (2020). Feasibility of controlling COVID-19 outbreaks by isolation of cases and contacts. *The Lancet Global Health, 8*, 488–496.

Hethcote, H. W. (2000). The mathematics of infectious diseases. *SIAM Review, 42*(4), 599–653.

Hochreiter, S., & Schmidhuber, J. (1997). Long short-term memory. *Neural Computation, 9*(8), 1735–1780.

Huang, N. E., & Qiao, F. (2020). A data driven time-dependent transmission rate for tracking an epidemic: A case study of 2019-ncov. *Scientific Bulletin, 65*(6), 425–427.

Huang, C., et al. (2020a). Clinical features of patients infected with 2019 novel coronavirus in Wuhan, China. *The Lancet, 395*(10223), 497–506.

Huang, C., et al. (2020b). Clinical features of patients infected with 2019 novel coronavirus in Wuhan, China. *The Lancet Respiratory Medicine, 8*(5), 475–481.

Infectious diseases data explorations & visualizations. [Online]. Available https://epistat.wiv-isp.be/covid/.

Jin, Y., Yang, H., & Ji, W. E. A. (2020). Virology, epidemiology, pathogenesis, and control of COVID-19. *Viruses, 12*(4), 372.

Jin, G., Yu, J., Han, L., & Duan, S. (2020). The impact of traffic isolation in Wuhan on the spread of 2019-ncov. *medRxiv.*

Jurisic, V., Radenkovic, S., & Konjevic, G. (2015). The actual role of LDH as tumor marker, biochemical and clinical aspects. *Advances in Cancer Biomarkers*, 115–124.

Kırbaş, İ. E. A. (2020). Comparative analysis and forecasting of COVID-19 cases in various European countries with ARIMA, NARNN and LSTM approaches. *Chaos, Solitons & Fractals, 138*, 110015.

Korolev, I. (2020). Identification and estimation of the SEIRD epidemic model for COVID-19. *Journal of Econometrics.*

Kucharski, A. J., et al. (2020). Early dynamics of transmission and control of 2019-nCoV: A mathematical modelling study. *medRxiv.*

Lauer, S. A., Grantz, K. H., et al. (2020). The incubation period of coronavirus disease 2019 (COVID-19) from publicly reported confirmed cases: Estimation and application. *Annals of Internal Medicine.*

Li, Q., Guan, X., Wu, P., Wang, X., Zhou, L., Tong, Y., & Ren, R. (2020). Early transmission dynamics in Wuhan, China, of novel coronavirus–infected pneumonia. *The New England Journal of Medicine, 382*, 1199–1207.

Li, X., et al. (2020a). Risk factors for severity and mortality in adult COVID-19 inpatients in Wuhan. *Journal of Allergy and Clinical Immunology, 146*(1), 110–118.

Li, X., et al. (2020b). Clinical characteristics of 25 death cases with COVID-19: a retrospective review of medical records in a single medical center, Wuhan, China. *International Journal of Infectious Diseases, 94*, 128–132.

Lin, Q., et al. (2020). A conceptual model for the coronavirus disease 2019 (COVID-19) outbreak in Wuhan, China with individual reaction and governmental action. *International Journal of Infectious Disease, 93*, 211–216.

Linton, N. M., Kobayashi, T., Yang, Y., Hayashi, K., Akhmetzhanov, A. R., Jung, S.-M., Yuan, B., Kinoshita, R., & Nishiura, H. (2020). Incubation period and other epidemiological characteristics of 2019 novel coronavirus infections with right truncation: A statistical analysis of publicly available case data. *Journal of Clinical Medicine, 9*(2).

Liu, Y., Funk, S., & Flasche, S. (2020). *The contribution of pre-symptomatic transmission to the COVID-19 outbreak*. [Online]. Available: https://cmmid.github.io/topics/covid19/pre-symptomatic-transmission.html.

Liu, Y., et al. (2020). Neutrophil-to-lymphocyte ratio as an independent risk factor for mortality in hospitalized patients with COVID-19. *Journal of Infection, 81*(1), e6–e12.

Lorencin, I. E. A. (2021). Automatic evaluation of the lung condition of COVID-19 patients using x-ray images and convolutional neural networks. *Journal of Personalized Medicine, 11*(1), 28.

Malik, P., et al. (2020). Biomarkers and outcomes of COVID-19 hospitalisations: Systematic review and meta-analysis. *BMJ Evidence-Based Medicine, 26*(3), 1–12.

Mei, X., et al. (2020). Artificial intelligence-enabled rapid diagnosis of patients with COVID-19. *medRxiv*.

Milovanovic, D. R., Jankovic, S. M., Ruzic Zecevic, D., Folic, M., Rosic, N., Jovanovic, D., Baskic, D., Vojinovic, R., Mijailovic, Z., & Sazdanovic, P. (2020). Lečenje koronavirusne bolesti (COVID-19). *Medicinski Casopis, 54*(1).

Musulin, J., Baressi Segota, S., Stifanic, D., Lorencin, I., Andjelic, N., Sustersic, T., Blagojevic, A., Filipovic, N., Cabov, T., & Markova-Car, E. (2021). Application of artificial intelligence-based regression methods in the problem of COVID-19 spread prediction: A systematic review. *International Journal of Environmental Research and Public Health, 18*(8), 4287.

Pan, F., Ye, T., & Sun, P. (2020). Time course of lung changes on chest CT during recovery from 2019 novel coronavirus (COVID-19) pneumonia. *Radiology, 295*(3).

Panteghini, M. (2020). Lactate dehydrogenase: an old enzyme reborn as a COVID-19 marker (and not only). *Clinical Chemistry and Laboratory Medicine (CCLM), 1* (ahead-of-print).

Peng, L., Yang, W., Zhang, D., Zhuge, C., & Hong, L. (2020). Epidemic analysis of COVID-19 in China by dynamical modeling. *arXiv Preprint*.

Pereira, R. M., Bertolini, D., Teixeira, L. O., Silla, C. N., Jr., & Costa, Y. M. (2020). COVID-19 identification in chest X-ray images on flat and hierarchical classification scenarios. *Computer Methods and Programs in Biomedicine*, 105532.

Piccolomini, E. L., & Zama, F. (2020). Monitoring Italian COVID-19 spread by an adaptive SEIRD model. *medRxiv*.

Poletto, C., Scarpino, S. V., & Volz, E. M. (2020). Applications of predictive modelling early in the COVID-19 epidemic. *The Lancet Digital Health, 2*(10), 498–499.

Quilty, B., Clifford, S., Flasche, S., & Eggo, R. M. (2020). Effectiveness of airport screening at detecting travellers infected with novel coronavirus (2019-nCoV). *Eurosurveillance, 25*.

Read, J. M., Bridgen, J. R., Cummings, D. A., Ho, A., & Jewell, C. P. (2020). Novel coronavirus 2019-ncov: early estimation of epidemiological parameters and epidemic predictions. *medRxiv*.

Ruan, Q., et al. (2020). Clinical predictors of mortality due to COVID-19 based on an analysis of data of 150 patients from Wuhan, China. *Intensive Care Medicine, 46*(5), 846–848.

Sanche, S., Lin, Y. T., Xu, C., Romero-Severson, E., Hengartner, N., & Ke, R. (2020). The novel coronavirus, 2019-nCoV, is highly contagious and more infectious than initially estimated. *MedRxiv*.

Sarzi-Puttini, P., et al. (2020). COVID-19, cytokines and immunosuppression: What can we learn from severe acute respiratory syndrome? *Clinical and Experimental Rheumatology, 38*(2), 337–342.

Schmidhuber, J. (2015). Deep learning in neural networks: An overview. *Neural Networks, 61*, 85–117.

Shaban, W. M., Rabie, A. H., Saleh, A. I., & Abo-Elsoud, M. A. (2020). A new COVID-19 patients detection strategy (CPDS) based on hybrid feature selection and enhanced KNN classifier. *Knowledge-Based Systems, 205*, 106270.

Sustersic, T., Blagojevic, A., Cvetkovic, D., Cvetkovic, A., Lorencin, I., Segota, S. B., Car, Z., & Filipovic, N. (2021). Epidemiological predictive modelling of COVID-19 spread. In *8th International congress of the Serbian Society of Mechanics, Kragujevac, Serbia*.

Šušteršič, T., Milovanović, V., Ranković, V., & Filipović, N. (2020). A comparison of classifiers in biomedical signal processing as a decision support system in disc hernia diagnosis. *Computers in Biology and Medicine, 125*, 103978.

Šušteršič, T., Ranković, V., Peulić, M., & Peulić, A. (2019). An early disc herniation identification system for advancement in the standard medical screening procedure based on Bayes theorem. *IEEE Journal of Biomedical and Health Informatics, 24*(1), 151–159.

Tan, L., et al. (2020). Lymphopenia predicts disease severity of COVID-19: A descriptive and predictive study. *Signal Transduction and Targeted Therapy, 5*(1), 1–3.

Tang, B., et al. (2020). Estimation of the transmission risk of the 2019-ncov and its implication for public health interventions. *Journal of Clinical Medicine, 9*.

Tindale, L., et al. (2020). Transmission interval estimates suggest pre-symptomatic spread of COVID-19. *MedRxiv*.

Wang, D., et al. (2020). Clinical characteristics of 138 hospitalized patients with 2019 novel coronavirus–infected pneumonia in Wuhan, China. *JAMA, 323*(11), 1061–1069.

Wilder-Smith, A., Chiew, C. J., & Lee, V. J. (2020). Can we contain the COVID-19 outbreak with the same measures as for SARS? *The Lancet Infectious Diseases, 20*(5), e102–e107.

Woelfel, R., et al. (2020). Clinical presentation and virological assessment of hospitalized cases of coronavirus disease 2019 in a travel-associated transmission cluster. *medRxiv*.

World Health Organization. (2020). *Novel coronavirus (2019-nCoV). Situation report - 6*. Geneva: World Health Organization. 26 January [Online]. Available https://www.who.int/docs/default-source/coronaviruse/situation-reports/20200126-sitrep-6-2019- -ncov.pdf?sfvrsn=beaeee0c_4.

Wu, Z., & McGoogan, J. M. (2020). Characteristics of and important lessons from the coronavirus disease 2019 (COVID-19) outbreak in China. *JAMA, 323*(13), 1239–1242.

Wu, G., et al. (2020a). Development of a clinical decision support system for severity risk prediction and triage of COVID-19 patients at hospital admission: an international multicenter study. *European Respiratory Journal, 56*(2).

Wu, J., et al. (2020b). Estimating clinical severity of COVID-19 from the transmission dynamics in Wuhan, China. *Nature Medicine, 26*, 506–510.

Wynants, L., et al. (2020). Prediction models for diagnosis and prognosis of covid-19: systematic review and critical appraisal. *BMJ, 369*.

Yan, L., et al. (2020a). An interpretable mortality prediction model for COVID-19 patients. *Nature Machine Intelligence*, 1–6.

Yan, L., et al. (2020b). A machine learning-based model for survival prediction in patients with severe COVID-19 infection. *MedRxiv*.

Yang, E. A. Z. (2020). Modified seir and ai prediction of the epidemics trend of COVID-19 in china under public health interventions. *Journal of Thoracic Disease, 12*(3), 165.

Yang, X., et al. (2020). Clinical course and outcomes of critically ill patients with SARS-CoV-2 pneumonia in Wuhan, China: A single-centered, retrospective, observational study. *The Lancet Respiratory Medicine, 8*(5), 475–481.

Yi-Cheng, C., Ping-En, L., Cheng-Shang, C., & Tzu-Hsuan, L. (2020). A time-dependent SIR model for COVID-19 with undetectable infected persons. *IEEE Transactions on Network Science and Engineering*, 1.

Zeng, T., Zhang, Y., Li, Z., Liu, X., & Qiu, B. (2020). Predictions of 2019-nCoV transmission ending via comprehensive methods. *arXiv*.

Zhou, F., et al. (2020). Clinical course and risk factors for mortality of adult inpatients with COVID-19 in Wuhan, China: A retrospective cohort study. *The Lancet, 395*, 1054–1062.

CHAPTER

13

Economic analysis for in silico clinical trials of biodegradable and metallic vascular stents

Marija Gacic

University of Kragujevac, Institute for Information Technology, Kragujevac, Serbia; BIOIRC Bioengineering Research and Development Center, Kragujevac, Serbia

1 Introduction

Classical clinical trials are designed to answer specific research questions relating to the efficacy and safety of a new intervention by measuring defined end points, including diagnostic biomarkers in real patient (Deloitte, 2021). Today, clinical trials can be commercialized after approval by a regulatory authority and an ethics committee review of the preclinical regulatory submission (Clinical Trials, 2021). Research start from the basic assumption that data collected in the clinical study are relatively small but representative selection of subjects and they have to generalize the results to the larger patient population. If the sample is too constrained or poorly selected, it hinders the broad applicability of the results. This is not only a statistical concern, but also an ethical and medical one (Properzi et al., 2019). Today, it takes 10–12 years on average to bring a new drug to the market, with limited change over the past decades in the linear and sequential process used to assess the efficacy and safety of drugs. Currently, drug discovery, which is the initial phase of R&D, takes 5–6 years, followed by around 5–7 years for clinical trials. According to the US Food and Drug Administration (FDA), approximately 33% of drugs move from Phase II to III, while around 25%–30% move from Phase III to the next phase (Taylor, 2019). The success rate in the traditional clinical trials is 10 percentages (Fig. 1).

Any medical device such as coronary or peripheral stent in the process of commercially released in the market, an appropriate level of scrutiny and rigorous testing must be undertaken. This testing is achieved through the clinical trials, a process that is carried out in three

https://doi.org/10.1016/B978-0-12-823956-8.00012-2

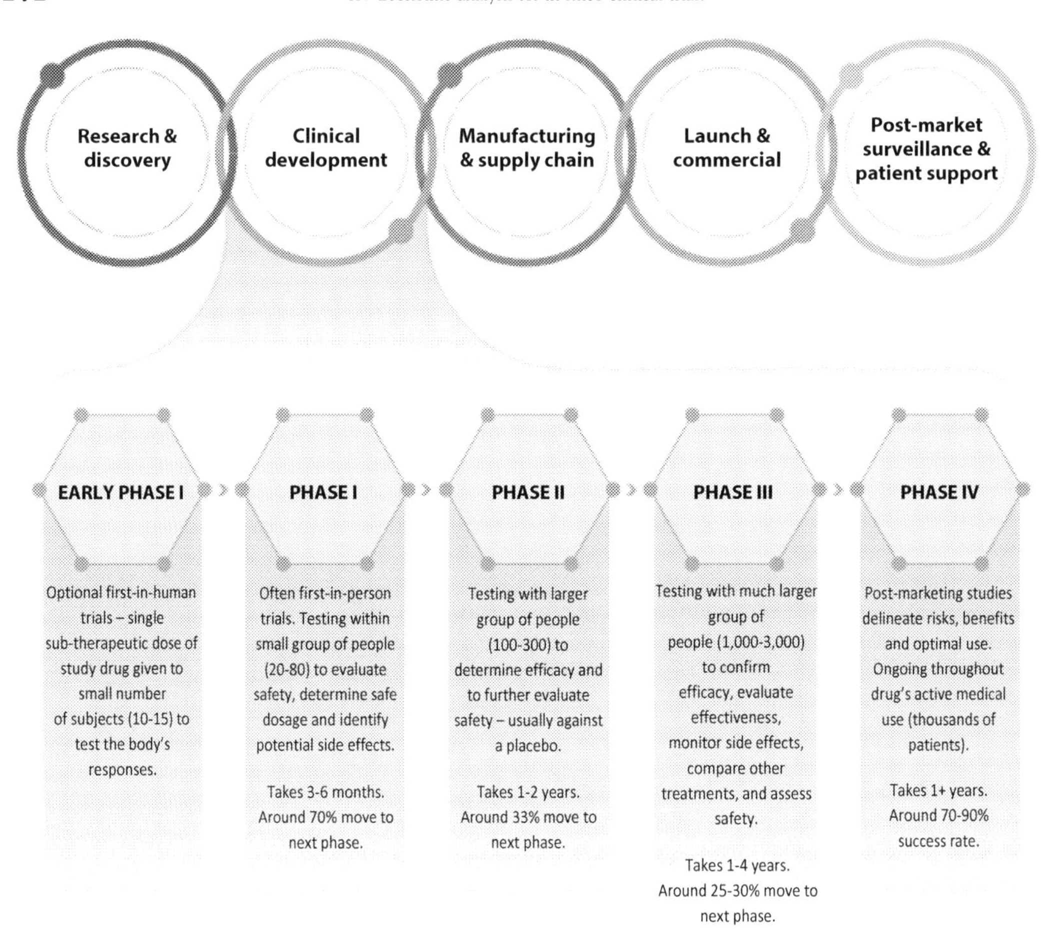

FIG. 1 The traditional approach for today's clinical trials with only 10% success rate.

phases and targets the evaluation of the safety and efficacy of stent. The difference between the three phases is the number of the enrolled patients, as well as the variables of interest in each of these phases. In the first phase, only a small number of patients were enrolled, while the ultimate objective is to ensure the safety of the medical device. In the other two phases, the medical device is tested on a larger number of patients toward evaluating its effectiveness and potential side effects (Phase II), and in multiple hospitals and countries (Phase III) to demonstrate the efficacy in a larger population. After this testing and the approval of the stent, postmarketing multicenter studies are performed to estimate and assess the effectiveness of the new stent compared to already available in the market.

InSilc project aims to develop an in silico clinical trial platform for designing, developing, and assessing drug-eluting bioresorbable vascular scaffolds (BVS), by building on the comprehensive biological and biomedical knowledge and advanced modeling approaches, to simulate their implantation performance in the individual cardiovascular physiology (H2020 InSilc Project, 2021).

In accordance with Directive 2010/63/EU, the principle of the 3Rs (Replacement, Reduction and Refinement) needs to be considered when selecting testing approaches to be used for regulatory testing of human and veterinary medicinal products. Testing of new models of vascular stents, scaffolds, and balloons in real clinical trials is time consuming, expensive, and highly inconvenient for the patients included in the study. Therefore, the intention is to replace, reduce, and refine the real clinical study with the in silico clinical study and in silico testing of the innovative models of stents in order to decrease the costs and the time required to perform real clinical study. In this chapter, we are presenting the innovative solution for designing, developing, and assessing coronary stents, which are developed within the EU funded project InSilc. The main question that has been raised and answered in this work is what the benefits of using in silico trials are. Analyzing market of similar solutions has shown that there is no similar integrated solution in the market and that potential savings are significant.

2 Market analysis

2.1 Stent biomedical industry

Intended users of InSilc cloud platform and separate modules are companies that develop new, innovative models of coronary stents. Significant resources that these companies invest in R&D could be reduced by using in silico testing to reduce the number of patients enrolled in the clinical study and time for performing the study. Huge investments in research and testing innovative designs of stents make the market of stent devices limited for new entrants, because small innovative companies often cannot invest in the development phase and bare the risk of failure. In silico clinical trials can significantly reduce the time and money for clinical trials and besides that we must consider no risk for patients and their life. Decreased costs of R&D activities can lower the barriers for entering the market for small companies, but also decrease the price of the final product and make it more affordable.

Obstacle-related challenges focus toward the approval process and translation of an R&D to a "product." Faced with budget constraints and the desire to offer patients access to most effective innovations, policy makers should think anew about the health innovation model. Leveraging the power of Big Data to optimize the current system, reviewing technologies that bring only limited health benefits, and thinking through novel approaches to manage areas where the current model does not work are just a few of the needed solutions.

The current approval process for medical devices and clinical trials in Europe is fragmented and requires improvement. The obstacles for approval of coronary stents can be classified within two major categories:

(1) Obstacles related to the complexity of the approval process and.
(2) Obstacles related to obtain evidence on safety and efficacy of devices through clinical trials.

However, medical use is not the only difference in the variety of stents.

InSilc modules have a great potential of creating new market opportunities since stents may vary also by the material: (a) metal stents, (b) standard polymeric stents, and (c) biodegradable polymeric stents.

Among the different types of stents, drug-eluting stents (DES) are considered to be the gold standard of treatment. Since the development of the first generation of DES, manufacturers have focused on optimizing and developing innovative low-profile platforms, stent coatings, drug-eluting components, and materials. Fig. **2** illustrates different types and uses of stents. A key issue is that in Europe the processes of device approval and clinical trial conduct are regulated by two collaborating agencies; namely notified bodies and competent authorities; in the United States, on the other hand, both processes are regulated by a single agency, the FDA (Escardio, 2014).

Despite the vague growth of the stent market and industry, the assessment of the risk of stent rupture is of primary importance to ensure its effectiveness during stenting procedure. At present, this is commonly performed through in vitro tests or computational approaches.

There are international standard regulations, which provide an immediate assessment of durability of the device subjected to a particular cyclic loading. For instance, ISO 25539 (ISO 25539, 2012), FDA Guidance (FDA, 2010), and ISO 5840 (ISO 5840, 2013) are some of those.

The experimental tests meet a number of disadvantages such as: (a) high costs and long duration, (b) difficulty in reproducing in vivo environment, and (c) difficulty in assessing biomechanical quantities.

On the other hand, numerical simulation methods have become a well-recognized and widely adopted tool to investigate biomechanical issues (Dordoni et al., 2015; Petrini et al., 2016). These so called in silico tests allow easy and quick modification of the parameters, for instance, geometrical features, boundary conditions, and external loadings toward virtual stent deployment. Moreover, numerical models provide direct and continuous access to many quantities (such as stress/strain field through the device) during the whole test simulation.

InSilc mechanical module reduces the time for classical mechanical testing **twice** and cost **three** times using specific innovation of in silico mechanical testing. Also, innovation of in silico stent deployment in anatomical patients' database will give a unique opportunity for industrial stent producers to reduce cost of clinical trials at least **five** times.

The world stent market has an estimated value of €6.4 billion, of which 37% is generated in the United States and 10% in the EU. Coronary stents are now the most commonly implanted medical devices, with more than 1 million implanted annually. By 2020, sales of coronary stents, including bare metal and drug-eluting stents, will grow to €5.62 billion, at a compound annual growth rate (CAGR) of 2.0%. Among the emerging markets, China has captured significant market share, which is expected to increase in the future. The emerging markets, including Brazil, China, and India, are expected to demonstrate the greatest growth in the market over the forecast period and will serve as an outlet of expansion for stent manufacturers to increase their global presence (Marketresearch, 2014). Major stent markets globally (until 2020), according to the international forecasts are: the United States, France, Germany, Italy, Spain, the United Kingdom, Japan, Brazil, China, and India.

The market for technologies and products in the treatment of coronary heart disease is forecast to grow from €12.2 billion in 2014 to €22.5 billion in 2021 (Fig. 4), according to a new study from Smithers Apex (Smithersapex, 2015).

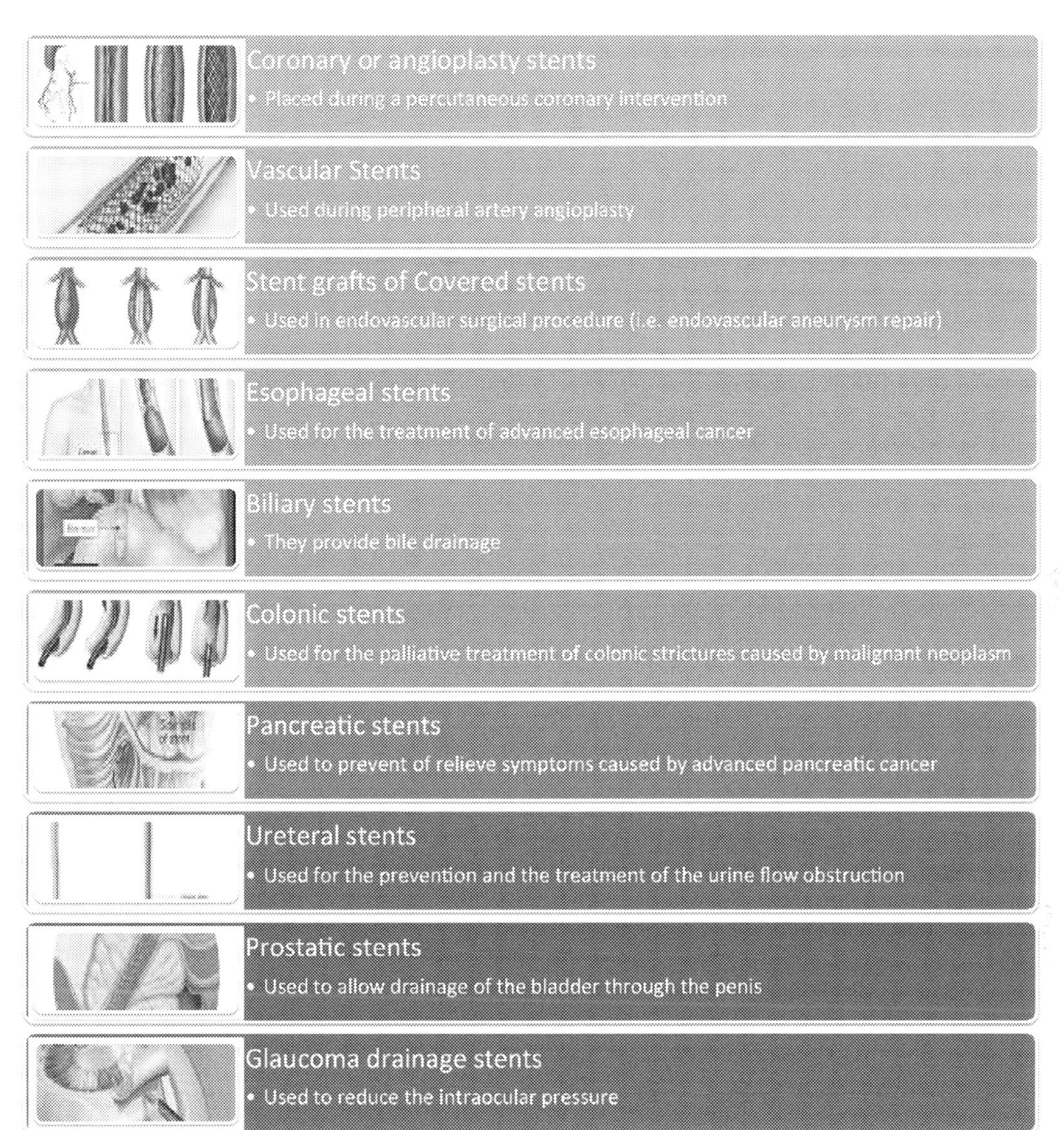

FIG. 2 Different types and uses of stents.

Increasing demand for minimally invasive surgeries globally will drive the market of vascular stents. In addition, aging population will further drive the market growth of vascular stents. According to a report published by the World Health Organization (WHO), the number of people aged 65 or older is expected to increase from 605 million to 2 billion by 2050. Thus, increasing geriatric population will fuel the demand of vascular stents and, hence, will

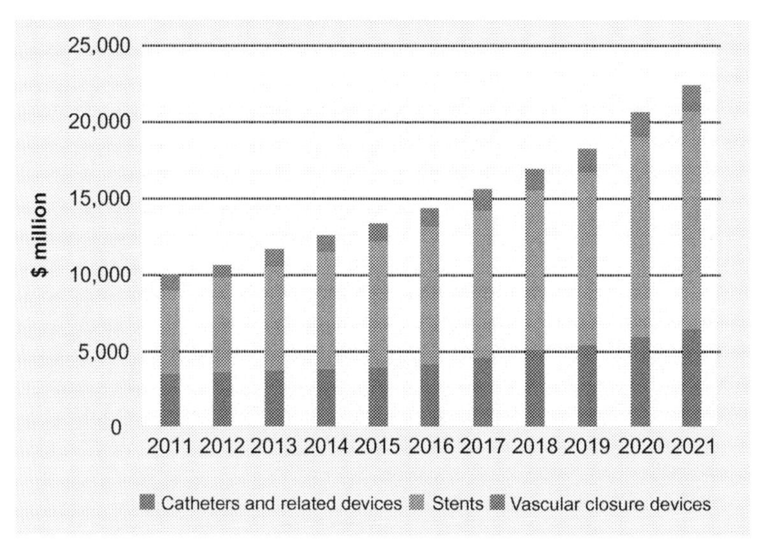

FIG. 3 Global market for coronary artery disease treatment devices, 2011–2021.

drive the market growth. In addition, increasing number of cardiovascular diseases drives the growth of vascular stents market. According to the WHO, about 17.3 million people died from cardiovascular disease (CVD) in 2008 and the number is expected to increase by 2030. This increase in number of patients suffering from CVD across the world will boost the global vascular stents market. The Future of Coronary Heart Disease Medical Devices to 2021 shows the continued demand for these devices as clinically and cost-effective solutions to coronary artery disease (Fig. 3).

In the largest product area, coronary stents, despite an upwards of 90% penetration of interventional cardiology procedures by the use of stents, forecasted stent sales will grow at nearly double-digit rates as a result of continued innovations, increasing coronary artery disease prevalence, and robust emerging market uptake.

Our market analysis includes companies with experience, know-how and resources that enable them to develop, manufacture, and market innovative products that can compete effectively in the global markets and influence the real R&D as well as the business trends.

Here are the 10 companies that are expected to lead the cardiology space by 2020, along with descriptions of how their businesses are evolving (Table 1).

The coronary stent market is fragmented with several global as well as local players. Players offer several products across different subsegments of this market. Key players in this market include Medtronic plc (Ireland), Abbott Laboratories (the United States), Boston Scientific Corporation (the United States), Biosensors International Group, Ltd. (Singapore), BIOTRONIK SE & Co. KG (Germany), B. Braun Melsungen AG (Germany), TERUMO CORPORATION (Japan), STENTYS SA (France), MicroPort Scientific Corporation (China), Meril Life Sciences Pvt. Ltd. (India), Vascular Concepts (India), and Translumina GmbH (Germany).

TABLE 1 Leading companies in cardiology by 2020 (Mddionline, 2014).

Rank	Company	2020 Sales (projected)	2014 Sales	% Annual growth
1	Medtronic	€11,120b	€9.361b	+3.6
2	St. Jude Medical	€6889b	€5.185b	+3.4
3	Boston Scientific	€5942b	€5.046b	+3.3
4	Abbott Laboratories	€3661b	€2.289b	+7.4
5	Edwards Lifesciences	€3050b	€2.844b	+2.1
6	Johnson & Johnson	€2959b	€2.208b	+2.9
7	Terumo	€2827b	€2.061b	+3.4
8	Getinge	€2573b	€1.778b	+4.6
9	W. L. Gore	€	€1.559b	+3.7
10	Asahi Kasei	€1864b	€270 m	+29.2

New developments in drug-eluting stents, such as innovative biodegradable materials and bifurcated stents, are expected to prompt an increase in market value.

InSilc platform or separate modules will be offered to key players in the stent industry as a service for competitive price, compared to real clinical trials, which also reduces time needed for in vitro experiment or clinical trial. Direct contact and presentation of InSilc will be performed in the next period. Mechanical module only can save a significant amount of money and time needed to perform all mechanical stent tests in silico. Also, other modules such as three-dimensional (3D) reconstruction and plaque characterization tool, deployment module, fluid dynamics module, drug delivery module, degradation module, myocardial perfusion module, and virtual population can be exploited by stent industry as partly replacement of clinical trials.

2.2 Contract research organization

According to a Grand View Research Inc. (Grandviewresearch, 2021) study, it is expected for the global health-care Contract Research Organization (CRO) market size to reach €54.7 billion by 2025.

CROs provide drug development services, regulatory and scientific support, and infrastructure and staffing support to supplement in-house capabilities or to provide a fully outsourced solution. In the 1970s, CRO industry started from providing limited clinical trial services, and has developed to a full-service industry encompassing the entire medical device development process. Today, CRO's clinical services include protocol design and management and monitoring of Phase I through Phase IV clinical trials, data management, laboratory testing, medical and safety reviews, and statistical analysis. Also, they provide services that generate high quality and timely data to support applications for regulatory approval of new devices. There are two major classifications of CRO—preclinical CROs and clinical trial CROs. Their global revenue share in 2015 was 19.95% and 80.05%, respectively.

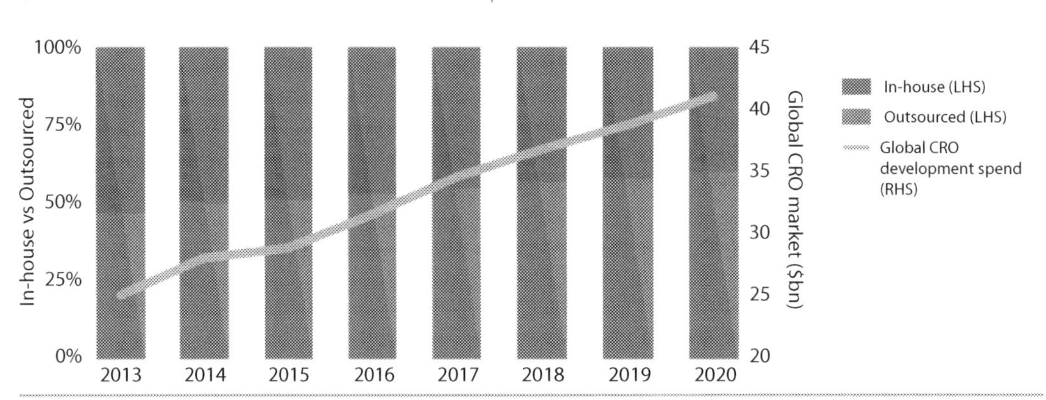

FIG. 4 Global CRO market and penetration rates. *Source: Baird, 2015; CRO Industry Primer, Credit Suisse, 2016; 2017 CRO Market Size Projections 2016–2021, ISR Reports, 2017.*

Today, the world's large service suppliers are mainly concentrated in the United States (Fig. 4) with the top three being Quintiles, LabCorp (Covance), and PPD with the global revenue market share of 20.06%, 9.72%, and 19.19%, respectively. The global CRO market was valued at 31 billion EUR in 2018. The estimations suggest that it will reach 50 billion EUR by the end of 2024, growing at a CAGR of 8.2% between 2019 and 2024 (Fig. 5) (Arcognizance, 2019).

2.3 Interventional cardiologists/hospitals

Interventional cardiology includes implanting of cardiology catheters, guide wires, balloon catheters, coronary stents, fractional flow reserve (FFR), intravascular ultrasound, etc.

This industry, mainly concentrated in Europe, the United States, Japan and China, will maintain more than 3% annual growth rate in the next 5 years. Chinese Interventional Cardiology market is predicted to be growing fast taking into account the increasing number of people suffering from cardiovascular diseases. Average industry gross margin is between 70% and 80%, suggesting that Interventional Cardiology Project is a good choice of investment. On the other hand, one should always consider negative factors and threat such as

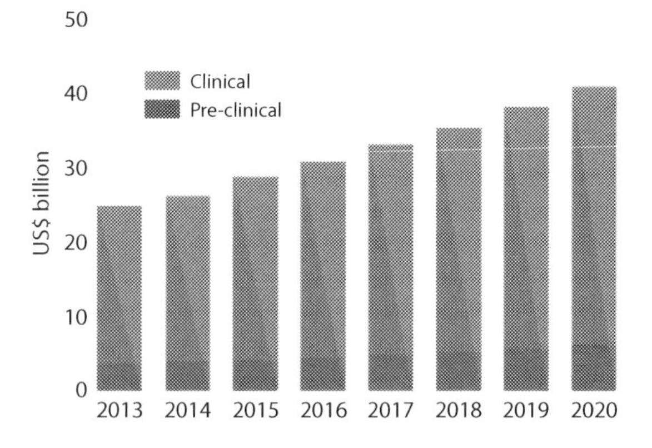

FIG. 5 Preclinical vs clinical outsourced development expenditure. *CRO Industry Primer, Credit Suisse, 2016.*

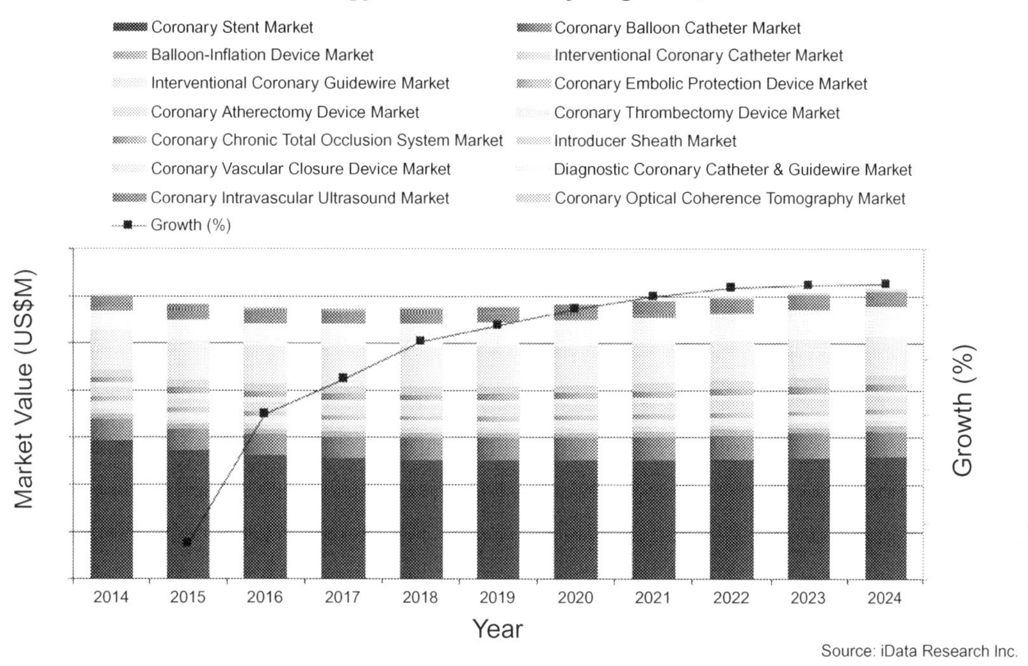

FIG. 6 Interventional cardiology market 2014–2024 (Idataresearch, 2021).

serious competition in Interventional Cardiology Industry. According to a new study (Fig. 6) (Marketreportsworld, 2014), the global market for interventional cardiology is expected to grow at a CAGR of roughly 3.1% and reach 3 million EUR in 2024.

In the United States, interventional cardiology procedures are broken into four segments: angiography procedures, angioplasty/PCI procedures, coronary atherectomy procedures, and catherization procedures. The number of catherization procedures is the largest, about four million per year. After catherization procedures, the numbers are the largest for angiography procedures. The European market has these same four segments and an embolic protection device procedure segment. The number of catherization procedures is also the largest in Europe, at over five million annually.

2.4 Animal testing organizations

Animal testing and experimentation organizations spend a lot of money and resources every year on drug and medical device testing. The salaries of researchers and technicians, and the income obtained from patents for the drugs and medical devices provide financial incentives and career advancement to the staff who chooses to conduct their research on animals. Universities, institutes, and other research institutions profit from "overhead" they receive from the funds for animal testing from different agencies. Both the Environmental Protection

Agency and the Food and Drug Administration require massive amounts of animal testing for the marketing of industrial chemicals, vaccines, and drugs.

Animal breeders obtain substantial amount of profit from breeding such animals, from mice to primates, in order to satisfy the demands of researchers. Recent prices from one animal supply company state the price of New Zealand white rabbits at €352 each, purebred beagles for €1049, and some primates costing more than €8000 each (Navs, 2021). InSilc platform can offer alternative to these animal testing and experimentation organizations in order to save financial resources and animal welfare. There is a huge society pressure for reducing animal testing because of protecting animal welfare, so besides obvious financial benefit we must not neglect the societal impact that reducing animal testing will have by using in silico testing.

2.5 Education and researchers

InSilc platform and separate modules could be used for further research and education. Numerous education courses related to biomedical engineering and medical informatics could use, for instance, different InSilc modules for educational and research purposes. The price for academic institution could be reduced in comparison with industry and commercial users. Also, an open module could be offered for mechanical modeling module, where users can change their own material model and easily build the stent model. Researchers can benefit from performing their research based on the InSilc platform and/or separate modules in cardiovascular topics.

If we analyze courses at universities for Biomedical Engineering (BE) and Medical Informatics (MI), there are number of compatible programs which need some of InSilc modules for research purposes. These programs are:

Biomedical Engineering, Medical Engineering, Medical Informatics, Computational Biology, Computational Physics, Computational Chemistry, Applied Mathematics, Electrical/Electronic Engineering, Mechanical Engineering, Medicine, Dentistry, Sports Engineering/Physical Education and Sports, Computer Science/Computer Engineering, and Life Sciences.

The courses appear under different titles, depending on the focus within the broad field of interest. The frequency analysis is based on generic titles that reflect the core (essence) of the disciplines and needs. For instance, Medical Information Systems, Information Systems and Databases, Information Systems in Healthcare, Biomedical Information Systems, Information Systems in Biomedicine, and similar titles appear more or less for the same content. Medical Information Systems is chosen as generic course title. The same approach is applied for all courses that are identified as similar in content.

The analysis of BE and MI courses was performed on 221 European universities, in over 30 countries. The selection was done according to the indices provided by world known ranking studies and portals (Leidenranking, 2021; Shanghai, 2021; Studyportals, 2021; University rankings, 2021). Not all European universities were analyzed for study programs containing BE&MI-related courses. The main lead in universities selection is based on Leiden Ranking list of European universities (Leidenranking, 2021), due to their indicators, but the Shanghai list (Shanghai, 2021) and search results of portals were used for additional selection. Leiden ranking list contains 214 universities (ranking 2013), and not all of them offer BE&MI-related

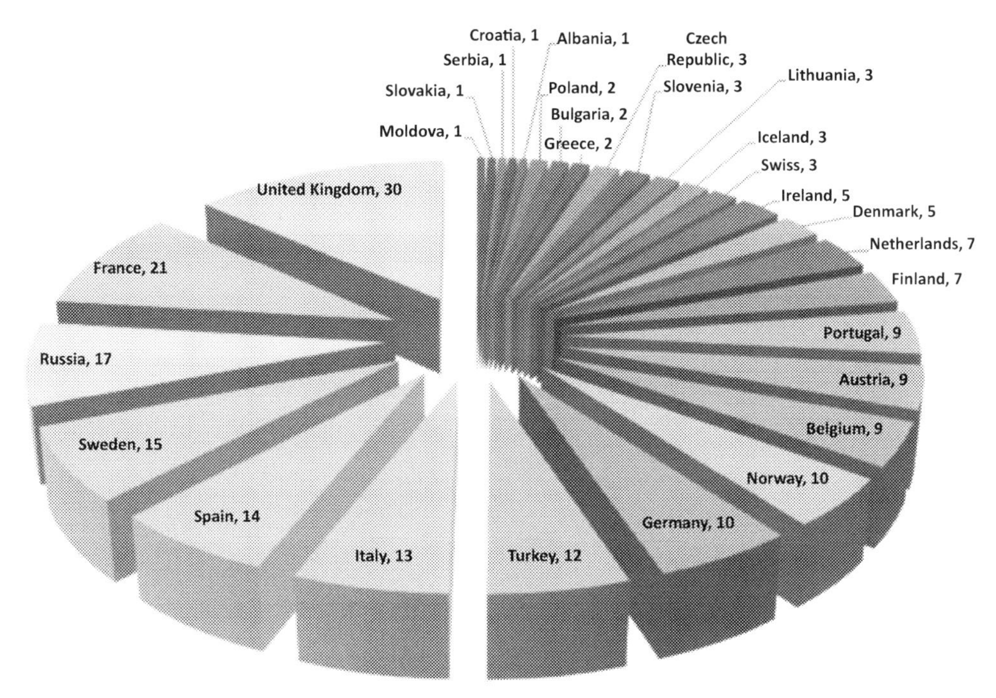

FIG. 7 Distribution of Biomedical Engineering or Medical Informatics-related courses (Navs, 2021).

study programs. The results include 94 Leiden ranked universities. The distribution of analyzed universities is shown in Fig. 7.

3 InSilc cloud platform

The stent manufacturer provided the average results of uniaxial tensile tests performed on a number of dog-bone samples with a different gauge length, width, and thickness. Tests are conducted at different temperatures (Bostonscientific, 2020). For each temperature, different curves were available referring to different strain rates: results are in accordance with typical PLLA behavior: at each temperature, the curves show a common initial elastic response, a strain rate-dependent yield point, and plastic behavior ending with a strong hardening. At higher temperature or lower velocity, stress values decrease despite the increasing final strains (Filipovic et al., 2021).

The InSilc platform is based on the extension of existing multidisciplinary and multiscale models for simulating the drug-eluting BVS mechanical behavior, the deployment and degradation, the fluid dynamics in the micro- and macroscale, and the myocardial perfusion, for predicting the drug-eluting BVS and vascular wall interaction in the short- and medium/long term.

The developed InSilc platform consists of different simulation modules/tools—some of which can be considered as stand-alone modules and, therefore, can be used separately if

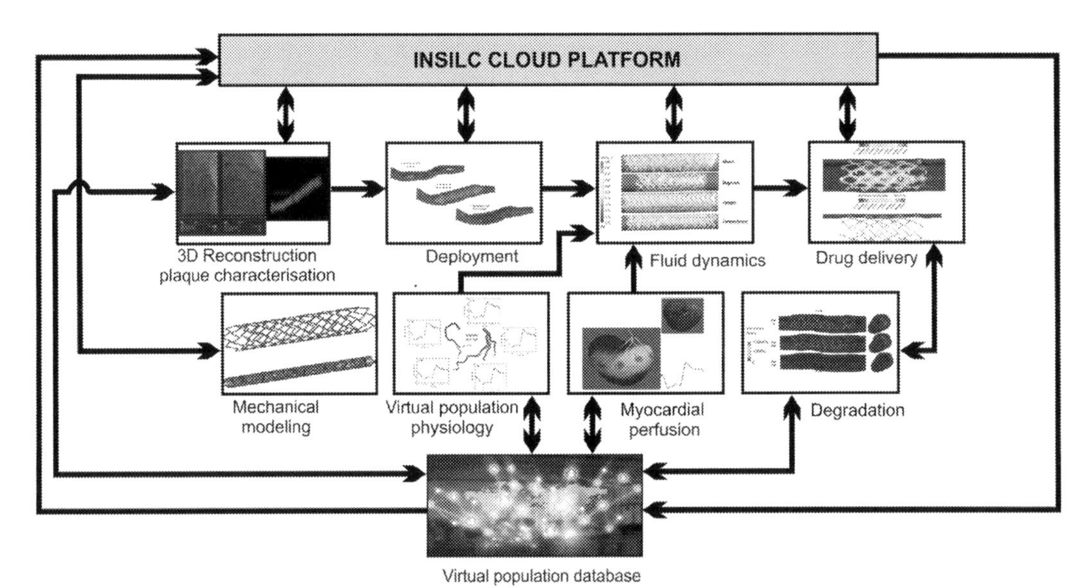

FIG. 8 InSilc cloud platform.

there is such demand from the targeted users. These modules integrated in the InSilc platform are: mechanical modeling module, 3D reconstruction and plaque characterization tool, deployment module, fluid dynamics module, drug delivery module, degradation module, myocardial perfusion module, virtual population physiology, and virtual population database (Fig. 8). These tools are applicable for all types of coronary and peripheral stents, such as bare metal stents (BMS), drug-eluting stents (DES), and bioresorbable stents. This is a great advantage of InSilc allowing this way the penetration of InSilc platform and modules to a wide range of market and interested stakeholders (Fotiadis & Filipovic, 2021). The chapter presents the detailed comparative analysis of the costs and time required for the real clinical trial and in silico clinical trial performed using the presented solution.

4 Description of the modules

The purpose of the deployment module is the simulation of the coronary stent (DES or BVS) implantation within stenotic coronary artery models. This simulation provides detailed information of the short-term outcome after stenting, in terms of deployed stent and vessel configurations as well as the stresses and strains in the two elements. These data are useful in predicting the in vivo performances of a new device.

The stent industry follows standard mechanical stent testing in the whole process of stent evaluation, i.e., according to the ISO test standards. Mechanical tests are very time consuming, expensive, and require many cycles/iterations, while in some cases total redesign of the stent are required or even the examined stent design is abandoned. The mechanical modeling module assists in reducing the required number of real mechanical tests and the associated costs. In brief, the module provides the ability of the following mechanical tests to be simulated in silico: simulated use—pushability, torquability, trackability, recoil, crush resistance, flex/kink, longitudinal tensile strength, crush resistance with parallel plates, local compression, radial force, foreshortening, dog boning, three-point bending, inflation, and radial fatigue test. The risk of fatigue failure is also predicted using fatigue criteria for metal stents with polymer.

The whole process for the mechanical module development includes the design, set up, and implementation of several finite element simulations performed with the advanced and beyond the state-of-the-art in-house BIOIRC's solver PAK (PAK Finite Element Program, 2020). The solver achieves the simulation of nonlinear material and geometry problems, nonlinear contact problems, dynamics and statics with residual stress, and strain analysis. The process that is followed, in general, includes the following steps: (i) creation of the 3D stent geometry (in case this is not available directly in a 3D format from the manufacturer), (ii) mesh generation, and (iii) application of appropriate boundary conditions (depending on the test a variety of boundary conditions are applied). BIOIRC has developed a nonlinear material model that is applied in the finite element solver PAK for prescribing material property from uniaxial stress-strain experimental curves. It is an open module used only in the mechanical modeling module (Fig. 9).

The deployment module requires detailed information about the delivery systems to be simulated to create reliable and realistic virtual FE models of the devices involved in the stenting procedure. In silico simulations of the stenting procedure consists of the following steps, to be repeated for each device (stent or balloon): (i) device positioning, (ii) balloon inflation, and (iii) stent deployment. Most of the computational steps are automatized and this allows a significant reduction in preparing and performing the simulations. In turn, this allowed a reduction of the process to be sustained by the users of the deployment module (Fig. 10).

The drug delivery module (Fig. 11) has been developed to model the in vivo release kinetics of the drug from the coating and its spatial distribution within the tissue over the course of weeks to months. Pharmacokinetics has been separately examined for the coating and the tissue. First, a mathematical empirical-trained model was developed to simulate release and extract the drug flux out of the drug-eluting surface, validated with the manufacturer's

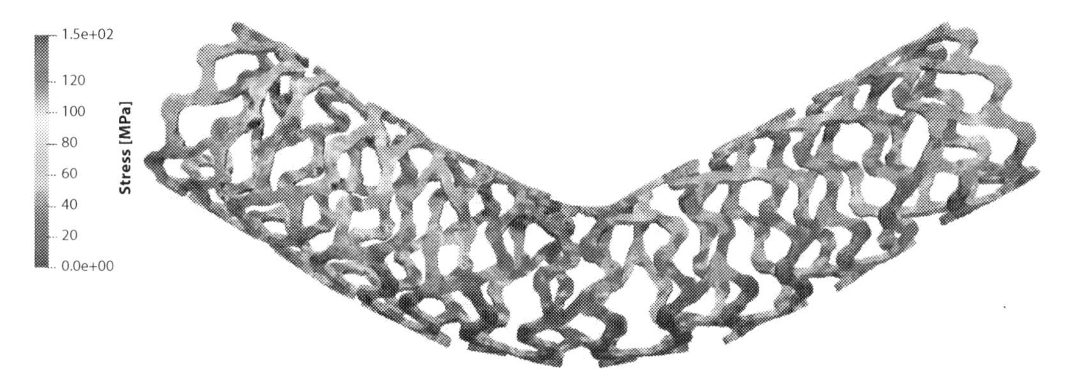

FIG. 9 Mechanical modeling module: three-point bending stent testing.

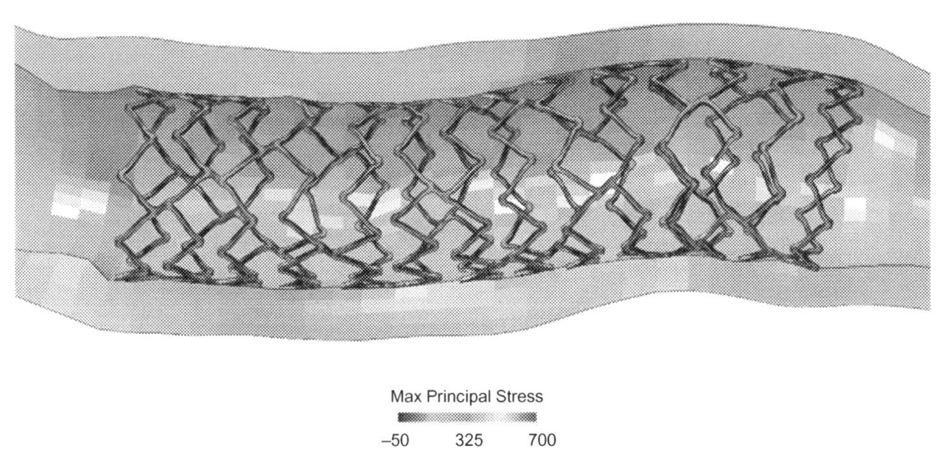

FIG. 10 Deployment module.

experiments. Then, a physics-based 3D advection-diffusion-reaction model was developed wherein using continuum mechanics equations the convection of drug by the plasma infiltration, diffusion of the drug within the tissue, and binding/unbinding of the drug to the extracellular matrix and specific receptors have been considered. Drug delivery is modeled for a sustained period of time to monitor both the early burst of drug as well as long-term retention and ultimate clearance rate.

The degradation module (Fig. 12) simulates the degradation pattern of implanted BVS. The InSilc degradation framework has been implemented within both Johnson-Cook and Parallel Rheological Framework (PRF) constitutive models, which have been found to form the basis for the mechanical behavior of several commercial BVS.

The InSilc degradation module depends on detailed input from the deployment module, whereby the implanted configuration of the relevant device and artery has been predicted. The postdeployment stent-artery configuration and the material stress-strain history at all model integration points are imported and these form the starting point for the InSilc degradation module. This approach ensures that model parameters remain consistent between the deployment and degradation modules, with continuity maintained in the discretization/mesh, element type, underlying constitutive model, and many of the numerical parameters that control the solution process (e.g., step times, mass scaling, etc.) allowing for a consistent predictive mechanical framework. The InSilc degradation module predicts the spatiotemporal progression of degradation. Based on this, the predicted long-term biomechanical performance can be related to several clinical end points relevant to implanted stents, including, minimal stent area, malapposed stent struts, stent fracture, or dismantling.

Fluid dynamics module (Fig. 13) is developed to compute the velocity, pressure, and shear stress patterns in stented segments of human coronary arteries. The fluid dynamics module requires two main inputs: geometrical information and flow boundary conditions. The geometrical information consists of two STL files, one describing the lumen of the vessel wall and the other the surface representation of the stent. These two STL files are combined to form the mesh of the fluid domain by using a commercial platform (ICEM, ANSYS). The boundary conditions consist of time-dependent inflow and outflow curves. These data are used to feed

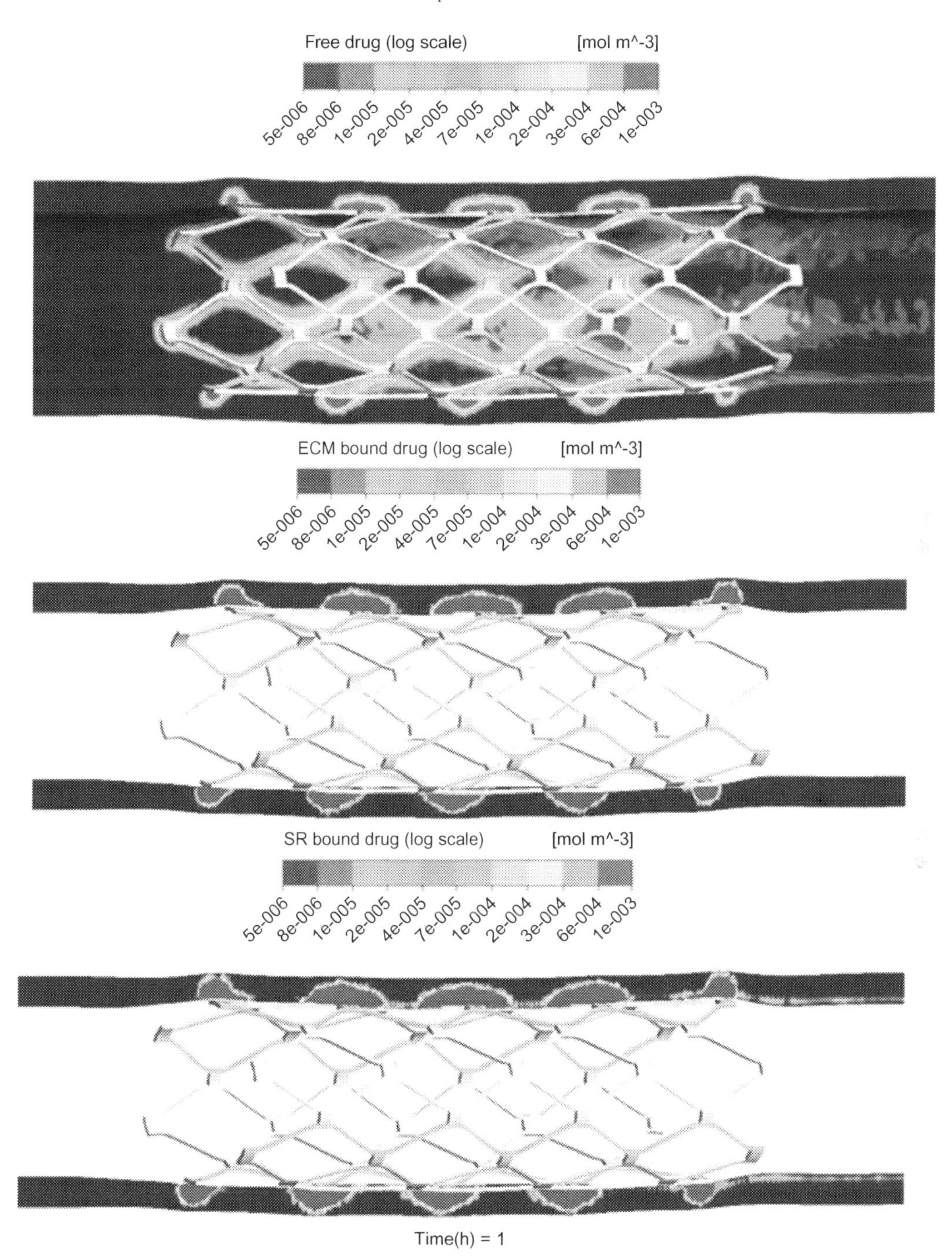

FIG. 11 Drug delivery module.

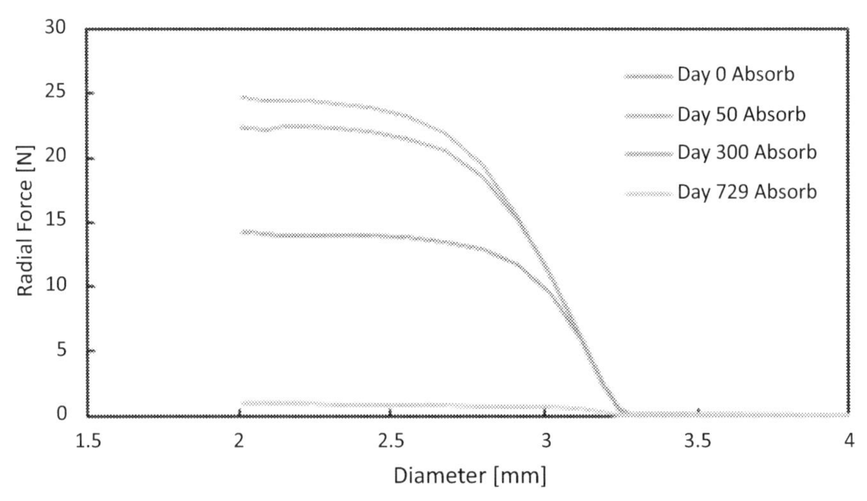

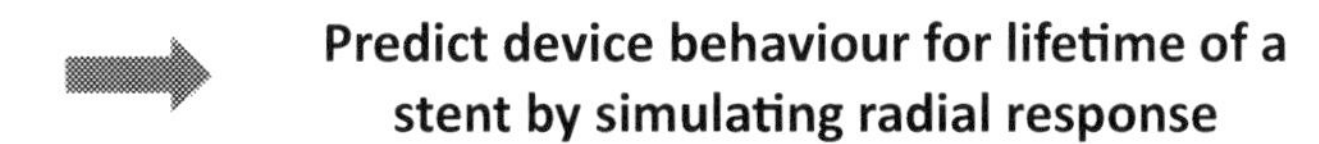

FIG. 12 Degradation module.

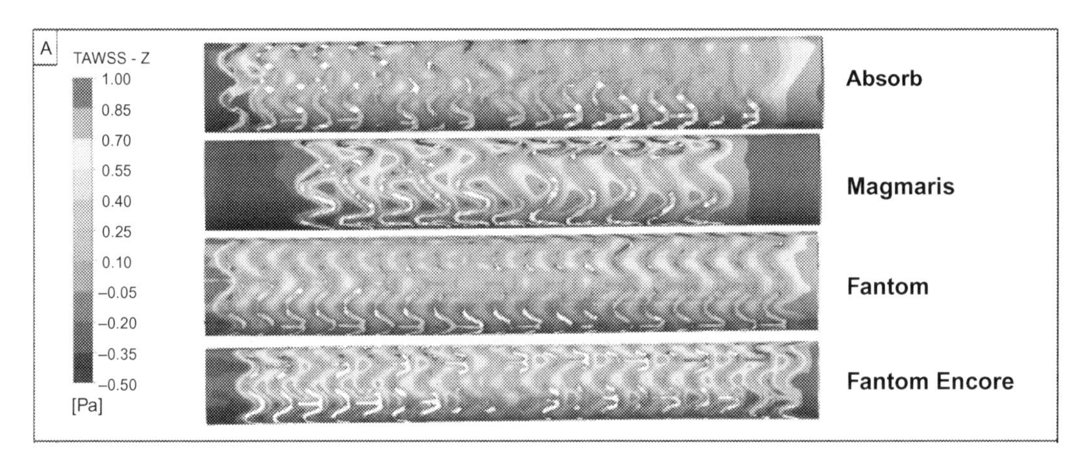

FIG. 13 Fluid dynamics module solution.

in a commercial solver (FLUENT, ANSYS) to compute velocity, pressure, and shear stress patterns. The output is formed by two-dimensional (2D) maps of pressure and shear stress derived parameters in the stented region (Fig. 13).

The myocardial perfusion module (Fig. 14) simulates the posttreatment performance of the drug-eluting BVS in improving myocardial perfusion distal from the treated vessel. The myocardial perfusion module predicts the whole-heart perfusion in the cardiac muscle, and generates virtual myocardial perfusion maps. The module takes as inputs CT coronary

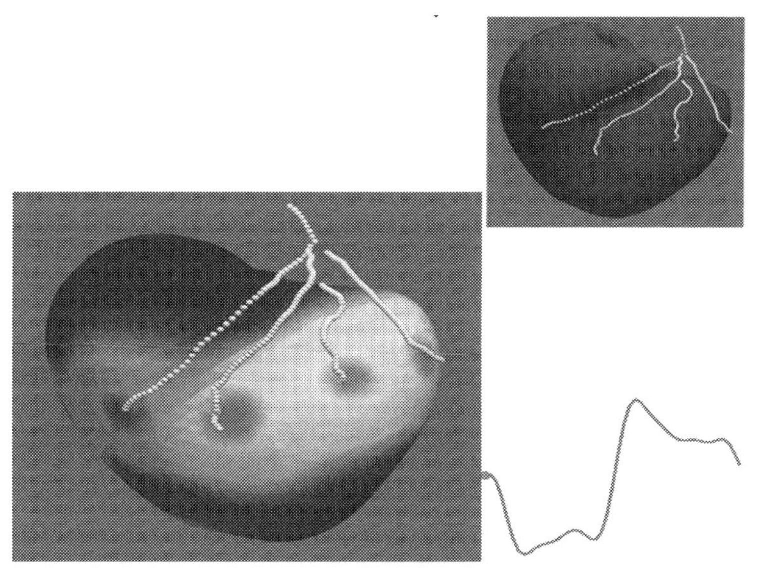

FIG. 14 Myocardial perfusion model solution.

angiography (CTCA) images, the model-generated pressure boundary conditions, and the outlet flow conditions from the fluid dynamics module. Prediction of postoperative perfusion is then provided by solving a multicompartment poroelastic flow model, from which the perfusion maps are estimated. By combining the myocardial perfusion module with the boundary condition variability model, it is possible to simulate perfusion differences under both rest and stress. By varying the boundary conditions between rest and stress and computing the summed difference score (SDS), a threshold value of $SDS > 4$ can be used to gauge whether the virtual patient is at risk of postoperative myocardial infarction and other major adverse cardiovascular events (MACE).

5 Economic analysis

In Table 2, detailed prices of each module have been presented.

Parameters for calculations of costs of real clinical trials based on literature analysis and/or Italian fees have been described in Table 3.

All the other costs that characterize a clinical trial (coordination, CRO, protocol, data management, statistical analysis, reporting, etc.) have not been considered as part of sensitive information that the industrial partner is not allowed to disseminate and as part of both the considered strategies. The average cost per patient for the execution of a real clinical trial is calculated according to the following equation (Meredith et al., 2012, 2013):

TABLE 2 Detailed prices of each module.

Module	Partner	Price €	Base for calculation in silico trial
Virtual database module	FORTH	**200**	Per simulation
Deployment module	POLIMI	**1.000**	Per simulation with the deployment of one stent in a single vessel. The price of simulations of more complex procedures (such as stent overlapping, treatment of bifurcation) is defined case by case. In all the cases it is supposed that the 3D geometry of the stent is already available
Fluid dynamics module	ERASMUS	**900**	Per simulation
Drug delivery module	CBSET	**1.900**	Per simulation. It is expected to receive the geometry of artery and device from the Deployment Module. The pricing assumes that the pharmacokinetics model of the release for the device has already been developed
Degradation module	NUIG	**500**	Per simulation
Myocardial perfusion module	UNIVLEEDS	**300**	Per patient with some additional automatization
Virtual population physiology module	UNIVLEEDS	**200**	Per patient
InSilc platform	FEOPS	**150**	Per simulation
TOTAL		**5.150**	

TABLE 3 Parameters for calculations of costs of real clinical trials.

Acronym	Parameter	Value	Unit	Source/note
$C_{DRG\text{-}557}$	Cost of the intervention (DRG-557)	8128	€	Italian fees—Decree of Health Ministry 18/10/2012
$C_{DRG\text{-}558}$	Cost of the intervention (DRG-558)	6434	€	Italian fees—Decree of Health Ministry 18/10/2012
C_{IVUS}	Cost of IVUS	650	€	Alberti et al., (2016)
C_{QCA}	Cost of QCA	2508	€	Italian fees—Decree of Health Ministry 18/10/2012
C_{Clin_FU}	Cost of Clinical follow-up	31,67	€	Cost calculated as average from the following Italian Regions: Friuli Venezia Giulia, Emilia Romagna, and Umbria Health Tariff 2012, Code 89.7A.3 (it includes cardiologic visit and ECG)

Cost for the real execution of the modules included in the InSilc platform (RTC)

$$CRCT = \frac{C_{DRG-557} * N_{Int557}(T_0) + C_{DRG-558} * N_{Int558}(T_0) + C_{IVUS} * \sum_{i=0}^{2} N\_IVUS(T_i) + C_{QCA} * \sum_{i=0}^{2} N\_QCA(T_i) + C_{Clin\_FU} * \sum_{i=0}^{2} N\_FU(T_i)}{N\_S} = 10.544€$$

The number of patients in real clinical trials and in silico changes according to the specific strategy (in the model, clinical trials after in silico trials are performed with a reduced number of subjects).

Two trials are needed to achieve a positive result for regulatory approval and release to the market (attrition rate 50% (Pammolli et al., 2011)), the perfect discrimination capacity of the InSilc platform in identifying the right trial to perform (sensitivity and specificity = 1) and the ability to reduce the number of patients needed for the real clinical trials by 30%, the strategy adopting the InSilc platform determines an overall cost of 2,652,120€ instead of 3,166,200€ for the strategy with only real trials (i.e., a 16% reduction). Moreover, a breakeven point (in terms of service price per patient) has been calculated (Fig. 15). It was found that everything below 6.8536 EUR for in silico clinical trials is very cost effective from economical point of view.

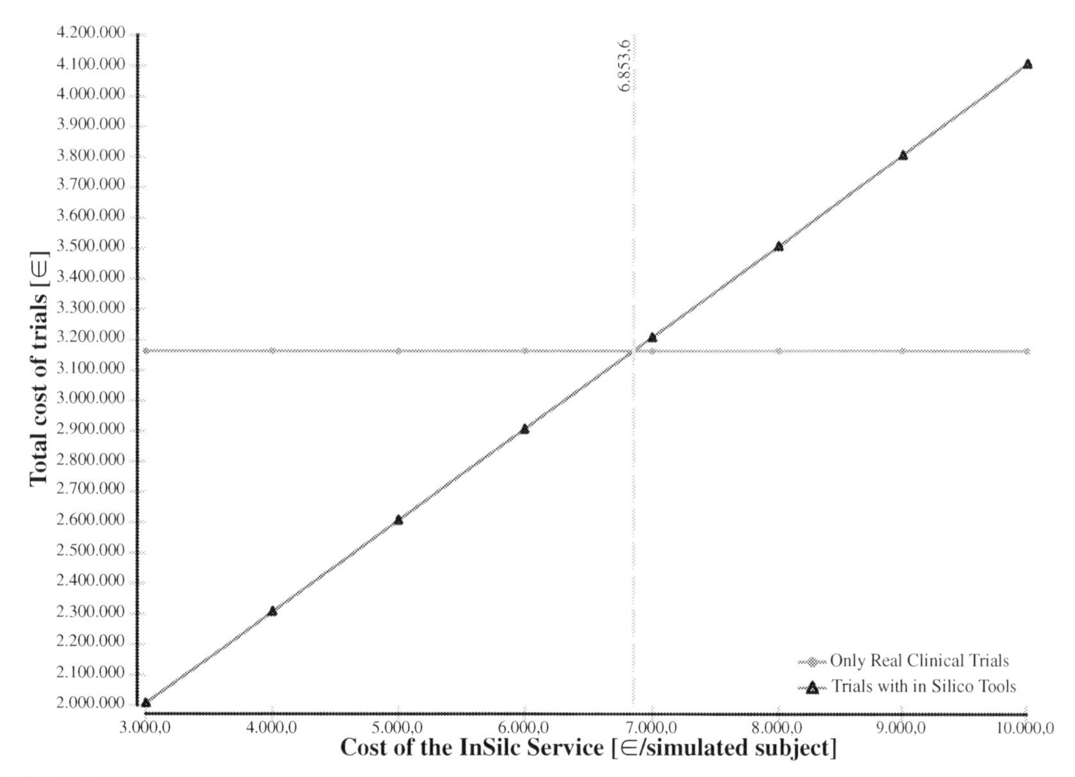

FIG. 15 Breakeven point for InSilc clinical trials modules based on InSilc service price variation.

6 Conclusions

The main exploitable products of InSilc are the mechanical modeling module, the 3D reconstruction and plaque characterization tool, the deployment module, the fluid dynamics module, the drug delivery module, the myocardial perfusion module, the degradation module, the virtual physiology module, the virtual population database, and the integrated InSilc cloud platform (Gacic et al., 2021; Miković et al., 2019; Stefanovic et al., 2015).

Coronary stents are currently the most widely used for treating symptomatic coronary disease. The traditional approach for today's clinical trials with only 10% success rate has been given. Detailed market analysis of the major users such as Stent Biomedical Industry (stent designers and producers), to Contract Research Organizations (CRO), Interventional Cardiologists/Hospitals, Researchers, and Animal testing organizations was described.

Economic analysis has been taken into account the prices per each InSilc module as total of 5150 EUR. The average cost per patient for the execution of a real clinical trial is calculated as 10.554 EUR. It was found that breakeven point for InSilc clinical trials modules based on InSilc service price variation is 6.8536 EUR. It means that calculated price for in silico trials is below breakeven point.

Acknowledgments

This chapter is supported by the projects that have received funding from the European Union's Horizon 2020 research and innovation program under grant agreements no. 777119 (InSilc project). This chapter reflects only the author's view. The commission is not responsible for any use that may be made of the information it contains.

References

Arcognizance. (2019). Gobal-contract-research-organization-cro-market-2019-by-company-regions-type-and-application-forecast-to-2024, [Online]. Available: https://www.arcognizance.com/report/global-contract-research-organization-cro-market-2019-by-company-regions-type-and-application-forecast-to-2024.

Bostonscientific. (2020). Bioabsorbable-polymer-stent-costs, [Online]. Available: https://www.bostonscientific.com/en-US/products/stents- -coronary/bioabsorbable-polymer-stent/bioabsorbable-polymer-stent-costs.html.

Deloitte. (2021). www.deloitte.co.uk/centreforhealthsolutions.

Alberti, A., Giudice, P., Gelera, A., Stefanini, L., Priest, V., Simmonds, M., et al. (2016). Understanding the economic impact of intravascular ultrasound (IVUS). *European Journal of Health Economics, 17*, 185–193.

Clinical Trials. (2021). *Learn About Clinical Studies, NIH U.S.* National Library of Medicine. https://clinicaltrials.gov/ct2/about-studies/learn#ClinicalTrials.

Dordoni, E., Petrini, L., Wu, W., Migliavacca, F., Dubini, G., & Pennati, G. (2015). Computational modeling to predict fatigue behavior of NiTi stents: What do we need? *Journal of Functional Biomaterials, 6*(2), 299–317. https://doi.org/10.3390/jfb6020299.

Escardio. (2014). Stent evaluation document_revision, Escardio, [Online]. Available: https://www.escardio.org/static_file/Escardio/Subspecialty/EAPCI/Documents/ESC_EAPCI_TF_StentEvaluationDocument_Revision_final-1.pdf.

FDA, 2010. Non-clinical engineering tests and recommended labeling for intravascular stents and associated delivery systems, FDA, [Online]. Available: U.S. Food and Drug Administration (FDA); Silver Spring, MD, USA.

Filipovic, N., Nikolic, D., Isailovic, V., Milosevic, M., Geroski, V., Karanasiou, G., Fawdry, M., Flanagan, A., Fotiadis D., Kojic, M., In vitro and in silico testing of partially and fully bioresorbable vascular scaffold, Journal of Biomechanics 2021;115:110158. doi:https://doi.org/10.1016/j.jbiomech.2020.110158. Epub 2020 Dec 2.

Fotiadis, D., & Filipovic, N. (2021). In-silico trials for drug-eluting BVS design, development and evaluation. *The Project Repository Journal, 8.*

Gacic, M., Karanasiou, G., Fotiadis, D., & Filipovic, N. (2021). Insilico clinical trials for bioresorbable vascular stents. In *11th International conference on information society and technology, Serbia, Mar 7–10.*

Grandviewresearch. (2021). Global-healthcare-cro-market, [Online]. Available: https://www.grandviewresearch.com/press-release/global-healthcare-cro-market.

H2020 InSilc Project, 2021: In-silico trials for drug-eluting BVS design, development and evaluation, www.insilc.eu.

Idataresearch. (2021). Interventional-cardiology-market, Idataresearch, [Online]. Available: https://idataresearch.com/product/interventional-cardiology-market/.

ISO 25539, 2012, Cardiovascular implants—endovascular devices, ISO 25539, [online]. Available: ISO 25539-2:2012 Part 2: Vascular stents. ISO; Geneva, Switzerland.

ISO 5840, 2013. Cardiovascular Implants—Cardiac Valve Prostheses, ISO 5840, [online]. Available: ISO 5840-3:2013 Part 3: Heart valve substitutes implanted by Transcatheter techniques. ISO; Geneva, Switzerland.

Mddionline. (2014). Who-will-be-10-largest-cardiovascular-companies-2020, [Online]. Available: https://www.mddionline.com/who-will-be-10-largest-cardiovascular-companies-2020.

Leidenranking. (2021). [Online]. Available: http://www.leidenranking.com/.

Marketresearch. (2014). [Online]. Available: http://www.marketresearch.com/product/sample-8538829.pdf.

Meredith, I. T., Verheye, S., Dubois, C. L., Dens, J., Fajadet, J., Carrié, D., et al. (2012). Primary endpoint results of the EVOLVE trial: A randomized evaluation of a novel bioabsorbable polymer-coated, everolimus-eluting coronary stent. *Journal of the American College of Cardiology, 59*, 1362–1370.

Meredith, I. T., Verheye, S., Weissman, N. J., Barragan, P., Scott, D., Chávarri, M. V., et al. (2013). Six-month IVUS and two-year clinical outcomes in the EVOLVE FHU trial: A randomised evaluation of a novel bioabsorbable polymer-coated, everolimus-eluting stent. *EuroIntervention, 9*, 308–315.

Marketreportsworld, 2014, [Online]. Available: https://www.marketreportsworld.com/purchase/13055860.

Miković R., Arsić B., Gligorijević Đ., Gačić M., Petrović D., Filipović N., The influence of social capital on knowledge management maturity of non-profit organizations—Predictive modelling based on a multilevel analysis, IEEE Access Journal, 2019 P. 1–15, ISSN: 2169-3536, Online ISSN: 2169-3536, DOI https://doi.org/10.1109/ACCESS.2019.2909812.

Navs. (2021). The-animal-testing-and-experimentation-industry, [Online]. Available: https://www.navs.org/the-issues/the-animal-testing-and-experimentation-industry/#.XbWX6K-VPHY.

PAK Finite Element Program, 2020 BIOIRC, Research and Development Center, 34000 Kragujevac, Serbia http://www.bioirc.ac.rs/index.php/software/5-pak.

Pammolli F., Magazzini L., Riccaboni M. The productivity crisis in pharmaceutical R&D. Nature Reviews. Drug Discovery [Internet]. 2011;10:428–38.

Petrini, L., Trotta, A., Dordoni, E., Migliavacca, F., Dubini, G., et al. (2016). A computational approach for the prediction of fatigue behaviour in peripheral stents: application to a clinical case. *Annals of Biomedical Engineering, 44*(2), 536–547.

F. Properzi et al., 2019 Intelligent Drug Discovery: Powered by AI, Francesca Properzi et al., 2019. See also: https://www2.deloitte.com/us/en/insights/industry/life-sciences/artificial-intelligence-biopharma-intelligent-drug-discovery.html.

Stefanovic, M., Nestic, S., Djordjevic, A., Djurovic, D., Macuzic, I., Tadic, D., & Gacic, M. (2015). An assessment of maintenance performance indicators using the fuzzy sets approach and genetic algorithms. *Proceedings of the Institution of Mechanical Engineers Part B Journal of Engineering Manufacture, 231*(1), 15–27. ISSN 0954-4054 https://doi.org/10.1177/0954405415572641.

Shanghai. (2021). Shanghai ranking, [Online]. Available: http://www.shanghairanking.com/.

Smithersapex. (2015). Coronary-artery-disease-to-drive, [Online]. Available: https://www.smithers.com/resources/2015/feb/global-market-for-coronary-artery-disease.

Studyportals. (2021). [Online]. Available: http://www.studyportals.eu/.

K. Taylor, 2019. Why improving inclusion and diversity in clinical trials should be a research priority. See also: https://blogs.deloitte.co.uk/health/2019/08/why-improving-inclusion-and-diversity-in-clinical-trials-should-be-aresearch-priority.html.

University rankings. (2021). Top universities, [Online]. Available: http://www.topuniversities.com/university-rankings/world-university-rankings.

Index

Note: Page numbers followed by *f* indicate figures and *t* indicate tables.

D

E

Q

R

S

T

U

V

W

Printed in the United States
by Baker & Taylor Publisher Services